Harry J. Stone

GUIDE FOR SAFETY IN THE CHEMICAL LABORATORY

Second Edition

MANUFACTURING CHEMISTS ASSOCIATION

VNR **VAN NOSTRAND REINHOLD COMPANY**

NEW YORK   CINCINNATI   ATLANTA   DALLAS   SAN FRANCISCO
LONDON   TORONTO   MELBOURNE

Van Nostrand Reinhold Company Regional Offices:
New York   Cincinnati   Chicago   Millbrae   Dallas
Van Nostrand Reinhold Company International Offices:
London   Toronto   Melbourne
Copyright © 1972 by Manufacturing Chemists' Association, Inc.
Library of Congress Catalog Card Number: 70-176320
ISBN: 0-442-05667-2
All rights reserved. Certain portions of this work copyright © 1954 and 1969 by the Manufacturing Chemists' Association, Inc. No part of this work covered by the copyrights hereon may be reproduced or used in any form or by any means—graphic, electronic, or mechanical, including photocopying, recording, taping, or information storage and retrieval systems—without written permission of the publisher.
Manufactured in the United States of America
Published by Van Nostrand Reinhold Company
135 West 50th Street, New York, N.Y. 10020
Published simultaneously in Canada by Van Nostrand Reinhold Ltd.
15   14   13   12   11   10   9   8   7   6

Library of Congress Cataloging in Publication Data

Manufacturing Chemists' Association.
    Guide for safety in the chemical laboratory.

    Original ed. published in 1954, prepared by the
General Safety Committee of the Manufacturing Chemists'
Association.
    Bibliography: p.
    1. Chemical laboratories—Safety measures.
I. Manufacturing Chemists' Association. General
Safety Committee. Guide for safety in the chemical
laboratory. II. Title.
QD51.M349   542'.1   70-176320

FOREWORD

The second edition of *Guide For Safety in the Chemical Laboratory* has been completely rewritten by the Safety and Fire Protection Committee of the Manufacturing Chemists Association. The latest methods and equipment for protection of laboratory workers have been incorporated and modern facilities and procedures have been illustrated. The table of flammability properties found in the 1954 edition has been replaced by a section in the appendix listing hazardous properties and referring to recommended methods of waste disposal. Another section in the appendix lists sources of information on a large number of chemicals and a third collects hazardous reactions which have been reported in the literature, supplementing reactions published in NFPA 491 M.

The book is quite properly called a Guide since it is intended to offer only a starting point in the solution of laboratory safety problems. However, it will be found of considerable assistance in setting up safety programs for schools and industrial or institutional laboratories.

ACKNOWLEDGMENT

Guide for Safety in the Chemical Laboratory is published as an activity of the Safety and Fire Protection Committee of the Manufacturing Chemists Association. The committee is grateful to the following members and guest authors who contributed.

Mr. R. H. Albisser, Merck & Co., Inc.; Mr. A. H. Christian, FMC Corporation; Mr. A. L. Cobb, Eastman Kodak Company; Mr. C. R. Eastman, Mobil Chemical Company; Mr. W. F. Elmendorf, Merck & Co., Inc.; Mr. H. H. Fawcett, National Academy of Sciences; Mr. G. G. Fleming, Celanese Corporation; Mr. J. T. Garrett, Monsanto Company; Mr. G. L. Gorbell, Monsanto Company; Mr. J. Guelich, Allied Chemical Corporation; Mr. W. H. Kistulenic, Mobil Chemical Company; Mr. F. E. Macauley, BASF Wyandotte Corporation; Mr. S. M. MacCutcheon, The Dow Chemical Company; Mr. A. L. Mossman, The Matheson Company; Mr. R. D. Poirier, Allied Chemical Corporation; Mr. J. S. Queener, E. I. du Pont de Nemours & Company; Mr. H. W. Snyder, The Dow Chemical Company; Mr. S. F. Spence, American Cyanamid Company; Dr. R. W. Van Dolah, U.S. Bureau of Mines; Mr. W. S. Wood, Sun Oil Company.

A special word of appreciation goes to Mr. W. F. Elmendorf for preparation of the list of hazardous reactions and to Dr. G. N. Quam of the University of Villanova for the research he directed in development of the section on laboratory waste disposal. Mr. W. S. Wood contributed greatly in final editing and coordinating of the manuscript, and Mr. A. H. Christian and Mr. F. G. Stephenson, secretary of the Safety and Fire Protection Committee, played essential parts in bringing this work to fruition.

CONTENTS

*See Contents on p. 325.

1

FUNDAMENTALS OF SAFETY IN THE CHEMICAL LABORATORY

Alchemists of earlier times went about their work in a most unscientific manner, creating hazards and endangering their lives and those of persons around them. They had neither adequate laboratory equipment nor enough prior knowledge of the hazards of chemical materials. Many paid the terrible price of injury and death from chemical exposures and accidents.

Today, extensive facilities for safe operation, and a vast and rapidly expanding wealth of knowledge, make it possible to operate a laboratory without injury to personnel. The attitude of the industrial laboratory director or the department chairman in the college or university laboratory toward safety and fire prevention is invariably reflected in the behavior of the entire staff. Safety results from an attitude of mind, and depends upon effort of each individual to eliminate unsafe acts and conditions that could cause accidents.

Continuing safety requires a genuine interest in safety from the top echelons to the bottom. Further, this interest must be continually demonstrated and transmitted—something more than just "Safety First" on a piece of paper tacked to the bulletin board. The development, adoption, and enunciation of a safety policy for the industrial plant or school are among the most direct methods for managements to express their attitudes and desires for accident-free operation. Similarly, the laboratory director must voice his approval and expectations in accord with a policy as follows:

1. All accidents are caused, and therefore can be prevented.
2. Supervisors or instructors at all levels are responsible for preventing injuries to those under their direction.
3. The environment, including equipment, must be safeguarded to prevent injuries.
4. Laboratory personnel must be trained to work safely.

Many universities and colleges have formally adopted such policies as cornerstones of their safety programs.

The ground rules for accident-free operation do not vary in basic organization and objectives whether they be applied to an industrial production facility or to a chemical laboratory. As one research safety supervisor stated, "There is no basic difference between the methods used to obtain a fine record of safety in research laboratory operations and the methods practiced to improve safety on the highway, in the home or in a production shop or an industrial plant. In each instance, the important ingredients are people, preparation, and prevention."

Public Law 91-596, the Occupational Safety and Health Act of 1970, places very direct responsibility upon all employers to assure safe and healthful work conditions. Existing consensus standards have become law by regulation, and inspections by federal or state agencies will be made to enforce compliance with the standards. Penalties are assessed for violations when cited by the inspectors. This law makes it even more necessary that laboratory administrations organize and implement active safety programs.

The member companies of the Manufacturing Chemists Association have proved that it is possible and practical to reduce the frequency and severity of accidental injuries by the application of sound accident-prevention controls. Safety engineers point out that the prevention of accidents is practical because few accidents are the result of unforeseen or unexpected circumstances. R. J. Crosby[1] has said that most accidents are the result of employees deviating from accepted, correct working methods. Accidents or "unwanted occurrences" must be handled similarly to any other interruption of efficient performance of the laboratory. Safeguards for prevention should become a normal part of the operation processes. Hence, an effective safety program seeks to develop and maintain such control over the activities as necessary to prevent accidents.

It is said that all the safe things in research have already been done, and that today the greatest opportunities in research lie in doing the things we were afraid to do yesterday. This is not true. On the other hand, we must point out the broad range of hazards encountered in today's laboratories—from simple corrosive reagents to the complexities of radioactive materials, cryogenics, and lasers—in order to indicate that there is still room for improvement in laboratory accident-prevention techniques. It is in the laboratory during the early stages of experimentation or research that the hazards must be identified and eliminated or brought under control. A hazard that is permitted to go uncontrolled can become disastrous in the eventual scale-up for quantity production.

Basically, the laboratory operates in two areas of possible hazard—those projects that have been done before and whose behavior is predictable, and those projects that have not been done before and whose behaviors are predictable only in theory. In the latter, precautions projected from available knowledge should suffice to guarantee safe results.

Prevention should be paramount in the safety effort of all laboratories. All too often the accidents that occur in the laboratory are simply human errors—errors committed by intelligent college- and university-educated persons who do not always apply their knowledge to the project at hand. Accidents in the laboratory have no respect for technical training or academic degrees. Highly trained researchers or instructors commonly assume that others in the laboratory know how to work safely. They cannot make such assumptions and still expect an accident-free laboratory. Those in charge must assume responsibility for the safety of associates or students in the laboratory.

Accidents and the resultant injuries in laboratories can be severe. *Case Histories of Accidents in the Chemical Industry*[2] indicates that flammable liquids, exothermic reactions, unstable materials, and toxic and corrosive materials play a large part in causing severe injuries. Severity of injury and damage can vary from minor to extreme, depending on the physical situation of the personnel and the design features of the laboratory and equipment. Although prevention is the initial objective of the program, it should be recognized that proper laboratory design and the provision of safe equipment will reduce the possibility of accidents, the severity of resultant injuries, and property damage.

Fires and explosions are major contributors to loss of life and property in laboratories. A study of a hundred significant laboratory fires by the National Fire Protection Association[3] provides some interesting facts: 71% of the fires originated in the laboratory itself; 56% of the laboratory fires originated between 6 PM and 6 AM; 67% of the fires were caused by:

electrical equipment (wire and appliances)	21%
misuse of flammable liquids	20%
explosions, miscellaneous	13%
gas	7%
spontaneous ignition	6%

Some of the factors influencing the spread of one hundred laboratory fires were:

structural features	
open stairwells	16%
no fire walls	15%
combustible interior finish	8%
undivided attics	7%
private protection—lack of automatic sprinklers	89%
contents features	
flammable liquids	17%
records	12%
public protection	
poor or inadequate fire equipment	13%
poor public water supply	8%
poorly situated hydrants	7%

Another basic reason, aside from the human value of accident prevention, for the company or educational institution to be interested in safety and prevention is that injuries to personnel and losses to property cost money. Simonds and Grimaldi[4] conclude that safety of operation is absolutely essential to any enterprise. "If that is lacking, operations are not under control and schedules and unit costs cannot be counted on." Safety must therefore also be handled in an effective manner if the laboratory is to take its rightful place in the successful operation of the company or enterprise. Also, there is a common tendency to look only on the surface of the cost of an accident—compensation, medical expenses, and property damage—and to ignore the many indirect costs involved. Similarly, the initial costs for losses in capital investment do not always reflect the loss of use and occupancy of the facility involved.

Simonds and Grimaldi also point out the intangible benefits of a good safety and loss-prevention program. These benefits are in the areas of the morale and attitude of laboratory personnel and the general goodwill or impression exerted on other elements of the company, government, school, or institution. Frequent accidents can result in loss of confidence by laboratory personnel in the technical ability of their supervisors or instructors. Managers or administrators who reflect a sincere concern for the welfare of their staff find it easier to secure staff acceptance of safety regulations and the need to use protective apparel and other devices so essential in safe operation.

Bibliography

1. Crosby, R. J., *A Management Guide for Accident Prevention,* Marsh and Mc-Lennan Insurance Company, 231 S. La Salle St., Chicago, Ill. 60604.
2. *Case Histories of Accidents in the Chemical Industry,* Manufacturing Chemists Association, Vol. 1, 1962; Vol. 2, 1966; Vol. 3, 1970.
3. Fire Record Bulletin, FR88-3, National Fire Protection Association.
4. Simonds, R. H., and Grimaldi, J. V., *Safety Management—Accident Cost and Control,* R. D. Irwin, Inc., Homewod, Ill. 1963.
5. *Safety in High Pressure Chemical Reactions,* National Safety Council, Congress Trans., Chemical Section, October, 1957.
6. Fawcett, H. H., and Wood, W. S., *Safety and Accident Prevention in Chemical Operations,* John Wiley & Sons, Inc., New York, 1965.

2

SAFETY ORIENTATION

"Because You're New, Be Especially Careful." How often are these few words the only safety instruction that a new employee gets? Or equally brief, safety orientation is limited to an afterthought remark from the boss, tossed over his shoulder, "Oh, and be safe around here."

Such casual initial introductions to safety are of little value, and are possibly detrimental. If a lasting impression of the importance of accident prevention is to be made, a much more positive and planned approach is required. This is true for students at all education levels, and basically the same approaches, fundamentals, and techniques should work in the academic world as in an industrial environment.

The new employee (or beginning student) is especially receptive to instructions and guides which will help him perform his job satisfactorily, and the time is ripe for a thorough safety orientation. This is the time to impress upon the new employee—or, for that matter, the recently transferred employee—that management considers safety a vital element of every job or operation.

A formalized approach to safety orientation assures optimum coverage of the most significant principles of accident prevention. These cardinal points can be dramatized and emphasized later by the continuing safety program.

Even the well-planned safety orientation program is, however, not a substitute for on-the-job safety training. It is an effective aid to the supervisor who is sincerely interested in accident prevention and eager to carry out his responsibilities for efficient laboratory operations.

Although there is no best way to orient new employees to the importance of safe job performance, programmed courses are becoming popular. Any program must be tailored to the industry, size of facility, location, turnover, heritage, and other pertinent factors. Here is one plan that has been used successfully in chemical operations.

BASIC APPROACH

All new employees should be introduced to accident prevention, even if briefly, on their very first day. They should be told of the general safety problems and the accident-prevention program. For uniformity and convenience, this can be done by the safety engineer before employees report to their working areas. At this time, too, appropriate literature, such as safety booklets, can be distributed.

In the case of experienced professional technical or administrative personnel, it would seem entirely fitting and proper that they also be given the orientation program, both for their personal benefit and to aid them in discharging their responsibility for the safety and well-being of those under their supervision.

FOUR MAIN AREAS

Safety Policy

The official policy on safety should be presented and explained. This can be simple or elaborate, but basically should state the organization's attitude toward accident prevention. If it is included in a safety booklet, all the better. At this time too, it would be well to emphasize that the individual *also* has a personal responsibility for his own safety and that of his co-workers. The safety booklet might say, for example:

1. XYZ Corporation recognizes safety as essential. It is important for the achievement of quality production, efficient operation, and worker satisfaction. No work is so important that it should be undertaken in an unsafe manner.
2. In our efforts to prevent injuries, our corporation provides safe working conditions, gives you full instructions on safe working methods, and makes available special equipment to protect you against particular hazards.

3. Freedom from accidents is of utmost importance to you, your family, and your fellow employees. To avoid injuries is to avoid suffering and financial loss.
4. Carry out your responsibility in accident prevention by following the safety instructions you receive from your supervisor and the safety suggestions in this booklet.

Hazards That May be Anticipated

The wide range of hazards that may be encountered in any operation, plus specific ones unique to the chemical industry or particular chemical laboratories, should be discussed generally.

Common Hazards.   These hazards exist in almost any industry, plant, or even home. They are mostly unrelated to any particular process or operation, and include cuts from broken glass, paper, knives, and fibers; burns from hot surfaces; foreign bodies (such as cinders) in the eye; falls on slippery floors or from makeshift ladders; hernias or back injuries from improper lifting; electrical shock, etc.

Hazards Unique to the Industry.   These include: corrosive chemicals, such as acid and caustic, which might cause serious body burns; flammable chemicals, such as process chemicals or those used for maintenance, which might cause a fire or explosion; toxic chemicals, which might cause serious injury or even death in the absence of proper precautions, e.g., through inhalation or absorption through intact skin; radioactive chemicals; process hazards, such as potential for runaway reactions, detonations, or rapid decompositions.

The employee should be informed that departmental supervision will subsequently explain to him the precise hazards of equipment he will have to operate, other hazards unique to his working area or his specific job, and how to safeguard himself and his co-workers.

Accident Prevention Program

In a well-run lab, just as in a well-ordered society, some rules and regulations are necessary. However, rules and regulations should be kept to a minimum especially in research laboratory environments and should be the ones management truly believes are essential and, accordingly, is prepared to enforce. Pertinent rules should be discussed and penalties for infractions explained. The following are examples:

1. Horseplay is strictly prohibited. Avoid pushing or poking fellow workers, unnecessary running, throwing objects, and all similar acts of so-called horseplay.
2. Smoke only in approved areas. (Explain the general system to control this source of ignition and source of housekeeping problems.)
3. Specialized equipment is to be operated only by trained and authorized personnel.
4. Good housekeeping is vital. (Explain general rules and standards.)
5. Follow departmental safety rules and regulations. (Discuss briefly, and mention that these will be distributed and explained by departmental supervision.)
6. Observe and obey safety signs and devices. Traffic lines, danger signs, flashing lights, alarms, and the like are provided only after careful consideration, and must be obeyed.

Certain *formalized* systems and procedures must be established and maintained to insure the safety of personnel and property. Brief details of these, with suitable illustrative slides, should be offered. Instruction and specific details should be given later, as needed. These systems include the following: (1) tank-entering procedure; (2) welding and burning permit system; (3) lock-out and tag-out system; (4) routine sampling, as for air contamination.

Although the prime objective is to prevent accidents, it is only sensible to use personal protective equipment at times as a second line of defense to prevent *injury* should an accident occur. In some cases, personal protective equipment is necessary because known hazards cannot be eliminated, and in other cases it is prudent where research or experimentation into the unknown is to be conducted.

As personal protective equipment needs may vary widely, only such general programs as eye protection, hard hats, and safety shoes should be discussed in detail at safety orientation meetings.

Other available equipment, such as gloves, aprons, boots, suits, respirators, and air masks, may be mentioned but training in their use, procedure for getting them, and instructions on when they are to be used are all departmental functions and responsibilities. A permanent exhibit of such gear may be maintained or shown on slides.

Slides may be used to stress the highlights of the entire safety program. They should be carefully selected to illustrate the wide range of activities and the effort expended with the safety of the individual in mind. They can illustrate the scope of the engineering effort and give examples of layout, design, special safety equipment, and the safety aspects of equipment. The following are some suggested topics that may be included:

1. Compatibility of chemicals.
2. Safety parameters of reactions and processes.
3. Aisles.
4. Machine guards, such as over gears and nip points.
5. Machine interlocks, such as to turn off electricity if protective doors are opened.
6. Ventilation systems.
7. Explosion-proof and vapor-proof electric equipment.
8. Three-wire polarized 110-volt electrical system with instruments and equipment grounded.
9. Solvent safety supply and disposal cans.

To provide training and instruction for safety, discuss such items as the following:

1. Industrial truck training program.
2. Rescue breathing training program.
3. Proper use of fire extinguishers.
4. Proper use of respiratory protective equipment.

To provide an organization for carrying on certain types of safety activities, including education, describe the organization and value of the following examples:

1. Safety council or policy safety committee.
2. Departmental committee.
3. Safety committeemen.
4. Inspection systems—safety and housekeeping—looking for both unsafe conditions and unsafe acts.

To provide for general promotion of safety, offer the following:

1. Safety slogan contests.
2. Safety calendars.
3. Safety posters.
4. Mailing of off-the-job material.

If an Accident Occurs

All personnel must be impressed with the need for reporting promptly all industrial injuries and sudden illnesses, no matter how minor, to insure proper treatment at the dispensary and to permit the supervisor to investigate. The accident investigation system should be explained briefly.

Outline the organization and function of the first aid crews or the system for getting qualified aid for seriously injured persons.

For chemical splashes and clothing fires, emphasize the need for immediate flooding with much water. Safety showers should be used; show picture of eye wash fountain also and give usual locations.

Discuss the theory of fire, demonstrate the use of fire extinguishers or have appropriate slides, and review the procedure for turning in a fire alarm.

Departmental Supervision

The detailed safety training for a specific job is ultimately the responsibility of the immediate supervisor. No one else can or should do it for him. To some supervisors, this comes easily and naturally. To others, a reminder might be helpful.

A control check sheet (see Fig. 2.1) is made out for each employee and sent to the department head or other responsible supervisor together with a brief reminder of the program. This is done for each newly hired employee, and may well be used for recalled or transferred employees.

Rather than assuming that each new member of the team gets off on the right foot in safety, set up a procedure for checking the effectiveness of the total orientation program. This might be done one month after employment. For impartiality, objectivity, and uniformity, one person—preferably the safety engineer—should do the job.

One approach is to give the person both a verbal and a written quiz, for often an informal chat doesn't reveal the level of his safety awareness.

The verbal part can be used for demonstration—for example, the proper use of fire extinguishers or respiratory protective equipment, or the proper method of lifting. A measure of safety stimulation for established co-workers can be achieved by having them either point out errors or agree on the correctness of the demonstration.

The written part can vary between questions requiring detailed answers or simply true-or-false. Typical of the first category is, "What must be done before entering a tank or other enclosed space?"

Although the proper answer to some true-or-false questions can become a controversial subject, the response by new employees can sometimes reveal the reason for the differences between theory and practice. For example, if horseplay on company property is forbidden, imagine the answer "True" to the statement, "Horseplay is permissible at lunchtime." Investigation might show obvious misunderstanding. More probably, the newcomer may have candidly accepted a departmental heritage, where, over the years, company rules have been interpreted to suit individual desires.

Important indirect benefits can be realized from a formalized safety

ACCIDENT AND LOSS PREVENTION

RESEARCH & DEVELOPMENT

SAFETY INSTRUCTIONS FOR NEW PERSONNEL

Section Leader or Group Head...... Date

Immediate Supervisor............. Date

Safety Coordinator............... Date   Name of Employee

Administrative Office............ Date

                                               Section or Group

(Please Initial and Forward)

       In addition to receiving from the Safety Coordinator copies of all safe practices memorandums issued to date, the above new employees should receive appropriate safety instructions from those directly responsible. To insure that this new member of Research and Development is fully acquainted with, (1) Research and Development safety policy and safety program, and (2) major hazards, together with precautionary measures, involved in the assigned area, please discuss the following applicable points as soon as employment commences, and check those discussed.

SECTION LEADER OR GROUP HEAD:

( ) 1. Research and Development safety policy, including brief review of Safety Committee organization. Responsibility self-protection and safeguarding others.

( ) 2. Necessity of reporting all occupational injuries, no matter how minor, to the Dispensary.

( ) 3. Necessity for thoroughly investigating hazards of all proposed reactions and operations and use of new chemicals, including information as to possible sources of precautionary information.

( ) 4. Eye Protection Program and necessity for wearing safety glasses at all times while in operational areas.

IMMEDIATE SUPERVISOR:

( ) 1. Correct method of operating specialized equipment and facilities.

( ) 2. Warning as to hazardous chemicals and reactions presently in the immediate area.

( ) 3. Location of emergency equipment and how to use it; e.g., respiratory protection, fire extinguishers, safety showers, etc.

( ) 4. Correct procedure for turning in fire alarm. Emphasize necessity for directing fire department personnel to exact location of fire. Review emergency evacuation procedure -- monitor system, etc.

FMC CORPORATION
AMERICAN VISCOSE DIVISION

Fig. 2-1. An example of a form giving safety instructions for new personnel.

orientation system since it includes a review of the overall safety program, and this in itself offers an opportunity for evaluation and improvement. Likewise, introduction of the plan to supervisors results in stimulation of safety supervision.

Bibliography

Am. Natl. Stds. Inst., Std. A 11.1, Industrial Lighting, 1965; Std. A 23.1, School
   Lighting, 1962; Std. Z 9.2, Design and Operation of Local Exhaust Systems, 1960.
Barrett, J. C., Laboratory Hoods Safety Maintenance, Jan. 1960.
Industrial Ventilation, Am. Conf. Govt. Ind. Hygienists.
Life Safety Code, 1966, Natl. Fire Protection Assoc.
National Building Code, Am. Insurance Assoc., 110 Williams St. New York, N.Y.

3

ORGANIZATION OF A LABORATORY SAFETY PROGRAM

Organization for safety in the chemical laboratory must establish a positive and sustained system of accident prevention in all operations of the facility. This can be accomplished only through a well-organized, well-defined approach to each facet of a prevention program. Of prime importance is the assignment of safety and loss-prevention responsibility to those having permanent functional responsibilities for administration and operation of the laboratory. Other factors affecting organization of a program are the permanency or transitory nature of the laboratory workers and the type of work or service performed by the laboratoy.

Certain elements necessary to a basic safety and loss-prevention program in the laboratory have been developed over a number of years. These elements range from administration, provision of a safe laboratory environment, education and training, and handling and storage of chemical materials, to emergency procedures.

In all school laboratories, effective safety-control responsibility must rest in the department chairman and his academic staff or in the laboratory director and his operational staff. The laboratory student or the laboratory worker cannot—nor should he be expected to—provide the coordinated effort and continuity necessary to educate other students or employees in the skills of accident prevention in the laboratory. It is in the faculty or functional management that formal organizational responsibilities are judged to be most important.

Direct control responsibilities by the faculty or management are particularly important where laboratory personnel, such as students, are transient. Safety meetings supervised by those having line or teaching authority offer the most effective means of providing pertinent instruction in accident prevention. Student participation is obviously lessened by this method, but it does give the specific instruction necessary to control unsafe behavior. In instances in which laboratory personnel are more permanent, safety meetings may be supplemented by further involvement in the program. Participation can take the form of committee assignments, self-inspection of work areas, project-hazard surveys, and development of safer techniques and standards for the laboratory.

Where the service performed by the laboratory is primarily one of instruction and demonstration, strong emphasis should be placed on pre-planning of exact safe procedures to be followed in experimental or project work. Reactants, toxicities, flammabilities, as well as other hazards should be considered by qualified technical personnel before release for routine experimentation by students or control laboratory personnel. Such considerations may also include a review of accident case histories, literature search, and gram-scale safety trials.

Special organizational features are necessary to control unusual hazards faced in research and development laboratories. Checks and double checks may be necessary to test safety devices; limits on quantities of reactants, calculations of energy potentials, and consideration of temperature and pressure extremes are some of the safety check systems to be established. These systems are intended to assure the success of the project without costly injuries and unnecessary delays caused by serious incidents. This philosophy is aptly expressed on the cover of every Manufacturing Chemists Association Chemical Safety Data Sheet: "Chemicals in any form can be safely stored, handled or used if the physical, chemical and hazardous properties are fully understood and the necessary precautions, including the use of proper safeguards and personal protective equipment, are observed."

There is some tendency to relax tight safety organization controls in research and development laboratories, particularly those controls which insist on a routine or methodical approach to accident prevention. Perhaps this tendency results from a certain attitude of independence on the part of some highly skilled technical people. Unfortunately, there is no defense for such a relaxation. It is well known that most laboratory accidents are caused by failure to observe primary safety techniques. Therefore, a basic laboratory accident-prevention program should be maintained but should be supplemented by more sophisticated procedures for handling unusual hazards. Those who insist on complete independence soon learn that no one can afford the luxury of assuming that he can control an incident once it is in the making. Maximum freedom for the researcher is of course a must, but not at the expense of serious danger to himself, his laboratory associates, or the project.

ORGANIZATION

There are certain common factors in the organizational patterns of safety and loss-prevention programs for the various types of laboratories whether they be educational, governmental, or industrial. The following briefly describes the elements considered essential.

Administration Participation in Program

The administration should issue a declaration of its safety and loss-prevention policy. This should be a clear and concise statement formalizing the administration's position in the eyes of all laboratory personnel. The following are excerpts from statements of safety policy used by a university and by an industrial chemical plant.

University: It is requested that the Deans and Department Heads be continuously cognizant of the safety needs of their personnel and initiate necessary preventive measures to control safety hazards associated with activities under their direction. It is essential that all supervisors, both academic and civil service, understand and accept this responsibility for the safety of all persons coming into their areas and for the safety of all personnel under their direction. It is further requested that safety be incorporated as an integral part of all courses in which there is a hazard of accidental injury in the classroom, laboratory, or shop.

Industrial Chemical Plant: The prevention of accidents is a part of everyone's job. Management at all levels has a prime responsibility for the safety of all employees working under its supervision and will be expected to conduct operations in a safe manner at all times. Every employee will be expected to follow plant and department safety rules and work to the best of his ability to prevent accidents and injuries to himself and his fellow employees.

Staff and Supervisory Level Participation

Each member of staff and supervision, including the instructor or first-line supervisor, is made responsible for safety and loss-prevention performance in his functional area. This responsibility cannot be delegated. Each laboratory worker is also held responsible for his own safety performance.

Functional Assistance Provided by Safety and Loss-Prevention Specialists

Safety, engineering, and fire-protection personnel should be given recognition and responsibility for providing adequate assistance, viz.,

1. By advising administration/management and staff/supervisors regarding safety, loss-prevention, and occupational health problems.
2. By correlating accident-prevention activities.
3. By promoting the entire program.
4. By providing assistance but not exercising line authority since this can be accomplished only by responsible administrators/managers and their staff/supervisors.

Building and Equipment Safety

Most industrial chemical companies develop guides for the design of their laboratories. Similar material has been prepared also by the Campus Safety Association and the National Safety Council for school laboratories. Provision should always be made for the safe storage of chemicals.

Education and Training

Education of supervisory/faculty personnel is necessary to assist them in carrying out their assigned safety and loss prevention responsibilities.

Objectives of the training program should include development of a positive attitude towards accident prevention. The laboratory instructor or supervisor has more than the usual responsibility of teaching safe experimentation or project work. Like all immediate supervisors, he must convince those in his charge of the importance that he personally attaches to safe operation. A student, or employee, is quick to determine the difference between a good example set by his supervisor or instructor and verbal admonitions.

Education of Students/Employees.   Laboratory personnel are usually of high caliber, fully capable of learning the most advanced techniques of accident prevention in a fairly short time. Programming should start early in the class year or period of employment. All personnel should be taught that accident prevention is of prime importance and that their success as students or employees is not complete until they have accepted and applied safety principles.

Laboratory indoctrination of new employees or students may be achieved through special lectures by supervisory personnel and/or by personal interviews with individuals. Specific project safety instruction should be given by the immediate supervisor or instructor.

Continuing safety educational sessions by means of safety meetings or personal contact systems are also important. One of the more effective educational methods used in the laboratory is the listing of key safe procedure points in a given project or in the use of laboratory equipment, referred to as the "Accident Prevention Key Points System." There should also be a demonstration of "Key Points" by a competent instructor or supervisor.

Records

Accident Investigation.   All accidents, whether or not resulting in injury should be reported. In addition to identifying information, three basic items of information should also be obtained: (1) a description of the accident (including injury, if any); (2) a list of causes of the accident; and (3) a description of measures taken to prevent a recurrence.

Standard methods of recording and measuring injury experience, available from the American National Standards Institute, as well as those of the U.S. Department of Labor are recommended.

Follow-Up After the Accident.   There is a strong tendency to file accident reports and forget them. The only value of an accident report is what can be learned in order to avoid a repetition of a similar incident. This can take the form of a change in procedure requiring education of those doing the operation, or a change in the design of the equipment. A definite system of review of accident cases should be established to insure positive results.

No-Injury Accidents.   Further refinement of record-keeping gradually leads to records of near-miss or close-call incidents in which no injury is involved. Each such incident represents a potential to be reckoned with in a preventive program. (See MCA publications *Case Histories of Accidents in the Chemical Industry,* 3 vols.)

Laboratory Standards and Manuals

The development of laboratory standards or a manual of safe operating procedures is the mark of maturity of a laboratory safety program, and serves the purpose of formalizing the best thinking on hazard-control information. These records must be kept up-to-date and reviewed on a regular basis to be effective. The content will vary considerably according to the service the laboratory renders. Some of the areas covered are the following: (1) listing of all chemicals used in the laboratory together with information on their hazardous properties; (2) instructions on acceptable methods of: storing, handling, and disposal of flammable liquids; radiation hazards, nonionizing and ionizing; ventilation in the laboratory; cylinders and regulators; eye protection, eye baths, and safety showers; storage of some chemicals in refrigerated cabinets; labeling procedures; cryogenics; pressure vessels, autoclaves; glassware; reaction bombs; fire protection—extinguishers, hose lines, and sprinklers; explosion protection; cubicles, shields, and relief vents.

Emergency Procedures

Chemical Exposures by Personnel. Standards directing the immediate use of copious quantities of water on affected part of the body are of the utmost importance.

Spills. Specific instruction should be given on what to do in the event of a spill of flammable, toxic, corrosive, or radioactive materials. When flammable liquids are spilled, immediate action must be taken to discontinue all sources of ignition, to remove concentrations of vapors by increased ventilation, and remove the spilled material. Corrosives may be removed by washing the area with water. Other contaminants, such as radioactive materials, require special decontamination procedures. For detailed recommendations on the handling of spills see Appendix 4.

Fires. Considerable judgment must be exercised in the event of a fire. Evacuation of personnel and calling the fire department should be the first order of procedure. Local control of the fire by extinguishers may then be attempted if the fire has not gained much headway.

Standard Alarm System. Standard signal devices and standard signals are advisable in laboratories in which the work may result in danger to personnel. Such signals must be audible in all areas where workers are situated, and must be recognizable and communicate their meaning.

Trial Runs of the Emergency Plan. Hypothetical problems or trial runs of the laboratory emergency plan are necessary to maintain a satisfactory performance of personnel under emergency conditions. Emergency plans rapidly deteriorate and become woefully obsolete unless trial runs are scheduled regularly. Most laboratories consider semiannual trial runs as a minimum requirement.

Inspection System: Hazards and Housekeeping

Visual, on-the-spot, inspection of laboratory facilities has proved a useful and necessary tool in the laboratory safety program. The objectives of inspection are to seek out and record hazards, both physical and operational, and to provide a positive means of follow-up and correction. Inspections are part of the proper administration of the laboratory in assuring hazard recognition, correction, and a clean laboratory. Supplemental inspections serve to provide an awareness of individual responsibilities in spotting hazards and keeping all areas clean.

Personal Protective Equipment

Adequate stocks of personal protective equipment should be maintained in locations easily accessible to all laboratory personnel. Some of the essential types of protective apparel that should be provided are:

eye protection (safety glasses and cover shields, louvered and un-louvered)
head protection (hoods and hard hats)
hand protection (gloves)
body protection (impervious clothing, aprons, and sleeve gauntlets)
respiratory protection (respirators and gas masks, cannister and air-supplied)
face protection (hoods and face shields)

Voluntary wearing of protective apparel "as needed" rarely proves adequate. A realistic preventive program should set forth standards for the wearing of protective equipment. Other types of protective equipment may include splash and explosion shields, blowers, exhausters, and other devices to prevent exposures to personnel.

Medical and Health Program

First aid treatment should be considered only as a stopgap, a means of emergency procedure to be followed until the physician arrives or until the injured person can be taken to a professional facility. Laboratory personnel selected to give first aid should be given special training by a competent physician having a clear understanding of the materials used in the laboratory. Work with extremely toxic materials may also require a special program of periodic physical examinations of laboratory personnel coupled with a monitoring of exposure atmospheres. In these circumstances, it is obvious that a close relationship should exist between the physician and the industrial hygienist to maintain a healthy work environment and to make frequent checks on the workers' health.

Fire Protection and Control Facilities

Laboratory fire protection and control equipment and facilities should be designed to prevent injuries to personnel and damage to the laboratory. Water, carbon dioxide, and dry-powder extinguishers, smothering agents, hose lines, and automatic sprinkler systems offer a wide range of protective devices. Special hazards inherent in flammable liquids and metal fires may call for special control applications. Laboratory equipment easily damaged by water may also present special problems best solved by a fire protection engineer.

A laboratory safety program founded on proved principles of accident prevention and fitted to the particular needs of the laboratory will bring about true accident prevention, which is essential to the successful operation of the laboratory.

Bibliography

"American Standard Method of Recording and Measuring Work Injury Experience", Std. Z 16.1, revised 1969, Am. Natl. Stds. Inst.
Breysse, Peter A., "A University Program for Laboratory Management and Safety Program", *Occupational Health Newsletter* **14** (April and May 1965), Univ. of Washington, Seattle, Wash.
Chemical Safety Data Sheets, MCA Inc.
Fawcett, H. H., and Wood, W. S., *Safety and Accident Prevention in Chemical Operations,* John Wiley & Sons, Inc., New York, 1965.
Heinrich, H. W., *Industrial Accident Prevention,* McGraw-Hill Book Co., Inc., New York, 1950.
Manual of Laboratory Safety, Fisher Scientific Co., 1969.
Steere, N. V., "Safety in the Chemical Laboratory," *J. Chem. Educ.* **41** (Jan. 1964).

4

SAFETY IN THE INSTRUCTIONAL LABORATORY: HIGH SCHOOL AND FIRST-YEAR COLLEGE

The teacher in a chemistry laboratory has a responsibility and often neglected opportunity to teach students how to handle chemicals and bring about reactions in a safe manner. This responsibility is largely ethical and humanitarian since the teacher is genuinely concerned about the well-being of the students in his charge. An injury, particularly if it is permanent, would be a reflection upon the teacher and his school since publicity relating to such an incident may be slanted to the teacher's disadvantage. One cannnot overlook the possibility of a suit against the teacher and other responsible persons in the school organization, particularly if there is any possibility of negligence. The law differs in various states, and the liability in such accidental injury should be discussed with the school's attorneys.

Laboratory manuals and chemistry texts are not noted for giving safety information. Some manuals do contain precautions where needed; but the teacher should review all experiments from this particular point of view in order to make sure that he gives the necessary safety instructions in advance. Another compelling reason for including safety instruction in chemistry teaching is that students continuing in a chemical career will be much more valuable employees in industry if they have had such instruction. Industrial safety standards are generally much higher than an average chemistry graduate has experienced during his training. The teacher should impress his students with the long-range benefits of learning safety in their work. The concept that safety is an integral part of every activity should be impressed upon them as early as possible.

Teachers in high school or first-year college chemistry laboratories have even more safety problems than do supervisors in research and control laboratories, and therefore even greater reasons for giving priority to safety promotion. Beginning chemistry students are largely unfamiliar with the potentials for injury which surround them in the laboratory. They are youthful, probably more impetuous than research students and plant employees, and more likely to indulge in playful experimentation with unfamiliar chemicals and equipment. And worse, the students are present in considerable numbers—often thirty or forty in a class.

Add to all this the possibility that the laboratory may not have been designed with student safety in mind. And also the frustrating circumstance not unknown in public schools, whereby school administrators, plagued by financial insufficiencies in other areas, have not recognized that the chemistry laboratory is a high-hazard, high-liability area, and have not given any significant support or leadership to a safety program. Such are the circumstances under which many beginning chemistry teachers must ply their profession.

Most accidents in the beginning chemistry laboratory result from the careless handling of glass or from exposure to acid or other caustic materials Serious incidents can occur, and the teacher should anticipate all the hazardous situations which might arise.

The following examples, taken from actual case histories, illustrate what can occur in a school laboratory when students either are not fully warned or fail to heed safety instructions.

Two students were each carrying two bottles of concentrated sulfuric acid down a corridor, holding the bottles by the necks. One student turned to talk to the other, and two bottles clanked together and broke. Both students slipped and fell into the broken glass and acid, breaking one more of the bottles. Another person coming to their aid also slipped in the acid and fell. Severe acid burns and lacerations were suffered by all three.

A student touched a match to the exit tube from a hydrogen generator without allowing enough time for air to be exhausted. The generator exploded inflicting serious cuts and burns.

Ignoring warnings to use tiny bits of sodium, a student placed a piece weighing several grams in a beaker of water. The resulting explosion caused an alkaline splash into his face. Failure to wash his face and eyes promptly and adequately nearly cost him his vision.

A student failed to use lubrication and other proper procedure for inserting a glass tubing through a rubber stopper. The tubing shattered, and he suffered severe cuts of the wrist, narrowly averting tendon damage.

A student connected a tightly stoppered Erlenmeyer flask to a vacuum pump. When the pressure was reduced, the flask imploded, scattering pieces of glass for approximately ten feet.

Whatever the cause of these accidents, the teacher must accept responsibility for them. Therefore, he should make plans to avoid their occurrence.

SAFETY PLANNING

The first planning tool the teacher needs is a formulated safety policy by the school administration, which with backing of the Board, should issue a clear, written statement, usually including declarations: that the safety of all concerned shall take precedence over all other considerations in the school activities; that structures shall be so designed and every activity and task so performed that the possibility of accidental injury is reduced to a minimum; that department heads shall make themselves aware of the safety needs of their personnel and shall initiate preventive measures to control safety hazards; that supervisors understand and accept responsibility for the safety of all persons under their direction; that safety be incorporated as an integral part of all courses in which there is a possibility of accidental injury; and that students must also accept the responsibility of following instructions and adopting safe practices established for their protection.

The teacher needs the support of such a formulation if he is to achieve a safe laboratory environment. He must have financial backing to provide minimal protective devices. This responsibility for student safety becomes all the more fixed on the teacher if the administration can demonstrate that it *has* fulfilled its policy of incorporating safety features in its laboratory structure and *has* furnished the equipment and facilities necessary to provide a safe environment.

The burden then rests with the teacher to train his students in their personal responsibilities to avoid accidents. So routine and persistent must be his insistence on safety conformity that safe laboratory practices ultimately become habitual. All laboratory work must be closely supervised, attention must be drawn to noncompliance with safety requirements, and determined disciplinary action must be taken against those who willfully and deliberately violate these requirements. In this, as well as in other phases of his responsibilities, the teacher must have the full support of his superiors; otherwise, his supervisory control of the laboratory will be lost. He should also have the assurance that the administration will take prompt action to correct unsafe conditions which are brought to its attention.

In planning his courses, the teacher should select the safest of chemicals and of laboratory techniques whenever such a choice exists in his laboratory work.

The demonstration lesson provides him with his best opportunity to illustrate the proper use of safety equipment and practices. He must wear the recommended protective equipment and follow safe operating procedures, laying proper stress on safety whenever the activity or discussion warrants.

Some teachers require students to include safety information in laboratory written work. This is a good idea; often a single paragraph will suffice to state what safety equipment was used and what safety practices followed.

Procedure for fulfilling a teacher's responsibilities for accident prevention follows a pattern similar to that adopted for most of his other teaching activities. He must know what the hazards are and must impart this knowledge, teach safe practices, supervise student activities, and, finally, correct unsafe situations and practices.

Eye protection is an absolute requirement in all laboratories at all times. Students need protection from their own mistakes as well as from those across the bench.

It is strongly recommended that regular inspections be made in order to reveal unsafe conditions and that the teacher report, in writing, to the school administration any unsafe conditions which he himself cannot correct. Student activities in a laboratory area containing unsafe equipment should be suspended until the unsafe condition has been corrected.

TEACHING LABORATORY SAFETY

When developing a laboratory course, teachers should check that the techniques and methods to be taught are based on sound principles of safety. They should take every opportunity to accentuate safety by explaining hazards and means of avoiding them. Experience has revealed many experiments or laboratory operations which involve dangers. Usually safe procedures have been developed. Teachers must thoroughly acquaint themselves with these safe procedures and, having brought them to the students' attention, insist on their adherence. In addition to specific procedures there are certain general aspects of laboratory routine which

Fig. 4-1. Testing the emergency evacuation alarm system.

contribute markedly to the avoidance of injury. The need for bench neatness and the avoidance of unauthorized experiments are typical examples.

The following partial check list should be expanded as necessary:

Emergency evacuation alarm system: Is the equipment in good working order (Fig. 4-1)? How often is it tested? Is an evacuation drill held twice a year?

In the event of fire or release of hazardous fumes or gases: What plans have been developed for speedy contact with professional fire fighters and police? Will air conditioning fans be shut off by the automatic or manual alarm signals? Are fume hoods and local ventilation equipment provided with manual shut-off devices?

Can all exit doors be readily opened in an emergency? Are the fire extinguishers clearly marked, easily accessible, and in good operating condition (Fig. 4-2)? Are they the correct type for the hazard? Are they inspected at least annually? Who is designated to use them? Are these persons trained? Are self-contained breathing masks available for emergency rescue? Are they located outside exposed areas so that they will be accessible when needed? Are key persons trained in their use?

Safety shower and eye wash facilities: Are these adequate and in good

Fig. 4-2. Are the fire extinguishers in good operating condition?

operating condition? Is there one of each in every laboratory? Are they tested at least annually (Fig. 4-3)?

Laboratory hoods: Are they operating? Have devices been installed to indicate that the exhaust fan motor has been turned on and that the fan is drawing air? Are the hoods equipped with safety glass sash? What is their open face velocity (50 lineal feet per minute minimum for nuisance vapors, 100 feet per minute for toxic gases or vapors). Is this flow checked periodically by maintenance engineers?

Storage: Is flammable liquid storage and dispensing done in a separate room? Are wires and the like provided for bonding metal containers during transfers? Is excessive storage prevented by purchasing containers of the smallest practicable size? Are all containers and materials properly labeled? Are highly reactive materials adequately separated? Are highly toxic materials, such as hydrogen sulfide, cyanides, phosgene, identified, stored, and handled so that no one will be exposed? Do you practice and insist on good housekeeping in the storage area? Do you use Underwriters' Laboratories approved fire-resistive storage cabinets and safety containers for storage of flammables within the laboratory? Have you

Fig. 4-3. Safety showers should be tested at least annually.

made provision for the collection and disposal of flammable liquid and hazardous chemical waste? Are compressed gas cylinders of a flammable or toxic nature stored outside the building in a secure manner and protected from the weather (Fig 4–4)? Are potentially explosive chemicals, such as organic peroxides, stored in separate areas away from flammable liquids?

Miscellaneous: Are hand-washing facilities available? Are the double benches (where persons work opposite each other) provided with safety shielding when necessary? Is all electrical equipment grounded? Are switches for shutting off electricity and gas available *outside* the laboratory? Are they marked (Fig. 4-5)?

Protective equipment: Are safety glasses worn in all areas of the laboratory? Are chemical goggles worn when handling caustic chemicals (Fig.

Fig. 4-4. Safe storage of compressed gas cylinders.

4-6)? Are dust respirators worn in addition to safety glasses or goggles when pulverizing hazardous or irritating powder? Are rubber aprons, face shields, and rubber gloves required when working with acids, alkalis, and other corrosives? Are safety carriers used for glass bottles? Are approved gas masks and self-contained breathing equipment readily available (Fig. 4-7)? When handling heated toxic flammable, or corrosive, materials in the fume hood, are the hood windows lowered? Is the blower running?

Housekeeping: Is everyone made responsible for keeping his own working area orderly and neat? Is there a storage place for everything, and is everything stored in its place? Are equipment and apparatus kept clean? Are cans available for disposal of broken glassware (Fig. 4-8)?

Spills and breakage: Clean up all spilled material promptly. Neutralize spilled caustic solutions with an acetic or hydrochloric acid solution.

Fig. 4-5. Marked electrical switches should be located outside the laboratory.

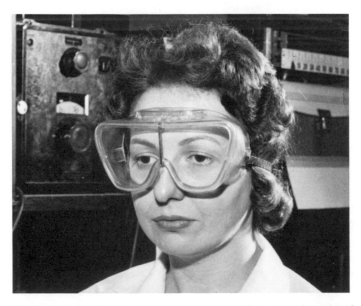

Fig. 4-6. Proper goggles to be worn when handling corrosive chemicals.

Fig. 4-7. Approved gas masks should be readily available.

Neutralize acids by sprinkling sodium bicarbonate on the spill area (Fig. 4-9). If the spilled material is flammable, shut off all electrical heating systems and extinguish flames. When cleaning up a spill of toxic material, wear self-contained breathing equipment or the proper canister mask and rubber gloves. When feasible, mop the spill area with water. For mercury, pick up all possible droplets with a suction tube and aspirator bottle. Redistill or return to vendor. Use a dust pan and brush to clean up broken glass.

Glassware: To insert glass tubing into a stopper or cork, first fire-polish the ends of the tubing. When boring rubber stoppers, lubricate the borer and stopper with water or glycerol to a depth of about ¼ in. Bore a hole large enough to accommodate the tube, starting at the narrow end of the stopper and using a block of wood as a backstop for the borer. Don't force the borer; twist it. Lubricate the glass. Protect the hands with gloves or a heavy cloth. Hold the stopper between the thumb and the forefinger with the palm parallel to the direction of force. Hold the tubing close to the end to be inserted. Insert with a twisting motion.

Avoid handling large open glass vessels when they are full of liquid chemicals. Use beaker tongs to move beakers of hot liquid (Fig. 4-10).

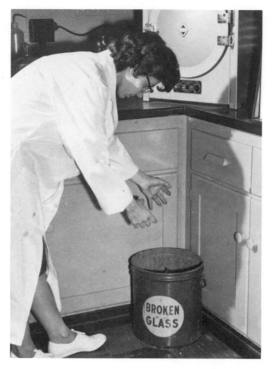

Fig. 4-8. Cans should be available for disposal of broken glassware.

Fig. 4-9. Sprinkling sodium bicarbonate on acid spills to neutralize the acid.

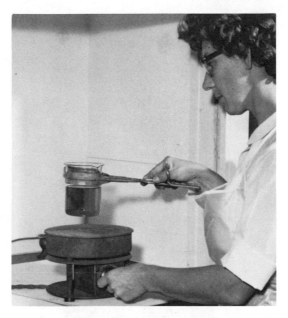

Fig. 4-10. Use beaker tongs to move beakers of hot liquids.

Store large or heavy glass containers on lower shelves. Keep all glass-ware at the rear of the shelves.

Miscellaneous procedures: When diluting acid, pour acid into water, *not* water into acid. To smell the gas content of a test tube, do so by waving some of the escaping gas toward the nose with the hand.

Use a pouring rod when pouring liquid reagents. Pipette by use of an aspirator bulb, *not* by mouth (Fig. 4-11). Before proceeding with a labora-tory experiment, find out the hazards of each of the chemicals to be used or prepared. Are any of them flammable, toxic, corrosive, or unstable? How does each react with water, acid, or alkali? Are any of them self-reactive, when contaminated, or when exposed to high temperature or shock?

Store flammable liquids in UL or FM approved safety cans. Label all containers with date and name of the chemical, and with hazard warn-ings and precautionary measures when applicable.

Gas cylinders are to be stored, moved, and used in accordance with procedures discussed in Chapter 20.

Gears, shafts, belts, and the like must be covered with suitable guards to prevent injury upon contact (Fig. 4-12). When handling fuming acid, wear gloves, apron, and face shield and provide adequate ventilation. When mixing liquids in a confined system, be alert to the possibility of

Fig. 4-11. Pipet by use of an aspirator bulb, not by mouth.

Fig. 4-12. Gears, shafts, and belts must be suitably guarded.

the mixture spurting from the vessel because of heat of solution or heat of reaction. A table shield or fume hood should be employed.

TOXIC AND CORROSIVE MATERIALS

The golden rule is to avoid direct contact with any chemicals and to avoid breathing vapors, particularly solvent vapors. Beware also of the increased vapor concentrations which develop when toxic materials come in contact with hot liquids; for example, hot water used for washing equipment.

The following are examples of common laboratory materials which emit toxic vapors or which may be absorbed through the skin: benzene, allyl alcohol, cresols, aniline, dimethyl sulfate, phenol, phenyl hydrazine, tetrachloroethane, acrylonitrile, carbon disulfide, carbon tetrachloride, hydrogen cyanide, methyl bromide.

In case of contact of such materials with the skin, immediately flood the affected parts with water and then wash with soap and water.

Before starting an experiment involving a toxic gas such as hydrogen sulfide, chlorine, or bromine, obtain the approved respiratory equipment and keep it handy. Review its use with the students.

With cyanides, make sure a cyanide antidote package is available. These first aid kits, prepared in the United States only by Eli Lilly and Company, Indianapolis, Indiana, are obtainable on medical prescription from many drug houses. They contain amyl nitrite pearls, which should not be used near any source of ignition, such as an open flame or a cigarette.

When you are using cyanides, it is wise to warn other occupants of the laboratory that you are doing so. These substances must always be used with adequate ventilation, such as in a fume hood.

RADIOISOTOPES

The Atomic Energy Commission and various state health departments have regulations governing the purchase, storage, use, and disposal of these materials. Anyone planning work involving radiation sources or radioisotopes must be properly informed and licensed. (See Chapter 23 for further discussion.)

PEROXIDES

Explosions have resulted from the formation of peroxides in ethers and other susceptible compounds. Among these are diethyl ether, isopropyl ether, dioxane, tetahydrofuran, glyme, and diglyme. Long-term storage, exposure to air, or any laboratory procedure that involves substantial contact with air may cause peroxide buildup. Such materials should be tested for peroxide content. For details, see MCA Chemical Safety Data Sheet SD-29, Ethyl Ether. Note that vinyl ether does not give a positive indication with the potassium iodide test.

EXPLOSIVE MATERIALS

The following are among the compounds which are potentially explosive; azo compounds, acetylides, compounds containing several nitro groups, peracids.

Before isolating pure compounds of the above types, examine the behavior of a small quantity when heated or upon impact. Such tests must be made by experienced persons utilizing effective shielding.

OPENING BOTTLES

Certain compounds—for example, acid chlorides, halides of aluminum, phosphorus, sulfur, and tin—evolve hydrogen chloride or hydrogen bromide when stored, and pressure may develop. When opening bottles of these materials or of bromine, nitric acid, or concentrated ammonia, protect the hands and wear a face mask; it is best to operate behind a screen in a laboratory hood. Take every precaution when handling any material which is aged or appears to have decomposed, or when the bottle stopper is stuck. An MCA accident case history records a fatality resulting from an explosion which occurred when a professor forcefully removed a stopper from a bottle of isopropyl ether in which peroxides had been generated.

DISPOSAL OF WASTE

Before putting aside for collection or washing, rinse out any containers which have held concentrated acids, alkalis, or any other toxic or corrosive materials.

Small amounts of water-miscible solvent may be flushed down a sink with cold water. All other solvents must be put into a clearly labeled UL-approved waste container.

Strong acids and alkalis must be carefully added to a large excess of water before flushing down a sink.

Malodorous materials should, if possible, be destroyed in the laboratory hood. Otherwise, they should be stored for special disposal.

For detailed information on the safe disposal of more than 1100 specific chemicals, see Appendix 4.

FIRST AID

Since accidents involving students occur even in the best-run laboratory, the instructor should be prepared to render immediate assistance in case of injury to a student or to anyone else in the laboratory. Chapter 29, on first aid, in this book gives reasonably complete guidance for such action. However, it is strongly recommended that the instructor receive first aid training, equivalent to the standard first aid course of the American Red Cross Society. He should be particularly informed as to first aid in case of bleeding, asphyxiation, or poisoning, since any of these three occurrences can result in loss of life unless immediate action is taken.

Hemorrhaging can usually be controlled by direct pressure on the wound, using a gauze compress, handkerchief, or other fabric. If the cut is so severe that bleeding cannot be controlled by this means, it will be necessary to apply digital pressure or, in extreme cases, a tourniquet. (Details on first aid will be found in Chapter 29.)

If a student or other person is asphyxiated, the instructor's first problem is to remove him from the contaminated area without himself becoming involved. This will usually mean wearing respiratory protection, preferably of the self-contained type. If the victim has stopped breathing, artificial respiration, preferably by the mouth-to-mouth method, also described in Chapter 29, must be started at the earliest possible moment. The assistance of fire or rescue personnel should be obtained to carry on resuscitation as long as is necessary. Medical help must also be obtained in all cases. A drill carrying out a simulated rescue will show what changes in procedure are needed.

In case of poisoning, liquids should be given immediately if the victim is conscious. Water or milk is excellent. Time should not be wasted searching for an antidote. Usually vomiting should be induced in order to remove the dilute poison from the stomach. If the victim lapses into unconsciousness, every effort should be made to place him in medical care at the earliest possible moment. It is important to identify the poison since this information can help the physician render effective treatment.

Lesser injuries, such as minor cuts, punctures, burns, and skin irritations, should be treated at once. The instructor should be alert to the possibility of infection or other complications. In case of fracture or suspected fracture, a doctor or ambulance should be summoned and the victim should not be moved except by trained experts or under medical guidance.

Complete records of all injuries or illnesses attributable to laboratory work must be collected and retained in the school office. Investigation of all accidents to determine cause factors will help prevent future accidents.

Bibliography

Accident Prevention for School Shops and Laboratories, National Safety Council, NSC Stock No. 421.52, 1969.

Quam, G. N., *Safety Manual for Chemical Laboratories,* Villanova Press, Villanova, Pa., 1961.

Safety in Schools, The Science Teacher, American Chemical Society, Committee on Chemical Safety, Sept. 1966.

Sax, N. I., *Dangerous Properties of Industrial Materials,* 3rd ed., Van Nostrand Reinhold Co., New York, 1968.

Steere, N. V., "Safety in the Chemical Laboratory," *J. Chem. Educ.* (continuing series).

5

BUILDING DESIGN AND FURNISHINGS

GENERAL CONSTRUCTION

The laboratory building, including walls, roof, drop ceiling if used, floors, and floor coverings, should be constructed entirely of noncombustible materials. A one-story building is preferred, since hood ventilation, gas services, accessibility for supplies, and waste removal are considerably enhanced. This, however, cannot always be accomplished, owing to lack of ground space. If a multi-story building is necessary, additional problems must be surmounted in order to provide proper fire protection and safety to personnel.

All buildings should have at least two standard exits for each floor as required by applicable building codes. Most state and city codes are based on the NFPA Building Exits Code. Each laboratory room should have two exits, one of which may be a connecting door between adjacent rooms where the individual rooms are of average size, or about 20 ft. × 20 ft. The main exit should be to the outside or into a corridor leading to the outside or to an approved stairway. Large rooms and those in which appreciable quantities of flammable or toxic materials are handled should have two or more exits leading directly to exit corridors or outdoors. Corridors should be wide enough to allow free passage, with individual room doors swinging out into line of travel (Fig. 5–1). Sometimes this is

Fig. 5-1. Corridors should be kept clear.

best accomplished by having door alcoves recessed in the rooms, with double swinging doors.

Corridors should not be used for the storage of any equipment except emergency devices such as fire extinguishers and breathing apparatus. These should be placed in clearly marked wall insets or cabinets. Clothing lockers are sometimes built into divider walls, if they do not obstruct passage and do not interfere with wall integrity.

Elevators will be required in large multistory laboratory buildings. These are usually the automatic type in modern buildings, and must meet applicable codes for safety and be cut off by approved masonary shafts. Laboratory quantities of chemicals, including flammable liquids, can be carried on the elevators if they are handled in suitable containers. If only a dumbwaiter is installed for handling chemicals to and from upper floors, it must also be constructed to meet applicable safety and fire-protection codes.

Stairway, ramps, and floors should be built of or covered with anti-slip material. Stair and ramp construction is regulated by several codes, usually the same as the NFPA Building Exits Code, which specifies width, slope, and construction details that should be followed. Corridors and individual laboratory floors are usually covered with plastic tiles of a composition resistant to most chemicals. The laboratory should be mopped daily.

LIGHTING

Modern construction practically precludes the use of large window areas, and many laboratory rooms are not located adjacent to outside walls, this space frequently being reserved for individual offices. It is therefore essential to provide adequate artificial lighting for all laboratory areas. Good general lighting is necessary in all corridors and individual laboratory rooms. Five to 10-candle intensity is recommended for corridors, and 50 foot-candles minimum for working areas. Fluorescent ceiling lights of the recessed type are preferred, with vapor- and dust-tight covers as best suited for safety and maintenance. Although special spot lighting may be required for some close-up work and instrument reading, general lighting should be such as to preclude the need for extension cords and drop lights. See ANSI Standards A11.1 and A23.1.

VENTILATION

General building ventilation with conditioned air will be required in most modern laboratory buildings. Such systems should be designed by qualified heating and ventilating engineers. The corridors, meeting rooms, and other areas where chemicals are not being handled should be furnished fresh air in sufficient quantity so that these areas are at a slightly positive pressure in relation to the laboratory rooms. Although recirculation of some air in office and similar areas should be satisfactory, corridors or rooms connecting directly to laboratory rooms should be supplied with once-through air only. Discharge from such areas may be into laboratory rooms.

The individual laboratory rooms must be supplied with sufficient air input to provide proper movement of air, fumes, vapors, and the like through the hood exhausts. There must be no recirculation of air, which puts a heavy load on air conditioning systems in hot climates. To compensate, auxiliary unconditioned air is usually introduced near, but not inside, the hood in sufficient quantity to keep the room at slightly negative pressure in relation to the corridor, and at the same time fresh conditioned air is added for normal control of the room. This is a design problem requiring the services of experienced air-conditioning engineers.

HOODS

Ventilated hoods are required in all laboratory rooms handling chemicals of any type, in order to remove toxic and flammable vapors and gases and to provide a safe place in which to perform experiments or routine analyses. Standard hoods, easily obtainable from several suppliers are preferred. Rooms should be sized around standard modules of hoods, benches, and laboratory furniture.

The proper placement and number of hoods in a room are important. These should be located so that exits from the room are not unduly exposed in case of fire or explosion within the hood, and should not face toward other hoods across relatively narrow rooms or aisles, so as not to interfere with proper movement of air from the room and through the hoods. Most modern laboratories are designed with only a few employees and relatively few hoods in each room. Large rooms usually promote confusion, are difficult to ventilate properly, and increase the exposure of individual employees to chemical hazards.

Fig. 5-2. Example of a hood with horizontal sliding doors.

Hoods should be ventilated by individual exhaust fans located on the roof with ducts leading as directly as possible from the hood to the roof. The duct should have a continuing slope from hood to discharge point, with no pockets in either duct or fan where solids or liquids can accumulate.

Although there is considerable difference of opinion regarding the ideal rate of hood air velocity, it is generally agreed that the minimum rate for ordinary laboratory work should be 50 lineal feet per minute at the face with all doors wide open. This should be increased to 150 feet or more for extremely toxic vapors, bearing in mind that the higher rates may blow out gas flames, requiring other means of heating. See ANSI Standard Z9.2 for ventilation details.

Hood doors may be either vertical or horizontal sliding. The horizontal sliding type (Figs. 5-2 and 5-3) are easier to maintain, provide more protection for the user, and will reduce the required exhaust volume by one-third for three-panel doors and one-half for two-panel doors. Vision openings in doors or entire doors should be made of safety glass (laminated) or heavy plastic (¼ in. minimum thickness) such as methyl methacrylate polymer or polycarbonate.

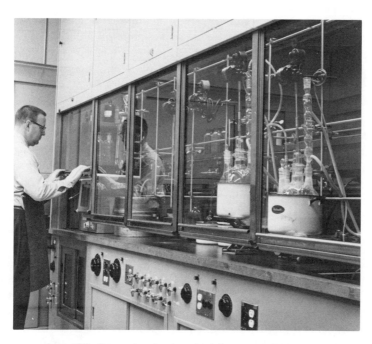

Fig. 5-3. Example of a hood with vertical sliding doors.

LABORATORY BENCHES

Conventional types of laboratory benches may be purchased from several sources. These should be arranged so that exit from any spot in the room may be accomplished in two directions; that is, there should be no built-in blind alleys. Back-to-back and wall benches are used, the arrangement and number depending on room size. For the back-to-back type, many companies have noncombustible separators between the benches to prevent equipment and material from protruding into the opposite work area.

BARRICADES

Shields or barricades, both fixed and portable, are needed for the protection of personnel performing routine or experimental work in which there is danger of explosion, rupture of containing apparatus and systems from overpressure, implosion due to vacuum, flash fire, or sprays of toxic or corrosive materials. Personal protective equipment should be worn as needed regardless of the protection provided by shields, including hoods.

Since the main hazard is from missiles, flames, and toxic or corrosive sprays, the probable intensity and volume of such occurrences can be controlled to a considerable degree by keeping quantities to the absolute minimum required for the individual experiment or analysis. The additional protection provided by the items described in the following paragraphs are imperative for full protection of personnel handling hazardous chemicals.

Steel barricades (Fig. 5-4) have been demonstrated to give better protection against missiles than do safety glass or plastic, but it is desirable to have direct vision in most instances. Indirect vision can be provided by means of mirrors, periscopes, and other devices when missile protection against metal container parts is required or heavy missiles of ceramics or glass may develop. Tests by several authorities have indicated that steel plate is about four times as effective as methyl methacrylate for shield protection; that is, use ⅛ in. steel for ½ in. methyl methacrylate, ¼ in steel for 1 in methyl methacrylate, and so on. The plastic shields are light in weight and easily shaped, but have the disadvantage of adding flammable material for fuel in a continuing fire. This is, however, relatively unimportant when considering personnel protection, since the plastics normally used are slow-burning and not easily ignited. Laminated glass shields give excellent protection for low-energy shielding, are noncombustible, and do not scratch easily. Ordinary plate or rolled glass, wire glass, and tempered safety glass are not recommended for shield

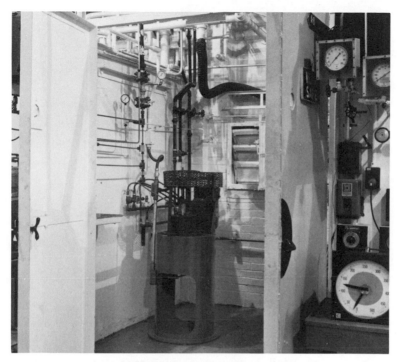

Fig. 5-4. Example of a reactor in a steel barricade.

use for various reasons. Ordinary plate glass should not be considered, because it breaks easily and provides missiles with sharp edges and slivers. Wire glass is undesirable when severe blast effect exists, because, if shattered, the wires may add to the missile damage. Tempered safety glass will shatter to small particles without cutting edges, but has the disadvantage of being easily broken by bending or impact stress if previously surface-scratched or the edge stress is upset by nicks or scratches.

Portable shields for low-order protection should be heavily weighted or, preferably, fastened at the bottom against overturning. Commercial models, usually of plastic, are available and are suitable for many uses.

Special tongs and manipulative tools can be used to avoid or reduce body, arm, and hand exposure when performing hazardous work. Considerable work along this line has been done in connection with radiation and high-energy chemicals.

One interesting safety device is a so-called tote barricade* for trans-

*Details may be obtained from Eastern Laboratory, E.I. du Pont de Nemours & Co., Gibbstown, New Jersey.

Fig. 5-5. Example of a tote barricade.

porting or storing small quantities of high-energy materials that may detonate outside barricade protection (Fig. 5–5). This consists of cast elastomer around glass laboratory apparatus and containers. A one-pint casting around a small glass vessel has contained without external rupture a 2-g nitroglycerine blast; a 1-gal size casting has similarly contained a 15-g nitroglycerine blast except for gases and particles expelled through the top. The casting is transparent and effective at temperatures from 0°F to approximately 212°F.

DESKS

Desks for chemists and other laboratory personnel are frequently placed in the laboratory rather than in a separate room. Care should be used in placing these desks so that personnel using them will not be exposed to toxic fumes, flash fires, or missiles from explosions and so that personnel will not be trapped in an emergency.

EMERGENCY EQUIPMENT

Laboratories handling much flammable materials should be equipped with standard automatic sprinkler systems. Fire extinguishers should be located in or near all laboratory rooms. The usual arrangement is to place one or more small extinguishers of the CO_2 or dry chemical type in each room, with larger units of the proper type placed in the corridors.

Emergency water showers should be available in or near all laboratory rooms. Commercially available deluge-type safety shower heads should be strategically located above an unobstructed floor space. Ball valves or other quick-opening type should be arranged for easy access. Supply lines should be capable of delivering 40 gallons per minute. Some laboratories also use eye wash fountains, which are commercially available.

An emergency alarm system should be installed in areas in which there is any danger of toxic gases or vapors escaping into the work areas. This system can be an elaborate wired one, or may consist of portable gas-pressured devices which emit a loud signal through a small horn. The price of the latter units is reasonable.

SERVICES

All services, such as water, steam, air, vacuum, and gases, which are piped into the laboratory should be clearly identified by name with stencils, adhesive labels, or similar means. Color coding, which may fade or change from chemical action, should not be relied upon.

Electrical

Electric service outlets will be required in the laboratory room and sometimes inside the hoods. One method of handling this problem is to install outlets, motors, and the like *under* laboratory benches or inside cabinets or hoods to meet the requirements of the National Electrical Code, Section 500, for hazardous areas. Outlets and electrical apparatus about bench-top level could be unclassified electrically, with open outlets and ordinary electric apparatus. This is predicated on the theory that flammable gases and vapors will always be confined to the hoods, and no handling of such flammables except in very small quantities will be done on the benches or in open containers. Of course, if flammables are handled in quantity, either in closed or open equipment, then all electrical wiring, apparatus, and equipment should be of the proper type as required for such areas by Section 500 of the National Electrical Code. Regardless of the wiring methods followed, the systems and service outlets should have an extra wire for grounding all parts not carrying current for such equipment as motors, portable and fixed tools, instruments, and so on.

Liquids and Gases

Service pipes are usually run into the laboratory to supply compressed air, vacuum, water, steam, natural or heating gas, nitrogen, and other commonly used gases. The gas lines should have outlets only inside the ventilated hoods, and should be supplied from headers or cylinders outside the building. If it is necessary to use other gases, small laboratory type cylinders, which can be placed under a hood, should be obtained.

Bibliography

Sax, N. I., *Dangerous Properties of Industrial Materials,* 3rd ed. Van Nostrand Reinhold Co., New York, N. Y., 1968.
Steere, N. V., *Handbook of Laboratory Safety,* Chemical Rubber Company, Cleveland, Ohio, 1970.

6

EQUIPMENT: APPARATUS AND PURCHASING

Many equipment hazards find their way into a laboratory as the result of poor judgment by the researcher in selecting equipment or because the manufacturer (who may even be the sole supplier) has inadequately engineered the apparatus. The researcher's greatest error in selecting a piece of equipment for purchase is *assuming* that the equipment is safe and adequately designed because it comes from a supposedly reputable manufacturer. This assumption can be fatal because either the engineering can be faulty or an error can be made during assembly.

There is only one way to control this problem: check the design, check the equipment for compliance with design, and check the intended method of usage for compatability with the manufacturer's design. The researcher will find that in some cases he must assume the guise of an expert to resolve his needs.

Sales personnel and literature should be considered and examined with some degree of doubt. They are naturally trying to sell the best features of their product. It is necessary to be critical in determining its liabilities or inadequacies. Research needs are difficult to satisfy. In many instances equipment is adapted to meet special needs and in so doing is stressed to its maximum capability.

We paint a black picture of purchasing practices because of the frequent cases of equipment failure and malfunction.

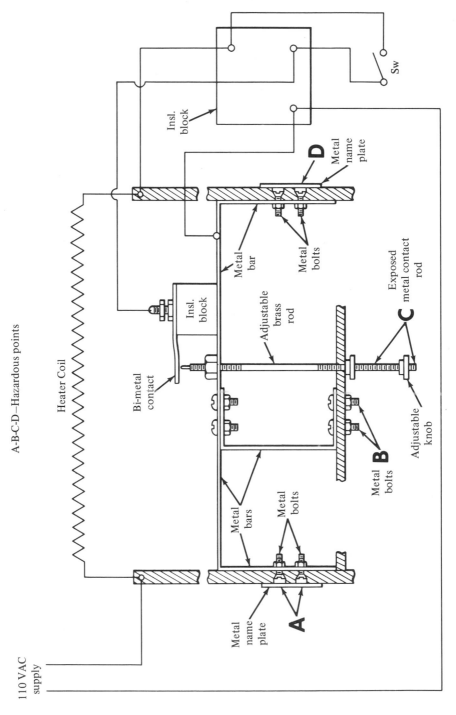

Fig. 6-1. Schematic wiring diagram of a hazardous electrical design.

For example, the schematic wiring diagram in Figure 6-1 shows how an apparatus was received in one laboratory. The chemist who purchased the unit assumed that the equipment was safely designed because he had purchased it from a reputable manufacturer. The unit arrived with an ungrounded plug; following normal practice, the chemist requested the maintenance department to change the cord and plug to the proper type. He also decided to have a pilot light installed so that he would know when the unit was turned on. As can be seen, the apparatus could have been a real shocker at points A, B, C, and D, which were exposed externally. The manufacturer was notified of the deficiencies, and recommendations were made to eliminate the hazard.

There is no reasonable way to discover faults of this type when buying equipment, but there is no reason why equipment cannot be inspected to determine compliance with good practice and proper assembly. Compliance with the electrical code is readily determined, and inspections should be routinely scheduled for all new apparatus on arrival.

Another instance shows the need for critical evaluation prior to purchase. A high-pressure cryogenic system was designed to study construction materials for compatability with a strong oxidizer. The engineer was sensitive to the need for an extremely clean system because he was using unusually small orifices. So he decided to use the most corrosion-resistant construction material available to avoid generating corrosion products which might flake off and plug the orifices. Figure 6-2 is a piping diagram of the system. Valves 2 and 3 failed simultaneously and sealed the system when the stems sheared in the closed position. The valves failed because a metal of low tensile strength was used, which is not compatible with the design requirements for valve stems on high-pressure valves.

This instance is a good example of inadequate design evaluation. The engineer didn't consider the high mechanical pressures developed when a valve stem is seated and high turning torques are applied to the handle.

The purchaser must be specific in describing the intended use to the manufacturer. He should ask his advice on the expected limitations of any change in the construction materials. He should discuss changes in design with his own engineering department, and get to know the materials engineer so that questions of this type can be resolved or the possible limitations anticipated.

Sometimes arrangements should be made to have an observer or inspector follow the manufacture of the item being purchased. In another high-pressure cryogenic system, a special valve stem assembly was ordered to repair a leaking valve. The purchasing instructions detailed a cleaning procedure to be followed because the assembly was to be

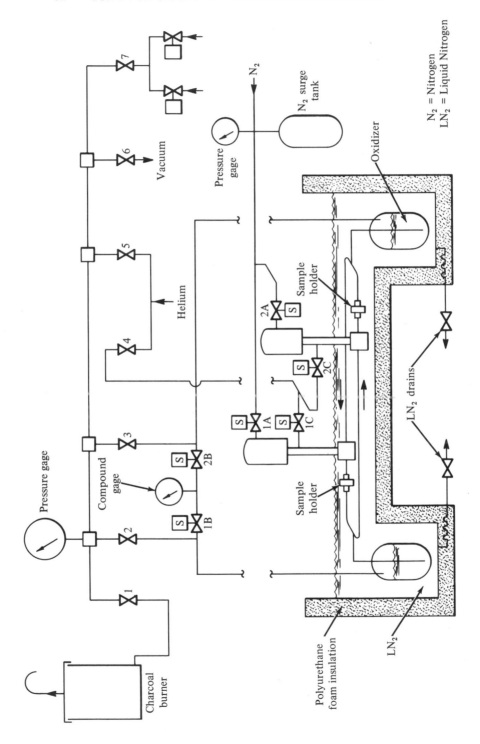

exposed to a strong oxidizer. The assembly arrived packed in excelsior. The engineer attempted to clean the assembly but, because of the tight convolutions of the bellows he wasn't sure of the cleanliness of the assembly. Owing to an extended delivery time, it was decided to use the valve. The valve was installed, and on the first run an explosive burnout occurred which was thought to have originated in the stem assembly of the valve.

When a component of a system needs special attention in assembly, serious consideration should be given to having an inspector observe the critical phases of assembly. This service is usually coordinated by the purchasing and engineering departments. This action must be considered for every component requiring long delivery time or when critical tolerances must be met or special assembly procedures must be followed.

An important source of information on laboratory equipment is the group of senior chemists and engineers who have years of experience. Many researchers seek out this knowledge. Unfortunately, not all do; nor do they gain the full advantage of this knowledge if they are not prepared to ask many detailed questions about their problem. Senior researchers must be careful to include their poor experiences as well as the good ones when advising someone about choosing equipment. A laboratory has two approaches to making use of this experience: to use a check list or outline of requirements in discussing or evaluating equipment for purchase; to establish a committee of senior researchers to prepare monographs on their experience in purchasing and using different types of equipment. A typical outline for use in evaluation equipment for purchase is shown in the Appendix to this chapter. Additional points for evaluation may be found which should be added to the list.

In every research laboratory, the level of common approach develops in the purchase and installation of process equipment. This is dictated by common fields of interest and observations by the researcher, of good and bad equipment arrangements used by fellow workers. The broad literature searching required by research work exposes the reader to the advertising of every conceivable piece of equipment available. The search for new and better equipment to foster laboratory work leads the researcher to purchase new items and materials. Once an item is used and meets the requirements of one researcher, its use begins to spread throughout the laboratory. Problems begin to develop because the item is being stressed to maximum performance. This important knowledge is usually lost with time.

To prevent this, researchers who have had broad experience in using a particular type of equipment should assemble a monograph of his experience. As an example, let's take the use of valves. Almost every laboratory procedure requires the use of one type or another. The evaluation could start with adjustable pinch clamps for hose lines, miniature needle valves, blunt needle valves, ball valves, plug valves, petcocks, valves for low pressure, high-pressure valves, construction materials for the different types, limits of usage for the different types of valves, valve mishaps of importance, and many other categories. Once prepared, the monograph can be updated or expanded for use by each researcher and particularly for use by new employees in evaluating their needs.

EQUIPMENT

To aid in the purchase of many common types of laboratory equipment and to give some ideas about potential problem areas, the following typical types of equipment are considered.

Refrigerators

Refrigerators for laboratory use are available in explosion-proof or explosion-safe design. If the refrigerator is to be used in an area where flammable vapors may be likely to develop, the explosion-proof type of unit should be purchased. If the unit is to be used in a normal laboratory where flammable atmospheres are unlikely to occur, then the explosion-safe type may be preferred.

If chemical spillage is likely to occur, stainless steel condensers should be used in preference to aluminum, particularly in the freezer section, since experience has shown that aluminum corrodes rapidly under the moist conditions of a refrigerator.

In addition, consideration should be given to installing a battery-operated bell as a high-temperature alarm. Be certain to have a sign made showing the purpose of the bell and listing clearly the last battery change. The internal box should be of seamless construction, with a pan as an integral part of the floor to catch spills and permit decontamination within the confines of the box.

Vacuum Pumps

Adequate guards must be specified on the purchase order. When a laboratory has many areas requiring explosion-proof equipment, it is advisable to purchase all pumps with explosion-proof electrics, thus permitting the use of any vacuum pump in all areas. All vacuum pumps should be vented to a safe area such as a hood or outside by piping from the exit of the pump.

Ovens

Careful analysis of potential uses is needed to evaluate specifications for purchase of an oven. If flammables may be released, operation must be such that an explosive atmosphere cannot be achieved and that a source of ignition is not available if a flammable atmosphere should be generated by mistake.

Latches for oven doors should release with internal pressure. High-temperature safety cutoffs should be provided. Integrated electrical circuits should be used, which will safely shut off the oven if a safety device fails or is actuated. NFPA Code 86-A, Ovens and Furnaces, should be consulted when flammables are to be handled in an oven. Exits from ovens should be piped to a safe location if the components to be placed in the oven can release toxic materials through evaporation or decomposition.

The insurance carrier should be consulted on the purchase of ovens and furnaces larger than bench size.

Oil Baths

There are three problems in using oil baths: flammability of oil to be used for high temperatures; safety cutoff for excessive temperatures; splash hazard from breakage of equipment during operation.

If temperatures within 20°C of the flash point are to be used, explosion-proof electrical accessories should be installed. Outer shields of heating elements should be grounded. High-temperature detectors and shutoffs should be provided. Metal catch basins should be provided for all glass bell jars or vessels holding hot oil. Splash shields should be provided for all oil baths. Chemical compatability with oil should be evaluated.

Low-Temperature Baths

The bath medium for low-temperature baths should be nonflammable if possible. If not, all equipment must be explosion-proof and ventilation provided to control flammable atmospheres.

Heating Mantles

Glass heating mantles can generate a shock hazard if not of the grounded type. This results from wear of the glass covering, thereby exposing the heating elements and allowing spillage of conductive liquids into the mantle. Whenever the mantle is wet, the power should be shut off before handling it.

The aluminum-covered mantles are preferred when purchased with the grounded receptacle. If a grounded receptacle is not provided, a separate grounding conductor can be installed and attached to the aluminum cover.

The use of glass heating tapes directly on metal equipment is not recommended without an electrical insulating shield between the tape and the metal. Glass electrical insulating tape will work well for this service. Glass tapes fray rapidly in service, and electrical leakages can occur on metal equipment to provide a severe shock hazard.

APPENDIX

The following items should be considered in addition to the specific requirements in either the design or purchase of laboratory equipment. Not all of these items will necessarily apply to all types of equipment, and other items will not have to be individually checked when buying from a known laboratory equipment-supply company. This list is presented in outline form so that it can be easily adapted to specific situations by additions, deletions, or expansion of certain items.

Material selection: metal and/or a borosilicate type glass (Pyrex and Kimax) are preferred to ordinary glass.

Corrosion: contamination may be a more important factor than the life of the equipment

Flammability: low-melting or easily ignitable equipment would quickly fail if the reactants were ignited

Explosive potential: brittle materials will become shrapnel in the event of an explosion

Toxicity: how serious will accidental breakage or leakage be?

Visibility: where is visibility required?

Cleaning: will reactants be easily removed from the surface?

Useful life: will it be needed for a week or a year? Some consideration should be given to longer than expected use

Pressures: ASME Pressure Vessel Code

Material stress: factors to be considered in specifying the required safety factor (ultimate stress/working stress).

Centrifugal force
    Normal
    Due to unbalance or shaft deflection
Impact
Fatigue: intermittent or continuous use
Pressures
Surface configuration and finish
Annealing required
Creep: time, temperature, and strength relationship
Temperature: maximum, minimum, duration, and cycling
Corrosion
    Surface
    Stress corrosion cracking
    Connecting equipment
Electrical
    Capacity
    Grounding
    Overload protection: adequacy and accessibility of fuses or breakers
    Type: National Electrical Codes
        Explosion-proof: this is frequently worth the additional cost because of the additional versatility
        Totally enclosed
        Nonsparking: induction motors, mercury switches
    Emergency shutoffs
    Interlocked with guards
    Lockout switch
    Static electricity control
Controls
    Overtemperature
    Overspeed
    Vibration
    Operating sequence: the order in which the controls function may affect the safety of this reaction
    Interlocking: to handle control failures
    Fail-safe design
    Accessibility: for operating and resetting
Guarding: isolation of hazard
    Nature of materials
        Flammability
        Reactivity, including explosion potential
        Effect of water (salt baths will explode)
        Toxicity
    Mechanical hazards: shafts, blades, nip points, and so on

Shields
    Strength and flammability
    Direction: protect front or all directions
    Stability
    Ease of use
Barricades: need should be reviewed for changes in use of equipment as well as for new equipment
Pressure relief
    Type: pop valves, rupture disc, fusible plugs, or a combination
    Piping: sharp turns and long runs reduce efficiency
    Vent: toxic or flammable materials vented into the same area or certain other areas could cause trouble
    Don't use: failure of devices in highly corrosive service may prove to be more hazardous than if left out of system

Maintenance
    Lubrication: frequency of, and access to lubrication points
    Periodic inspection by maintenance personnel
    Pressure testing frequency of vessel and pop valves
    Access to parts: lengthy disassembly for small items discourages prompt attention
    Standard parts: the use of standard parts will decrease downtime for repairs

Ventilation
    Local: adequacy and area of exhaust
    General: excess heat loads or leakage may require additional local exhaust

Lighting: poor lighting can cause accident through incorrect readings, failure to notice small changes, and fatigue

Noise: excessive noise causes fatigue, a breakdown of verbal communications, and injury to ear

Equipment arrangement
    Adequate working space
    Access to working parts, such as stopcocks, regulators, shutoffs
    Sample removal
    Access for maintenance and cleaning
    Rough edges and parts protruding in awkward positions
    Adequate handles on portable equipment
    Pedestrian safety: adequate passageways

Operating instructions
    Vendor's: check clarity, completeness, and ease of following . . .
    Special instructions for specific application

Approvals: by other personnel in your organization

7

SAMPLING TECHNIQUES

In the course of routine laboratory and pilot-plant operation it is usually necessary to transfer chemicals from their original containers to smaller secondary containers or to reaction vessels. This transfer of chemicals may be made for the ease of handling of a smaller container, for analytical purposes, or for use of the chemical in any of its intended applications.

SAFETY PRECAUTIONS

The infinite variety of chemicals used in the laboratory presents a variable hazard factor to the operating personnel. Of prime consideration in handling chemicals are the physical state and the physical and chemical characteristics of the product. Such factors as toxicity, flammability, corrosiveness, radioactivity, pyrophoric nature, and friction and shock sensitivity must be evaluated by the chemist or technician before sampling efforts are begun. Particular consideration must be given to those chemicals which exhibit more than one specific hazard.

The individual performing the sampling operation must provide not only for his own safety but also for the welfare of those persons working nearby in the same area or in adjacent laboratories. The use of personal protective equipment, such as safety glasses, acid goggles, gas masks or self-contained breathing apparatus, flame-proof wearing apparel, and rubber gloves, should be mandatory when indicated.

In addition to personal protection equipment, structural safety features and general safety equipment should be utilized when necessary. For example, exhaust hoods should be used where toxic gas or vapor may be liberated in a sampling procedure, and spark-resistant tools should be used when sampling flammable materials. Thorough preparation for sampling operations should take into account the known existing hazards and should also anticipate the latent dangers attendant to the procedure. In no circumstances should sampling of a chemical be initiated until all hazards have been appraised and proper remedial action taken.

CONTAINERS TO BE SAMPLED

In general laboratory practice, an infinite variety of containers may be encountered.

Liquids usually are packaged in small glass bottles, jugs or metal cans, five-gallon cans, ten-gallon carboys, and intermediate units up to and including fifty-five-gallon drums.

Dry solids are usually packaged in paper or fabric bags or fiber pack drums. Also used are wooden barrels or boxes and, in some cases, metal, glass, or plastic containers.

For compressed gases, the high pressure cylinder is the standard container.

SECONDARY CONTAINERS

The secondary container (the container into which the chemical sample is placed) must be at least equal to the original container in its ability to safely store the product. In addition, the secondary container must be properly labeled to indicate the name of the chemical, the source, the date of transfer and the general and specific hazards involved in the use of said chemical (see Chap. 17).

PROCEDURES

Specific sampling procedures differ widely. They are dependent upon the physical state and the hazardous characteristics of the products being handled, and also upon the size and type of containers being sampled and the amount of product required.

Liquids may usually be transferred by simple pouring, by pipette, by hypodermic syringe, by syphon, by spigot, or by transfer pump. Never use oral suction when pipetting or starting a syphon flow. Use only a suitable suction bulb or vacuum hose fitted with a trap to prevent contamination of the vacuum system. Do not pressurize containers to remove samples unless they have been specifically designed for pressure transfer of its contents. Occasionally, it may be necessary to warm a material to effect liquification of a frozen chemical or to reduce the viscosity of a highly viscous chemical. Open flame should not be used as the heat source. Electric heating tapes, steam coils, warm water baths, or approved safety ovens should be used as dictated by the situation. Care must be taken to properly vent the container being heated and to guard against overheating, which may cause ignition or violent decomposition of some products. Secondary containers should never be filled to capacity with liquid samples. Adequate outage (vapor space) must be allowed for expansion of the liquid. Limit filling of a secondary container to ninety percent of its capacity, and leave the remaining ten percent capacity as outage to compensate for liquid expansion.

Dry solids may usually be transferred by pouring and by scoop or shovel. Care must be taken when breaking up solid materials which have formed hard cakes or have fused together. When it is necessary to melt a dry solid material, the same safety precautions should be observed as for liquids.

Compressed gases must be transferred by suitable hose, tubing, or pipe assembly to the secondary container or reaction vessel. A pressure regulator should be used for proper control of pressure and flow of gas. Caution must be exercised to prevent the escape of toxic or corrosive gas into the laboratory atmosphere during sample transfer.

The chemist or technician should closely follow any sampling directions issued by the manufacturer or supplier of a chemical. These directions may be printed on the original container label or supplied as a separate bulletin. Also available to operating personnel are a number of specific chemical safety bulletins published by the Manufacturing Chemists Association.

Bibliography

ASTM Standard on Sampling Petroleum Products.
MCA Safety Guide, *Liquid Chemicals—Sampling of Tank Car and Tank Truck Shipments.*

8

LOCK-OUT AND TAG-OUT

NEED FOR LOCK-OUT

In laboratory and pilot-plant operations, it is important that an effective lock-out system be built into the procedures for carrying out maintenance work as well as such operations as clearing plug-ups or adjusting certain equipment.

Any equipment that can be operated, or moved by a remote force, is subject to the need for lock-out. If electrical or electrically driven equipment were to move unexpectedly while being repaired, adjusted, or cleaned, serious injury could result. Many cases of amputation and death are on record because no positive lock-out system was developed or, if developed, not properly used.

The only positive method of preventing such cases is by means of a mechanically opened disconnect, properly locked out, in the electrical power supply line. It should be pointed out that equipment in laboratory and pilot plant operations is sometimes driven by steam or air. Here again, the only good protection is to disconnect the steam or air supply line.

NEED FOR TAG-OUT SYSTEM

Tagging out equipment has been used on occasion as the sole protection for men working on equipment. A tag-out system is of positive value only when used with a lock-out system. The tag has a definite purpose: to point out the name of the equipment locked out, the name of the man who locked it out, and such other pertinent information as date of lock-out. Although the tag has a definite purpose in isolating equipment, it should be used only in conjunction with a positive disconnect.

WHY A POSITIVE DISCONNECT IS ESSENTIAL

Many times in industrial operations, particularly in pilot-plant and laboratory operations, personnel have been known to operate power-driven equipment wth only the stop-start button in the off position. Sometimes a "Do Note Operate" tag has been applied as a protective device. This is a potential situation for a serious accident. Stop-start buttons or switches on equipment are normally in the control circuit, and are designed to control the flow of power to drive the equipment. They do not offer protection to the worker, for the following reasons:

1. It is possible to activate the equipment by ignoring the tag and pushing the button or switch.
2. Even if a lock is applied to the stop-start button (some equipment is designed in this manner), shorts can occur which will by-pass the button and cause the equipment to operate. (Water in lines and some types of dust, for example, can supply a path to close the control circuit.)
3. Sticky or broken switches are common. Equipment has been known to start when a slight jar or vibration closes an opened control circuit.
4. Remote or interlocked switches could possibly start equipment.

The answer then is a *mechanically opened disconnect switch in the power circuit* (which positively opens the circuit) and a lock applied to this disconnect.

TYPES OF LABORATORY EQUIPMENT REQUIRING LOCK-OUTS

The following list is indicative of the types of laboratory equipment which require positive disconnects when maintenance and/or servicing work is being done.

laboratory milling machines
calenders
hydraulic or electric presses
powered molding machines
centrifuge machines
ventilation and air-conditioning machines
compressors

electric or steam heaters
extruders
mixers
blenders
drill presses
vessels with powered agitation
electrical equipment generally

This list should be considered as only indicative of the types involved. The complete list would include any piece of mechanical or electrical equipment which can be started or could start up while someone is working to repair it.

METHODS OF APPLYING AND SUPERVISING LOCK-OUTS

Simplicity and safety are the keynotes of a good lock-out procedure. Wherever possible, it is a good idea to tie the lock-out procedure with a work-permit system (see Chap. 9). In setting up the procedure, the following items are of great importance.

Stress need for mechanically opened disconnect.
Have disconnect points as convenient as possible.
Do not stop and start equipment with disconnect.
Use proper stop and start switches if they are on equipment.
Supply suitable single-keyed locks and keys to each individual craft involved.
Bring the electrician into the procedure, if only as a reference man.
Have each craftsman or tradesman apply his own lock (not to be removed until passed by those applying lock).
Be certain that lock is on right disconnect (checking by electrican and checking start-stop button are good methods).
Enter all pertinent information on tag and attach tag to every lock applied (workman's name, trade or craft, department, and supervisor).
Identify both disconnect and equipment with same name and number (make these clearly visible).
If more than one trade or crew is involved, the use of scissors-type lock arrangements are advisable, so that locks can be applied.
Review lock-out and tagging procedures with all involved workmen and foremen on a frequent basis.

It is possible that the problem of the positive disconnect can be solved in some areas on some equipment by simply pulling fuses or plugs. If this procedure is followed, make sure that the control of the plug or fuse is at all times directly with the workmen concerned. Caution should be exercised by using only proper fuse pullers with properly trained personnel doing the pulling.

Bibliography

Accident Prevention Manual for Industrial Operations, 6th ed. National Safety Council, 1969.
Electrical Switch Lockout Procedure, Safety Guide SG-8 MCA, 1961.

9

WORK PERMITS

THE NEED FOR WORK PERMIT SYSTEM

A fairly common misbelief in some laboratories and pilot plants is that a standard work permit system is not essential. It must be assumed that this idea stems from the fact that such operations are reasonably small and that operators of such locations feel they can reasonably well control the safety of maintenance and servicing work without a permit system. This is a difficult position to justify, since, in the majority of cases, equipment used and chemicals handled are similar (although smaller in size and quantity) to those used in manufacturing operations. Common sense would dictate that every means possible should be explored to protect all employees working in laboratories and pilot plants. It should be pointed out that hot work frequently takes place close to or on equipment that has been in flammable liquid service. The repair of electrical equipment and the handling of equipment which has contained many varieties of dangerous chemicals is also frequent. In areas where one, or all, of the above types of work takes place, a work permit system is essential to guarantee safe working conditions.

TYPES OF WORK REQUIRING PERMITS

It is extremely difficult to list all types of laboratory and pilot-plant activities for which a work permit is required. The basic philosophy should be that all equipment be put in a "safe" condition before the work actually starts. Much laboratory and pilot-plant equipment requires no preparation whatsoever because the area involved is safe for almost any type of work. In cases of this kind, the work descriptions should be so written that persons are not bound by unnecessary and unwieldy procedures. It is extremely important that the permit system apply only where it is needed.

Jobs in areas where flammable liquids or vapors could exist; tanks which have handled corrosive or flammable liquids; rooms where fire hazards exist; electrically driven equipment or electrical equipment—all these can be put in condition for safe work if a suitable check list is used to point out all the steps necessary to achieve this end. This check list is most acceptable and workable in the form of a safety work permit.

Examples of equipment and jobs requiring work permit coverage include:

1. Entry to all tanks and vessels.
2. Work on equipment containing powered agitators.
3. Work on equipment that has contained hazardous chemicals (instruments, small reactors, and so on probably require steaming, flushing, and the like).
4. Welding, burning, and use of other spark-producing tools in areas where a fire or explosion hazard could exist.
5. Powered machinery (other than portable tools).
6. Work requiring special precautions to insure personal safety or protection, materials, and property.
7. Roof permits (where toxic vapors may be vented).

PREPARATION OF EQUIPMENT

The work permit system used in laboratories and pilot plants should be set up in such a manner that all of the items necessary to insure safe working conditions are listed. Examples of the types of items that should appear on the work permit form are the following:

Washing and steaming
Lock-out of power drive
Blanking of supply lines
Protective clothing necessary
Respiratory equipment necessary
Fire equipment necessary
Results of gas tests (toxic, explosive, or flammable, oxygen deficiency or excess)
Ventilation of equipment

Condition of surrounding areas (removal of rubbish, direction of wind)
What work consists of (hot work, cold work, vessel entry)
Special attention to units utilizing inert gases
Methods of removal of persons from a vessel, and escape methods

The above list is by no means complete; individual operations should include other safety precautions that might appear necessary. The permit system forms should include instructions to answer all items noting what has been done. In many types of work, some of the items would not apply and should simply be noted accordingly. On the permit forms, there should also be sufficient room to allow the entry of specific comments and notes which would help the workman perform the job safely.

METHODS OF APPLYING AND SUPERVISING PERMIT SYSTEMS

The procedure outlining methods of using a work-permit system should be simple and easily understood by the personnel owning and operating the equipment and the personnel who will be working under the permit's jurisdiction. It is desirable, first of all, to establish the types of jobs that require work permits and to insure that all concerned are fully aware of the hazards involved and what is to be done to eliminate them. The following points should be considered when developing the procedure and its applications:

List, if possible, all jobs requiring permits.

Establish responsibility for permit (usually the equipment operator or owner should issue, and the work crew should sign as understood).

Tie in the issuing of the permit with the request for work.

Be sure that gas, toxicity, radiation, oxygen deficiency or excess, and other tests are entered on permit form.

In the procedure, list guidance instructions for the issuer and receiver on steaming, washing, blanking, locking out, and so on.

Establish a system for the renewal of permits.

Tie in housekeeping with the permit system.

Each laboratory and pilot plant may have peculiarities in its operation which require special consideration in a permit system. What is needed in cases of this kind can be decided only by the particular operation concerned. Careful preparation, aided by an adequate check list, guarantees safe work.

SAMPLE WORK-PERMIT PROCEDURE

Nearly every laboratory and pilot-plant operation has developed a type of work-permit form that is suitable to its particular operation. An example of a permit form now in use is shown in Figure 9-1.

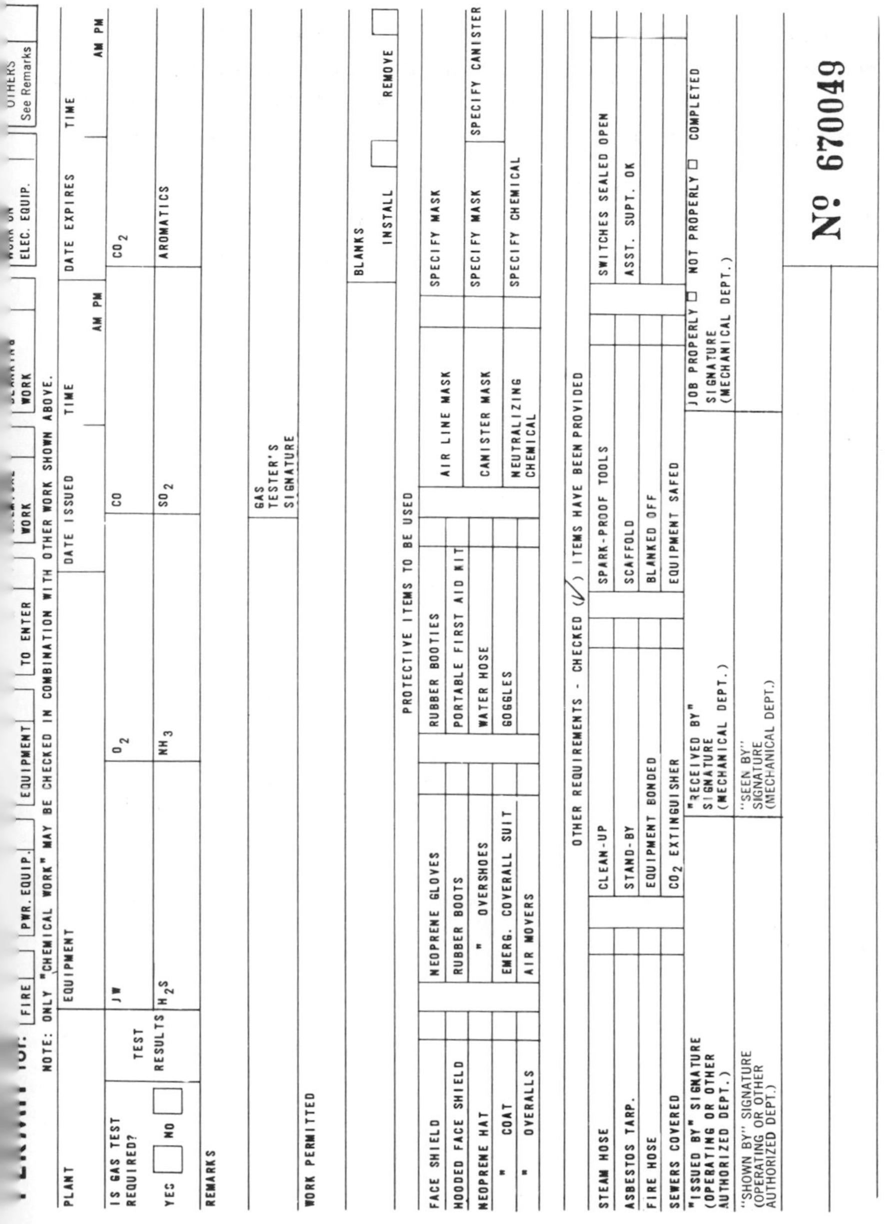

FIRE | PWR. EQUIP. | EQUIPMENT | TO ENTER | WORK | WORK | ELEC. EQUIP. | OTHERS See Remarks

NOTE: ONLY "CHEMICAL WORK" MAY BE CHECKED IN COMBINATION WITH OTHER WORK SHOWN ABOVE.

PLANT | EQUIPMENT | DATE ISSUED | TIME AM PM | DATE EXPIRES | TIME AM PM

IS GAS TEST REQUIRED? YES ☐ NO ☐ | TEST RESULTS | JW | O_2 | CO | CO_2 | H_2S | NH_3 | SO_2 | AROMATICS

REMARKS

GAS TESTER'S SIGNATURE

WORK PERMITTED

BLANKS INSTALL ☐ REMOVE ☐

PROTECTIVE ITEMS TO BE USED

FACE SHIELD	NEOPRENE GLOVES	RUBBER BOOTIES	AIR LINE MASK	SPECIFY MASK			
HOODED FACE SHIELD	RUBBER BOOTS	PORTABLE FIRST AID KIT					
NEOPRENE HAT	" OVERSHOES	WATER HOSE	CANISTER MASK	SPECIFY MASK	SPECIFY CANISTER		
" COAT	EMERG. COVERALL SUIT	GOGGLES	NEUTRALIZING CHEMICAL	SPECIFY CHEMICAL			
" OVERALLS	AIR MOVERS						

OTHER REQUIREMENTS - CHECKED (✓) ITEMS HAVE BEEN PROVIDED

STEAM HOSE	CLEAN-UP	SPARK-PROOF TOOLS	SWITCHES SEALED OPEN
ASBESTOS TARP.	STAND-BY	SCAFFOLD	ASST. SUPT. OK
FIRE HOSE	EQUIPMENT BONDED	BLANKED OFF	
SEWERS COVERED	CO_2 EXTINGUISHER	EQUIPMENT SAFED	

"ISSUED BY" SIGNATURE (OPERATING OR OTHER AUTHORIZED DEPT.) | "RECEIVED BY" SIGNATURE (MECHANICAL DEPT.) | JOB PROPERLY ☐ NOT PROPERLY ☐ COMPLETED SIGNATURE (MECHANICAL DEPT.)

"SHOWN BY" SIGNATURE (OPERATING OR OTHER AUTHORIZED DEPT.) | "SEEN BY" SIGNATURE (MECHANICAL DEPT.)

Nº 670049

Fig. 9-1. Hazardous work permit.

Manufacturing Chemists Association Safety Guide, SG-10, discusses the dangers connected with entering tanks and other enclosed spaces. It is recommended that this be used as the basis for preparing work-permit systems which include entry into tanks and dangerous enclosed spaces.

Bibliography

Accident Prevention Manual for Industrial Operation, 6th ed., National Safety Council, 1969.
Entering Tanks and Other Enclosed Spaces, SG-10, MCA, 1961.

10

AFTER-HOURS WORK AND OPERATION

Most chemical laboratory work can be carried out during normal daytime hours. However, the very nature of some investigations requires work after hours either occasionally or on a programmed basis. Most laboratory personnel who return after hours to write reports, work in the library, do calculations, or make mechanical or physical changes in experimental arrangements can safely perform these functions alone. The only risk involved may be the risk of blackout, fainting, or being incapacitated by stroke, heart attack, or other physical disorders. Obviously, if such an event does occur, help is needed immediately. However, the incidence of such attacks among the working population is so low that it is generally considered as an acceptable risk on the part of the individual and management.

On the other hand, experimental operations involving highly toxic, high-pressure, high-energy, or unstable materials, should not be performed unless two or more persons are present. Even then, one of the persons should be continually out of the immediate danger area. In other words, a "buddy" system is considered essential. Unfortunately, there is a gray area between the relatively problem-free safe operational area and the high-hazard area, where it may be difficult to justify the expense of necessary preventive measures.

A survey of the chemical industry made in 1962 by the Engineering

Committee of the Chemical Section of the National Safety Council (Volz, Heesman, Levens, and Cobb) developed some interesting facts. From responses received covering 265,000 employees, the companies involved indicated that "existing policies" covered 195,000 of these employees. In general, these policies specify conditions under which solitary after-hours work is acceptable and conditions under which it is not permissible and also specify some means of checking. On the other hand, approximately 50% of the sponsors indicated that there was no written policy, and 20% indicated that the increased tendency for persons to work alone is becoming a serious problem.

Fortunately, there have been relatively few reported incidents of injury or property damage sustained while an employee was working alone. However, it may be seen that the risk inherent in such operations requires the development and implementation of complete accident-prevention and loss-prevention policies and guidelines to keep this risk under reasonable control.

Some functions can be performed safely after hours by the solitary laboratory employee, including the following:

Night watchman's functions.

Desk work in the office or laboratory.

Janitorial or custodial functions.

Assembling of laboratory apparatus or modifications when chemical and/or electrical hazards are avoided.

Maintenance functions other than on active chemical process equipment.

Routine job functions which as part of an operating procedure have such long standing that the operations are routine and known to be safe.

On the other hand, some operations definitely should not be permitted if the investigator is unattended or unmonitored, including the following:

Experimental operations involving toxic gases, liquids, or solids.

High-pressure investigation.

Investigation involving the handling of cryogenic materials.

Work with high-energy or thermally unstable materials.

Operations involving moving equipment or machinery, particularly that which is not completely guarded.

Investigations involving electrical work.

Work involving transfer of quantities of flammable materials, particularly in open systems.

In the area between these extremes, the phases of the work must be reviewed in order to identify those which can be safely performed by a person working alone. This analysis should be a joint effort involving the individual investigator, the responsible laboratory supervisor, and the laboratory's accident-prevention specialist.

There should be an established procedure for routine checks upon personnel working after hours, both for the person working alone, or in high-risk operations, even if two or more persons are present. One system which is fairly common requires that the individuals involved call at regular specified intervals to a control desk, watchman station, fire station, or some other attended point. The guard or fire watchman can also be instructed to call the individuals at some specified interval. It must be recognized that these periodic calls cut into the investigator's time. One research executive, in objecting to this approach, points out that once an hour is too frequent unless something happens, in which case it would not be frequent enough. Consequently, more sophisticated systems are finding application, particularly on high-risk operations. These may involve a personal check on the area by a co-worker or watchman, or the installation and use of intercom monitors or two-way radio equipment. In the latter cases, the base station would be an attended location, such as the watchman's office or fire station, and the field set would be placed in the experimental area. If the field instrument is set to transmit continuously, the investigator can call for assistance or abnormal noises in the area can be heard by the monitoring station. Of course, the most complete system uses closed-circuit television monitors, thus supplementing audio monitoring with visual supervision.

There is an increasing trend for experimental equipment to be left in operation continuously with occasional attendance by a night technician, night chemist, or, in some cases, night watchman. Obviously such arrangements introduce the hazard of unanticipated abnormalities developing during the unattended periods. The presence, even if only occasionally, of a single night operator exposes him to the same risk as contemplated above. Consequently, there is a growing reliance among chemical research laboratories to equip continuous-basis processes with instrumentation and automatic controls which, under emergency conditions, may operate fail-safe devices, automatic cut-offs, safe venting or purge systems. Thus, the research investigator is freed from the necessity of working alone after hours. On the plus side, such instrumented controls also continuously safeguard the experiment during normal working hours when the investigator is absent from the laboratory.

Papers presented at the 1964 National Safety Congress by Fox, Lynch, and Richmond of Monsanto Company have indicated that this approach is a growing trend and, further, that the necessary instrumentation devices are readily available.

The great variety of laboratory experimental operations precludes the recommendation of any one particular system to solve all after-hour problems. It must be recognized that the problem exists and will continue to exist to some degree. The laboratory organization, as a part of its responsibility for personnel safety and property protection, must carefully assess the risks apparent in each after-hours operation and develop systems which will minimize exposure to operating personnel. The independent investigator, the laboratory safety staff, and the administrative management must all contribute to the development of satisfactory methods and equipment.

11

INSPECTIONS

Among the many types of inspections used throughout the chemical industry today, including laboratory operations, safety inspections remain the most effective means of detecting accident causes. Basically, accidents are caused by unsafe conditions, unsafe acts, or both. This chapter presents guidelines to aid those responsible for the formulation of safety inspection procedures individually tailored to detect these conditions in specific operating areas.

Detecting the unsafe condition is of course only the first step leading to its ultimate correction. Inasmuch as both unsafe environmental conditions and unsafe acts are entirely within the control of man, they can be corrected by man. If hazardous conditions can be detected by careful inspection, corrective measures can be devised which are both practical and effective.

Such inspections should include the following: fire prevention measures and methods; fire protection equipment; lifesaving devices; housekeeping; safe working methods and procedures; and special inspections indicated by experience.

Special attention should be given in inspections to the following: layout and design; ventilation; illumination; electrical equipment; personal protective equipment and clothing; glassware; chemical storage and handling; pressure vessel; emergency showers and eyewash fountains; ladders; and material handling equipment.

CONDUCTING AN INSPECTION

The assignment of inspection personnel depends upon the type of inspection and the area to be covered. Safety of assigned areas is inherently the responsibility of the supervisor. To detect unsafe conditions, he may assign inspection duties to subordinates or use available staff inspection groups. However, he must know the conditions of his department, must evaluate the soundness of recommended corrective safety measures, and must push for the elimination of accident causes; therefore he must be an active participant in any inspection within the area of his responsibility. In many laboratories, inspection teams are made up of operating employees. The members of these teams are rotated to afford everyone an opportunity to participate. In some cases, safety committee members may act as inspectors, with the area of inspection responsibility assigned on the basis of the most effective use of past or present experience. Participation by involved personnel is the keynote to effective accident prevention.

The frequency of inspections depends on the type of inspection being made. Periodic inspections of condition are recommended for such fire-protection equipment as automatic sprinkler systems or hand extinguishers, and for such life-saving devices as cannister gas masks or self-contained breathing apparatus. Suppliers of these types of equipment will assist in establishing the frequency of inspection as well as procedures to follow. OSHA regulations should also be consulted.

The need for special inspections may be indicated as the result of accidents. Analysis of accident causes may indicate high frequencies of particular types of accidents, within certain departments, at given times of day, or with certain processes or equipment. Specific inspections may then be used to identify underlying causes and to indicate steps toward correction. Examples may be failure to wear personal protective equipment or misoperation of equipment.

WHAT TO LOOK FOR: UNSAFE CONDITIONS

Safety devices, signal devices, and guards
Inadequate or missing guards on shafts, gears, pulleys, belts, rolls, chains, agitators, grinders.
Bypassed safety devices, such as limit switches, pressure-relief valves, hold-downs, lock outs, overload switches, grounding connections.
Missing or faulty warning devices.
Blocked or hidden signal devices, such as fire and emergency alarms.

Structural defects and material characteristics
Sharp-edged, jagged, splintery conditions.
Worn, frayed, cracked, broken items, such as electrical cords, glassware, hoses.
Slippery, uneven, rough surfaces.
Decayed or corroded conditions.

Functional defects
Inoperative safety equipment, such as eyewash fountain, emergency showers, respiratory equipment, fire protection equipment, emergency exits, ventilation equipment.
Susceptibility to rupture; breakage; falling, tripping, or flying objects; spills.

Storage areas
Condition of stacking, heat, tipping, blocking automatic sprinklers, aisle blockage, secure piles.
Chemical storage.
Drum storage: contents flammable, fire protected, remote area, subject to temperature extremes.

WHAT TO LOOK FOR: UNSAFE ACTS

Failing to use personal protective equipment provided
Chemical goggles.
Safety glasses.
Appropriate gloves.
Escape masks.
Face shields.
Aprons.

Temporary installations
Hose instead of piping.
Glass containers for flammable liquids.
Electric and air lines on floor.
Electric lines not properly grounded.
Exposed hot surfaces.
Electric connections not covered.
Material or product left temporarily in aisle.

Operating or working at unsafe speeds

Running instead of walking.

Throwing instead of carrying or passing articles.

Jumping from moving vehicles.

R & D LABORATORY SAFETY INSPECTION SHEET

Division _____ Section _____ Date _____

Inspected by: _____   Reports to: _____   Div. Mgr.
              _____   _____   Sect. Chief
              _____
                                                            _____   Safety Coord.

Room No. or Location											
General											
Aisles obstructed											
Excess materials											
Excess equipment											
Floor slippery											
Tripping hazards											
Hot surfaces exposed											
Work space cluttered											
Unstable equipment											
Gas Cyl. unsupported											
Appearance-cleanliness											
Other _____											
_____											
Safety Equipment											
Needs maintenance											
Not accessible											
Other _____											
_____											
Special Hazards - Specify											
_____											
_____											
Electrical Hazards											
Repairs needed											
Exposed wire or term.											
Other _____											
_____											
Mechanical Hazards											
Guards needed											
Sharp points or edges											
Other _____											
_____											
Chemical Hazards											
Too much reactive chem.											
Pans & carriers needed											
Other _____											
_____											
Flammable Hazards											
Excessive flammables											
Safety cans not used											
Other _____											

Notes:_____   Recommendations:_____
_____   _____
_____   _____

Fig. 11-1. Example of a laboratory safety inspection sheet.

Unsafe positions
Lifting with the back.
Lifting too heavy a load without help.
Insecure footing.
Improper method of descending or ascending ladders.
Exposure to suspended loads.
Overextended reaching from ladder or scaffold.

Distracting, teasing, startling
Calling, talking, or making unnecessary noise.
Throwing material.
Practical joking.
Horseplay.

Working on dangerous or moving equipment
Cleaning, oiling, or adjusting moving equipment.
Welding improperly purged containers.
Working around machinery with loose clothing, rings, and the like.
Working on electrically charged equipment.

CHECK LISTS

Check lists play an important part in inspections, and care must be taken in their preparation. Check lists are best suited to a special inspection where specific items are to be checked. Making a check list for general inspections is difficult because some items may be overlooked; then too, a check list may tend to cause an inspector to check only the items listed. Any check list should be reviewed periodically to verify that it serves its purpose. Figure 11-1 is an example of an inspection sheet which has been successfully used in a large industrial research and development laboratory.

12

PERSONAL PROTECTIVE EQUIPMENT

The laboratory safety program must be concerned with eliminating or controlling hazards by establishing safe methods, procedures, and the necessary applied engineering. It must also be concerned with providing laboratory personnel with personal protective equipment that will increase their chances of avoiding even minor injuries. The provision, storage, maintenance, and proper use of personal protective equipment, as in any other phase of the safety program, means increased proficiency in the overall performance of the laboratory work load.

The quantity and types of protective equipment a chemical laboratory should have available will depend upon its size and type of work.

It is most important that personnel be familiar with the various types of equipment they will use. They should know how it can be obtained; why and when it will be needed; its manner of use and care; and, above all, its specific limitations. This information should be reviewed on a planned periodic basis—at least every six months. Training and review sessions may include visual aids, demonstrations, group discussions, and review of pertinent case histories and experiences.

The orientation program should make each person knowledgeable and self-sufficient in the use of the equipment.

Convenient storage of personal protective equipment is important. Respirators, face shields, aprons, gloves, and sleeve protectors are examples of equipment that can be issued to the workers for a particular job or on an as-needed basis, with an adequate back-up supply being immediately available. Fire extinguishers and emergency self-contained breathing apparatus may be stored in doorway exits or strategic locations where it will be readily available in emergencies. Rescue equipment may be stored in a preplanned rescue station or in close proximity to the supervisor's office outside the immediate laboratory work area. The main idea is to have the equipment so located that an emergency condition does not limit or prevent its availability. Some industrial laboratories have found it feasible to locate storage cabinets in the exterior walls—the cabinets being equipped with double doors—thus making the contents available from either the interior or the exterior of the laboratory.

With the provision of personal protectors comes the inherent need for a maintenance plan. This means periodic inspection and cleaning, sanitization, and repair services. Since the equipment is of a personal nature, it should be cleaned and sterilized before use by another person. It should always be kept clean and in good repair so that it will be dependable and give maximum efficiency. When protective equipment is used to protect against sensitizing chemicals, obviously it must be kept scrupulously clean.

Soap and warm water have generally been used in cleaning personal equipment. Sterilizing methods will depend on the nature of the equipment. Ordinarily, face pieces and other items made of rubber or plastic will not withstand temperatures of sterilization. There are, however, a number of synthetic detergents, such as those used for sterilizing glassware in restaurants, which are satisfactory. Rubbing alcohol has also been used for sterilizing respiratory personal protective equipment. If there is any doubt, a request should be made to the respective supplier as to the proper maintenance methods, cleaning materials, and practices to follow for effective sanitization.

EYE PROTECTION

Eye protection is of paramount importance, and experience indicates that workers should always wear suitable eye protection when in the laboratory.

Use of contact lenses in the laboratory is discouraged for the following reasons. They are not a substitute for safety glasses; sudden displacement of the lenses can create visual problem; the lenses are difficult to remove when chemicals get into the eyes; they have the tendency to prevent natural eye fluids from removing contaminants (solids and liquids) which may get under the lenses.

Many styles and types of safety glasses and face protectors are available. The following are representative of the principal styles which have been found satisfactory for laboratory use:

1. Metal-rimmed safety glass spectacles with colored or clear lenses, with or without side shields.
2. Plastic-rimmed safety glass spectacles with colored or clear lenses, with or without side shields.
3. All-plastic (chemical) safety goggles, which can be worn over corrective glasses.
4. Rubber-framed goggles with plastic or hardened glass lenses. (These latter two offer more complete eye protection than the first two mentioned.)
5. Cup-type safety goggles with glass or plastic lenses, available in several styles, some of which are suitable for wearing over corrective glasses.
6. Nitrometer masks, for use where complete face protection is required.
7. Plastic face shields, which may be worn over the spectacle type safety glass or chemical goggle for additional face protection.
8. Special protective type lenses are needed for laser, welding, burning, ultraviolet, and infrared exposures.

Properly fitted spectacle-type safety glasses are comfortable and offer adequate protection for most routine laboratory work. It is desirable, however, when doing work known to be potentially dangerous to the eyes, to wear more complete eye protection, such as is offered by chemical goggles and/or the combination of glasses and plastic face shields. The proper care of safety glasses includes the following:

1. Keep the lenses clean. Dirty glasses obscure vision and may lead to eye fatigue. Never clean lenses with abrasive hand soap, since it will scratch them. When cleaning plastic lenses, any abrasive dirt which may be on the surface should be flushed off by holding the lenses under running water; otherwise the lenses will become scratched by the abrasive matter being rubbed into the lenses. Lenses with surface scratches should be replaced, since the hardened glass will thus be weakened.
2. Safety glasses should never be left with the lenses in contact with hard surfaces such as table tops.
3. Safety glasses should not be carried unprotected and in the same pocket with other objects, such as pencils, files, and depth gages. It is good practice to keep glasses in a case when they are not being worn.
4. Goggle cleaning cabinets located in the immediate laboratory work area have proved beneficial and efficient.

The eye protectors provided should meet the requirements of the American National Standard Z87.1–1968, Practice for Occupational and Educational Eye and Face Protection.

HEAD PROTECTION

In general laboratory work, hard hats are not normally used. Head protection can be required for special work or when laboratory personnel work assignments take them into certain areas.

Hard hats should be worn to protect against falling, sliding, flying objects, and other causes of head injuries.

Many such hats are available. All are constructed to protect against high impact, electrical, and bump type hazards. Caution should be used in selecting the proper one to be certain that it offsets the hazard. See ANSI-Z89.1

HAND PROTECTION

Many minor injuries in laboratories can be prevented by the proper utilization of effective hand protection.

Knit cotton, leather, and leather-faced gloves are available in wrist length and gauntlet style, and afford protection for general handling of abrasives, sharp objects, and glassware.

Knit cotton gloves are also available dipped in various materials, including rubber, synthetic rubber (which is particularly oil resistant), and various plastics which are resistant to solvents.

Where hand protection is desirable but finger dexterity is essential, a surgeon's type glove of either natural rubber or various synthetic rubbers may be used. One type has roughened surfaces to make it particularly suitable for handling wet glassware safely.

Heavy rubber gloves, available in different sizes up to elbow length, are recommended for handling concentrated acids or other corrosives. These are especially recommended for washing glassware using chromic acid cleaning solution.

Fleece-lined asbestos or leather wool-lined gloves should be available for handling hot objects.

FOOT PROTECTION

Leather or rubber safety shoes with built-in steel toe caps are recommended where heavy objects are customarily handled or there are other foot hazards. Leather safety shoes should prove satisfactory for general wear, but rubber safety shoes are recommended where there is considerable amount of water, acid, or other chemicals present on the floors. Rubbers may be worn over leather safety shoes. See ANSI-Z41.1.

Conductive, safety-type shoes are not necessary for general laboratory work. For special situations when there can be an explosive atmosphere or condition, it is recommended that the hazard of static electricity be mitigated by the use of safety shoes with conductive soles. However, static charges will not be dissipated as they are generated unless the floor is concrete or of some conductive covering, properly grounded.

BODY PROTECTION

Laboratory coats, aprons, smocks, coveralls, pants, jackets, hoods, and similar garments need to be used, when indicated, for protection of the body and clothes from corrosive chemicals. Some of the garments made from synthetic fibers are highly resistant to corrosive chemicals. Since it is possible that none of the garments will possess all the protective qualities required, it may be necessary to select garments made of the material that will be most resistant to the chemicals most frequently handled in the laboratory. Air-supplied suits, helmets, and masks should be provided for decontamination or work with highly toxic materials.

Certain synthetic fibers and plastic materials used in personal protective clothing can, owing to friction during use in very low humidity areas, generate static electricity; therefore, their use in fire restricted areas or when working with flammables could be a possible fire ignition source. The suppliers should furnish information about the equipment and its static potentialities and provide suitable ways to overcome this hazard.

RESPIRATORY PROTECTION

Self-contained breathing apparatus provides the user with either a supply of air or oxygen. The compressed air type is equipped with a cylinder containing a thirty-minute supply of air. The oxygen generating unit uses a special canister to provide a supply of oxygen from thirty minutes to one hour.

The liquid air and liquid oxygen type contains a one- to two-hour supply of air or oxygen. Compressed oxygen types serve from one to two hours.

Filter-type canister masks offer limited protection since they depend on an absorbing and filtering material to remove contaminants from the air. They do not offer protection against possible oxygen deficiency (below 16% by volume) in air. The life of a canister varies, depending on the type

TABLE 12-1.   PRINCIPAL TYPES OF CANISTERS

Atmospheric Contaminants to be Protected Against	*Colors Assigned**
Acid gases	White.
Hydrocyanic acid gas	White with ½-inch green stripe completely around the canister near the bottom.
Chlorine gas	White with ½-inch yellow stripe completely around the canister near the bottom.
Organic vapors	Black.
Ammonia gas	Green.
Acid gases and ammonia gas	Green with ½-inch white stripe completely around the canister near the bottom.
Carbon monoxide	Blue.
Acid gases and organic vapors	Yellow.
Hydrocyanic acid gas and chloropicrin vapor	Yellow with ½-inch blue stripe completely around the canister near the bottom.
Acid gases, organic vapors, and ammonia gases.	Brown.
Radioactive materials, excepting tritium and noble gases.	Purple (Magenta).
Particulates (dusts, fumes, mists, fogs, or smokes) in combination with any of the above gases or vapors.	Canister color for contaminant, as designated above, with ½-inch gray stripe completely around the canister near the top.
All of the above atmospheric contaminants	Red with ½-inch gray stripe completely around the canister near the top.

*Gray shall not be assigned as the main color for a canister designed to remove acids or vapors.

NOTE: Orange shall be used as a complete body, or stripe color to represent gases not included in this table. The user will need to refer to the canister label to determine the degree of protection the canister will afford.

factor and the concentrations encountered. Table 12.1 lists the principal types of canisters and the protection they will provide.

Canister gas masks have been widely used for routine service work, but, because of their limitations, many laboratories are installing the more reliable self-contained types of breathing apparatus. A person's sense of smell is the principal test for the need of and usefulness of a canister-type mask, but it must be remembered that most toxic gases, although their odor is pronounced at first, soon affect the sense of smell and become barely noticeable. It is important, therefore, that a person wearing a canister-type mask leave the area immediately upon detecting the odor of any gas. The odor may mean that the concentration of the gas encountered is dangerously high; that the face piece is not properly adjusted, and is leaking; that the canister is no longer serviceable; or that the mask is equipped with the wrong type of canister.

Air-line masks obtain their air supply through a nonkinking hose line from a remote source, such as bottled air or the laboratory air supply. This equipment can often be used to an advantage in nonemergency operations. These devices are available with either a half-mask or full-mask face piece. They are usually equipped with a flexible or corrugated air supply hose connecting the face piece to a control valve, which is worn on the belt. The control valve can be of the demand type, or it may permit the wearer to adjust the air to the desired amount for protection and comfort.

It is also important that the blower's air intake or air source be in an area which is free from air contaminants. Line filters containing activated charcoal may be used to advantage to remove moisture and certain other contaminants. Air from an improperly cooled reciprocating type compressor can contain carbon monoxide from the decomposition of the lubricating oils used.

Since air-line masks depend on a remote air supply, they should be used only where conditions will permit safe escape in the event that the air supply fails.

The following is taken from United States Department of the Interior, Bureau of Mines, Information Circular #8281, *Respiratory Protective Devices Approved by the Bureau of Mines as of October 1, 1965.* (Lists of respiratory protective devices approved after October 1, 1965, may be obtained from the Branch of Health Research, Bureau of Mines, 4800 Forbes Avenue, Pittsburgh, Pennsylvania 15213.)

To assist users in the selection of reliable and satisfactory respiratory protective devices, the Bureau of Mines has established five schedules of minimum performance requirements for the approval of respirators. The current revisions of these schedules include 13D for self-contained breathing apparatus, 14F for gas masks, 19B and the amendments to 19B for supplied-air respirators, 21B for dust, fume, and mist respirators, and 23B for chemical cartridge respirators. Copies of these schedules may be obtained from the Bureau of Mines, Publications Distribution Section, 4800 Forbes Avenue, Pittsburgh, Pennsylvania 15213.

Approval tests on respirators are performed only at the request of the manufacturer of the respirator and at his expense. If the device meets the approval requirements, a certificate of approval is granted on the complete respirator. This certificate states that as long as the respirator is manufactured as originally approved, or in accordance with any changes covered by an extension of the original approval, the manufacturer may advertise and sell his respirator as being approved by the Bureau of Mines. Many industries and regulatory agencies require that respirators approved by the Bureau of Mines be used where respiratory protection is needed. The Bureau of Mines, however, has no regulatory power over the use of the respirators that it approves. It acts as an impartial testing agency to insure the availability of suitable and safe respirators.

At frequent intervals, the Bureau of Mines purchases approved respirators and replacement parts on the open market. These are examined for conformity to the approved status, and are checked to insure that the manufacturer's production facilities are maintaining a satisfactory quality control.

Bibliography

Accident Prevention Manual for Industrial Operations, 6th ed., National Safety Council, 1969.

Sax, N. I., *Dangerous Properties of Industrial Materials,* 3rd ed., Van Nostrand Reinhold Co., New York, N.Y., 1968.

Schutz, R. H., Ferber, B. I., and Kloos, E. J., *Respiratory Protective Devices Approved by the Bureau of Mines as of October 1, 1965,* U.S. Department of Interior, Bureau of Mines Information Circular #8281, 1966.

13

WASTE DISPOSAL

All laboratories, regardless of their size or the nature of their work, must dispose of some waste chemicals daily, and occasionally of rather large quantities of accumulated materials. Depending upon geographical location and local ordinances, this can present a wide variety of problems. Improper disposal can not only be extremely hazardous to personnel involved but can result in poor public relations and sometimes expensive legal action. The following paragraphs will attempt to outline the basic methods which can be applied to both large and small laboratories. (The MCA Laboratory Waste Disposal Manual is reproduced as Appendix 4. It gives physical properties, hazard information and recommended disposal procedure for 1121 chemicals.)

RESPONSIBILITY

It is extremely important that the entire matter of waste disposal be under the control of a single individual or department specifically charged with that responsibility. The local safety organization, if not actually in charge of waste disposal, should act as consultant and have sufficient control of the operation to insure the development of safe methods and planned action for emergency situations.

METHODS

Some commonly used methods of waste disposal are discussed below.

Burning in the Open

Many materials may be safely disposed of by burning in the open. Ignition should be accomplished from a safe distance, and personnel handling flammable materials should be equipped with flameproofed clothing. The main objection to open field burning is the smoke resulting from incomplete combustion and, in many cases, to the appreciable amounts of irritant and toxic materials in the smoke. Smoke clouds not only create a severe nuisance in the neighborhood but may cause damage to plant life, and house and car paint, and thereby result in law suits. Recent legislation has discouraged and even prohibited any type of open burning of any kind of waste materials in many communities.

Burning in Incinerators

Although burning in incinerators under proper control can reduce most of the smoke nuisance, the hazards created by flammable vapors or explosive solids in the incinerator structure may be serious. Chemical plant incinerators require expert design and the provision of automatic controls to prevent the development of unsafe conditions.

It is usually possible through the burning of relatively harmless wastes, such as ordinary waste paper, to build up temperatures in the "combustion chamber" or the "after burner" which are high enough to ignite flammable liquid vapors immediately and to cause decomposition of irritant and toxic organic compounds. Sufficient temperature control and interlocking devices are required so that these materials cannot be introduced into the furnace unless a proper operating temperature, usually in the vicinity of 1700°F, is maintained. Great care is needed in the arrangement of flammable liquid tanks, piping, and control valves. Safe clothing for workers, adequate showers, and means of escape are positive requirements. Operations of this type must be under the responsible charge of competent operating personnel.

Disposal Through Sewers

While many water-soluble materials may be readily flushed down sewers, great care is needed to avoid placing material in sewers which will create flammable vapors, stream pollution problems, or upset the normal operation of sewage disposal plants. All disposal of wastes by this means must be in accordance with federal, state, and local laws relating to stream pollution and to the type of material acceptable in the sanitary sewer system.

In instances where wastes are discharged directly into streams, special sewage-treatment plants designed for the particular wastes may be necessary.

Great care must be taken to make sure that radioactive materials do not enter the normal waste disposal operation but are properly segregated and handled in accordance with recommendations of the Atomic Energy Commission.

REACTIVE WASTES

Reactive chemicals which may cause a violent reaction when in contact with water or other common materials require special treatment so that they may be made relatively harmless before disposal. The exact nature of this treatment will, of course, vary with the nature of the material.

Explosive Materials

Standard procedures should be set up for the careful removal of shock-sensitive materials from the laboratory to the disposal area, where they may be safely destroyed, usually by burning. In some instances, it may be desirable or necessary to destroy the material by exploding with a detonator as initiator. This procedure, however, requires the services or advice of an expert in the handling of explosives, and also requires the approval of local authorities. Never attempt to use rifle shots to explode such materials because of the danger of a wild shot or richochet traveling long distances, danger to the person firing the rifle, and the uncertainty of an explosion even if the material is hit as intended.

Disposal by Burying

Although many ordinary materials may be safely disposed of by burying or land-fill operations, it should be remembered that water-soluble materials may eventually leach into streams and wells and create a hazardous situation. Buried materials may also create a severe hazard when dug up during later construction or other earth-moving operations. Land-fill operations are acceptable for many ordinary materials of organic origin, but the use of this method should be carefully limited to those which will not cause hazards later on.

Commercial Disposal Service

Many cities and industrial areas can obtain the services of commercial disposal companies who will remove and destroy or recover valuable chemicals from waste materials. This should be investigated, since it may not only be the safest but also the most economical solution of the problem.

Bibliography

Air Pollution Abatement Manual, Manual Sheet P-1, Manufacturing Chemists Association.

Fire Hazard Properties of Certain Flammable Liquids, Gases and Volatile Solids, NFPA 325M-1969, 49-1969.

Hunter, D., *The Diseases of Occupations,* 4th ed., Little Brown and Co., Boston, Mass., 1969.

Laboratory Waste Disposal Manual, Manufacturing Chemists Association, Washington D.C., 1969.

Sax, N. I., *Dangerous Properties of Industrial Materials,* 3rd ed., Van Nostrand Reinhold Co., 1968.

14

MISCELLANEOUS HAZARDS

HANDLING GLASSWARE

Most chemical laboratory work involves the use of glassware, and glass breaks easily. Therefore laboratory personnel must guard against being casual in handling glassware and always treat it with the care it requires. Vigilance in the selection and use of glass equipment can prevent breakage and subsequent injury from broken glass or from the contents. In the case of unavoidable rupture of glass equipment, care can prevent the accident from having serious consequences.

Receiving, Inspection, Storage

Shipments of glassware should be examined and the contents of all packages checked against the vendor's marking for correct storage. Manufacturers' or suppliers' catalog numbers serve as satisfactory guides for identification. Packages should be piled so as not to impose excessive weight on the lower containers. Storage should permit removal without risk of breakage.

Each package containing glassware should be opened and inspected for cracked or nicked pieces, pieces with flaws that may become cracked in use, and badly shaped pieces. Such rejects, if not returned to the supplier, must be disposed of in an approved receptacle. All packaging materials should be removed from each piece before storing.

Glassware should be stored on well-lighted stockroom shelves designed for this purpose. The shelves should have a coping of sufficient height around the edges to prevent the pieces from falling off. The following rules should be observed:

1. Store heavy pieces on lower shelves, light pieces on upper shelves. Store tall pieces at the back, smaller ones toward the front of the shelf.
2. Store glassware no higher than a person can easily reach without standing on a stepladder or something even more dangerous, such as a box or shelf.
3. Store glass tubing and rods in a horizontal position separated according to length, with no piece protruding over the coping.
4. Store delicate pieces in separate cartons clearly marked for ready identification.
5. As in the laboratory proper, keep aisles between shelves in the storeroom clear of obstacles and debris at all times.

Selection

The piece of glassware that is designed for the type of work planned is the one to select. For pressures slightly above normal, pressure bottles should be chosen; for filtration with the aid of suction, vacuum flasks. Each piece should be inspected before use. Types of glass rods and tubing can be identified by refraction in a medium of carbon tetrachloride and benzene by comparison with approved standards. In a mixture composed of 59 ml of carbon tetrachloride and 41 ml of benzene, Pyrex glass is invisible while soft glass is clearly visible (Fig. 14-1).

Setting up Apparatus

Apparatus (a combination of two or more units) is set up with the units in line and adequately supported by clamps on stands so that no unit exerts a strain on another. The use of spherical joints in setting up complex systems diminishes the probability of breakage due to misalignment. Laminated safety glass or heavy plastic protective shields can be placed around the apparatus to protect workers on both sides of the bench if necessary.

Cutting Tubing and Rods

To cut glass tubing or a rod, make a straight, clear mark with a cutter or sharp triangular file at the point where the piece is to be severed. Use leather gloves and eye protection. Make the break by placing the fingers

Fig. 14-1. Method of identification of soft or Pyrex glass.

around the piece with the thumbs together opposite the cut, and bend toward the body (Figs. 14-2 and 14-3).

Large-size tubing may be cut by means of a nichrome wire looped around the piece after marking with a file or cutter at the desired point of severance. The wire should be heated electrically with a rheostat in the circuit. When the wire around the glass is hot, place a drop of water along the wire, thus causing the glass to break. Cut ends of glass tubing or rod are then fire-polished.

Large tubing may also be cut safely and accurately by means of a motor driven cut-off wheel.

Glass and Rubber or Cork Connections

When it is necessary to insert a piece of glass tubing or a rod through a perforated rubber or cork stopper, select the correct bore so that the insertion can be made without excessive strain. The glass and/or stopper should be wet (water or glycerine). Hold the stopper in the fingers of the

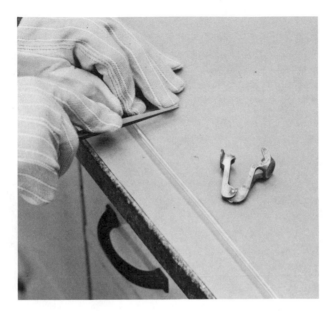

Fig. 14-2. Proper scoring for cutting of glass tubing.

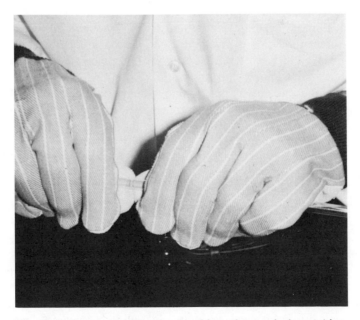

Fig. 14-3. Proper handling for breaking of scored glass tubing.

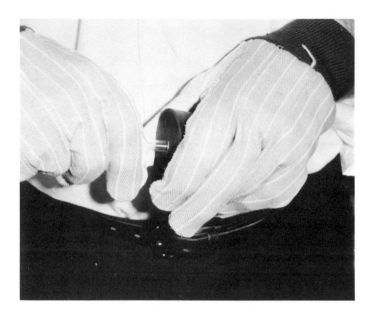

Fig. 14-4. Proper method of inserting glass tubing into rubber stopper.

other hand placed near the end being inserted, avoiding any leverage motion. Keep hands out of the line of the end of the glass being pushed through the hole. Pay careful attention to this procedure when working with bent glass tubing. It is always best to wear leather or heavy coated cotton gloves or to wrap a towel around the palms of the hands. Bear in mind that leather gloves are preferable (Fig. 14-4).

Another good method of positioning glass tubing in rubber stoppers is to use the borer as a guide. Merely insert the borer in the stopper, position it along the tube, and then remove the borer.

It is occasionally necessary to remove a glass tube or a rod that is stuck to a rubber stopper. Sometimes it can be separated by rolling the stopper with a block of wood under enough pressure to flex the rubber. If the glass tubing is straight, a brass cork borer that will fit neatly around the tube can be used to cut the rubber and free it. If it is difficult to free the tube, it may be preferable to scrap both the stopper and the tube.

Rubber stoppers or corks are appropriately selected to fit the opening into which they are to be inserted and of correct density to make a tight closure without undue pressure. Use the fingers and not the hands when making this insertion.

Choose rubber tubing to fit glass connections snugly, with the end of the rubber cut at an angle so that it will stretch easily. To make connections, grip the glass piece close to its end with the fingers, keeping the hands out of its line. Wetting of the glass will expedite insertion (water or glycerin). If possible, formation of a slight bulge followed by a taper on the end of the tube will facilitate making the connection.

Heating of Glassware

Flasks are usually best heated by electric mantles, particularly when flammable liquids are involved. If mantles are not available, a bath of liquid or sand heated with an electric hot plate is a second choice. If glassware must be heated with an open flame, it must be rested on a nichrome wire gauze or iron mesh with an asbestos center to prevent the flame from coming into contact with the glass above the level being heated. This will minimize the chance of cracking the vessel.

Confining

Pressure flasks which are to be heated under temperature control, as by a water bath or electric oven, must be shielded so that in case of breakage the contents and fragments of broken glass will be safely deflected. Similarly, any glass apparatus used at subatmospheric pressures must be shielded.

Pressure Relief for Glassware

Pressure should be slowly released from, or applied to, glass vessels. Provide a pressure control valve, a water column trap or other appropriate relief device installed between a cylinder of compressed gas and a glass vessel into which the gas is to be conducted. An empty flask, bottle, or suitable surge-equalizing container can be installed where there is likely to be a change in pressure that might cause liquid in one vessel to flow into another and create a hazardous reaction or condition.

Pressurizing Glassware

It is generally unsafe to put pressure on laboratory glassware to expel its contents. An exception is the use of the laboratory wash bottle, where only light pressure is applied. Liquids might be transferred by pouring or syphoning.

Small-scale reactions may be safely carried out at pressures somewhat above atmospheric in sealed glass tubes by heating the tube in a metal bomb (a pipe capped at both ends) with some of the same liquid in the bomb as is contained in the sealed tube. This equalizes the pressure on the sealed tube and prevents breakage.

Outage in Bottles

Fill reagent and other bottles to not more than 90% capacity at room temperature, leaving 10% of the capacity as outage to allow for their expansion.

Carrying Containers

Severe vibration of a vacuum container may result in a collapse equivalent to the shattering effect of an explosion. If it is absolutely necessary to move containers while under vacuum, vacuum desiccators may be transported in a wooden box or metal shield.

Beakers and all usual shapes of flasks and bottles should be carried with the fingers around the body of the vessel. Do not grasp or hold by the lip or edge.

Fig. 14-5. Safe method of mixing contents of a volumetric flask.

Volumetric and other long-neck flasks should be held at both top and bottom when their contents are being agitated (Fig. 14-5). When examining the contents of glassware by light from above, the laboratory worker should hold the vessel at a safe distance and height to avoid injury to the head and eyes in case some of the contents should spill or the vessel should slip out of control.

Transport all bottles in standard carriers; no more than can be carried safely should be taken at one time.

Special metal or plastic containers should be used to transport bottles of acids, alkalies, or other corrosive or flammable liquids; the capacity of the container should be sufficient to hold the contents of the bottles if they should break.

For construction of the container, select a material resistant to the chemical to be transported.

Reagent Shelves

Double-sided reagent shelves require a center partition high enough to prevent the pushing off of a bottle from one side by pressure from the opposite side.

Glass flasks, empty or containing clear liquids, should be placed where the sun's rays cannot fall upon them. This will avoid potential fires.

Large bottles of acids, alkalies, or other corrosive liquids should not be placed on shelves above head level. They also should be placed in pans or similar container of appropriate material such as lead, stainless steel, or rubber.

Reagent or solution bottles provided with delivery tubes should be secured so that they cannot be tipped when the contents are being withdrawn.

Cleaning Glassware

Before turning glassware over to another person to be thoroughly cleaned for storage, the user must see that each piece is free of any material that might cause a hazard.

The user must clean any apparatus which may require special treatment because of the material it may contain or because of its design and construction.

In washing glassware, no more than gentle pressure should be applied in wiping the inside. Excessive pressure may cause jagged sections to split off. The washer should wear rubber gloves to prevent cuts and skin damage from chemical irritants. The use of a mixture of trisodium phosphate and pumice or other mild cleaners or equivalent is preferred to strong acid or caustic.

Correctly designed brushes should always be used to clean glassware. Worn brushes should be discarded, because the metal portions of the brush may scratch the glass.

Broken Glass Disposal

To remove glass splinters from a sink or other area, use a whiskbroom and a dustpan. Small pieces can be picked up with a large piece of wet cotton. Do not allow broken glass to enter the sewer system.

Provide a separate specially marked receptacle for broken glassware, which must be kept out of baskets or containers of wastepaper, rags, and other discarded materials.

Glass Stoppers

In this manual, no great emphasis is placed on a safe procedure for removing stuck (frozen) glass stoppers in bottles. This was a frequent hazard years ago, but a commercially available dependable bottle stopper remover has now been developed and is recommended.

Special Hazards

Special hazards are encountered in handling explosives, oxidizing materials, poisonous substances, radioactive materials, and flammable solvents in glassware. In such instances, specific precautions should be formulated for the guidance of the laboratory technicians.

LABELING

Proper labeling of chemical containers is necessary to insure safe operations. The labels should identify[1] the contents and, at the same time, be instructive about handling precautions. *The Guide to Precautionary Labeling of Hazardous Chemicals*[2] Manual L-1, published by the Manufacturing Chemists Association, Inc., is a primary guide to proper labeling. The present manual offers only general guidance. Some laboratory operations are difficult to stereotype. The following suggestions may be found useful.

A laboratory uses both chemicals manufactured on the outside and also those prepared by its own personnel. Chemicals received from the outside are usually labeled in accordance with applicable federal and state regulations. It is highly important that each time a chemical is transferred from an original container to another one, this latter container should be

Fig. 14-6. Proper labeling of transferred chemicals.

properly labeled with a duplicate or facsimile of the original label (Fig. 14-6).

There is probably no greater hazard around the laboratory than that of unmarked or improperly labeled chemicals. All chemicals, especially those prepared in the laboratory, should have complete identification securely fastened to the container (Fig. 14-7). Chemicals of unknown stability and those which deteriorate with age should have a preparation date clearly indicated on the label.

The purpose of proper labels is multifold: (1) They indicate the source, supplier, or manufacturer of the chemicals. (2) They warn about the possible hazards. (3) They indicate the age of the chemical; this information will be extremely helpful in case of any unstable compound. (4) Labels throughout the laboratory will be uniform. Such uniformity will contribute not only to good safety practice but also to neater laboratory housekeeping.

While most of the foregoing is applicable mainly to bottles and con-

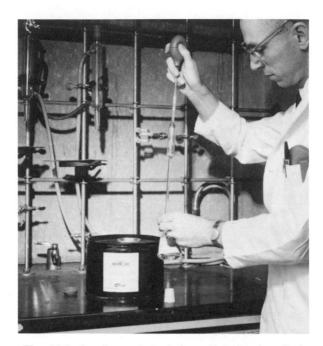

Fig. 14-7. Small-sample technique. Note label on flask.

tainers up to five gallons, the same guide is valid for larger containers, cylinders, and so on. If the material is to be stored outside, or if there is the possibility of spillage over the label, the label should be suitably over-coated or placed into a plastic envelope and wired to the container. Fire insurance carriers, who maintain competent staffs of engineers for advising and inspection, usually wish to be consulted regarding installations in hazardous areas.

The importance of proper labeling cannot be overemphasized, as has been long recognized by laboratory workers. In recent years, state legislatures and the United States Congress have passed regulations and laws governing labeling. Thus, proper labeling is not only necessary for safety[3] but also required by law.[4]

ELECTRICAL AND MECHANICAL HAZARDS

Electrical

The varied hazards met in chemical laboratory operations require careful selection, proper installation, and adequate maintenance of electrical equipment in order to avoid explosions and fires. The most serious problem is that incurred in areas where flammable liquids, gases, and dusts are involved.

The National Fire Protection Association in the National Electrical Code offers guidance as to classification of atmospheres and types of equipment approved for use for the class of location and the specific gas, vapor, or dust present. The reader is cautioned that interpretations of the code and complex installations are best left to competent individuals who are familiar with the requirements and will follow local, state, and national codes.

The National Electrical Code states: "For purposes of testing and approval, various atmospheric mixtures have been grouped on the basis of their explosive characteristics, and facilities have been made available for testing and approval of equipment for use in the following atmospheric groups":

Group A, atmospheres containing acetylene.

Group B, atmospheres containing hydrogen, or gases or vapors of equivalent hazard such as manufactured gas. (Manufactured gas refers to a mixture of coal gas, water gas, and petroleum hydrocarbons adjusted to 540 BTU per cubic foot. (It contains a high percentage of hydrogen.)

Group C, atmospheres containing ethyl ether vapors, ethylene, or cyclopropane.

Group D, atmospheres containing gasoline, hexane, naptha, benzene, butane, propane, alcohol, acetone, benzol, lacquer solvent vapors, or natural gas.

Group E, atmospheres containing metal dust, including aluminum, magnesium, and their commercial alloys.

Group F, atmospheres containing carbon black or coal or coke dust.

Group G, atmospheres containing flour, starch, or grain dusts.

Groups A through D cover Class I, gases and vapors, whereas the others cover Class II, dusts.

Class I locations are those in which flammable[5] gases or vapors are or may be present in the air in quantities sufficient to produce explosive or ignitable mixtures. Class I is then subdivided into:

Division 1. Locations in which hazardous concentrations of flammable gases or vapors exist continuously, intermittently, or periodically under normal operating conditions, in which hazardous concentrations of such gases or vapors may exist frequently because of repair or maintenance operations or because of leakage, or in which breakdown or faulty operation of equipment or processes which might release hazardous concentrations of flammable gases or vapors might also cause simultaneous failure of electrical equipment.

Division 2. Locations in which flammable volatile liquids or flammable gases are handled, processed, or used but which will be normally confined within closed containers or systems from which they can escape only in case of accidental rupture or breakdown, or in case of abnormal operation of equipment, in which hazardous concentrations of gases or vapors are normally prevented by positive mechanical ventilation, but which might become hazardous through failure or abnormal operation of the ventilating equipment.

The choice of electrical equipment in hazardous areas is thus firmly defined by the code. For example, an area involved in the use of hydrogen and defined as Class I Group B, must be provided with Group B approved equipment; Group C equipment cannot be used since it is not intended or approved for this application. Improper installations or misin-

Fig. 14-8. An explosion-proof electrical installation, including Variacs®.

terpretations for convenience on the basis of availability give a false sense of security and can lead to serious consequences.

Class II locations are those in which combustible dusts may be present. Dust-tight electrical equipment must be so located that the dust deposits will fall off, or that regular cleaning be maintained.

The complexities of interpretation and installation of the aforementioned are such that it is advisable to reiterate the importance of the use of competent and expert personnel in this area. The accepted standard for installation of electrical equipment is the National Electrical Code available from the National Fire Protective Association, Boston, Mass., the American Insurance Association, or the American National Standards Institute (Figs. 14-8 and 14-9).

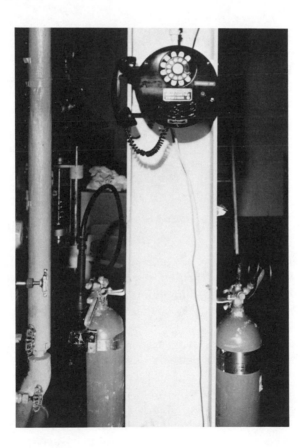

Fig. 14-9. Special telephone in an explosion-proof area. Note the tag on the right indicating the emergency telephone number for obtaining assistance in the event of an emergency.

Fig. 14-10. Grounding of electrical equipment.

General recommendations or counsel can be made as follows: Electrical distribution systems and their outlets should be provided with grounded connections, so that plugged-in laboratory equipment is automatically grounded (Fig. 14-10). If this is unavailable, special grounding connections should be provided.

Spillages of flammables in an area where electrical equipment is in use should be handled with extreme caution to avoid fires. Power to the area should be shut off, rather than being locally disconnected, to avoid the generation of sparks or arcs which might ignite the vapors. Power should not be restored until the area is clear and the equipment checked for possible damage. It is advisable, before the purchase of electrical equipment, to consult with engineers or electrical service personnel to determine whether the equipment is compatible with the area in which it is to be installed. Consideration should be given to whether the equipment is explosion-proof or not, its voltage and amperage, and its effect on existing circuits and equipment.

Certain types of operations are capable of generating static electricity; provisions should be made for grounding to avoid electrical arc or spark formation.

Care must be exercised in the use of electrically driven blenders which are in wide use in research laboratories. Explosion-proof or air-motor-driven blenders should be used for blending of flammable solvents. Non-explosion-proof blenders should be restricted to uses not involving flammable liquids.

Explosion-proof refrigerators should be used in hazardous areas. Explosion-safe or modified refrigerators may be used to store flammables if the general area is free of vapors.

Temporary wiring is poor practice, and should be avoided. If it must be used, it should be of limited duration; the actual installation should be neat, orderly, and safe.

All switches should be labeled to show the equipment they control.

Portable equipment should be inspected by an electrician on a quarterly basis, and repairing should be done as necessary.

Mechanical

Laboratory equipment—benches, hoods, auxiliary equipment—should be so installed that they provide prompt and easy access to valves, switches, and controls. It may be advisable, in individual laboratories that contain several benches and hoods, to install control valves in the utilities serving each bench or hood (Fig. 14-11). This will afford the opportunity to shut off individual units in the event of trouble and provide for convenient maintenance without the need for a complete shutdown of an entire laboratory. Certain utilities, such as steam, hot water, cold systems, should of course be properly insulated to avoid burns and frostbite. Aisles and passageways should be kept clear and uncluttered to provide ease of mobility and eliminate tripping hazards. Wet floors caused by spillages or leaks present hazards which should be corrected immediately. Floor waxes should be of the nonskid type.

Some laboratory operations require the use of standard tools such as used by mechanics. These should always be in good condition, and the proper size should be used for the application required; for instance, the use of an undersized wrench can result in the tool slipping and possibly injuring the user. Cutting or sharp-edged tools should be stored in proper containers and used in such a manner that the sharp edge is directed away from the body or its appendages.

Fig. 14-11. Readily accessible shutoff valves of services to a distillation table.

Fig. 14-12. Guard installed on a vacuum pump.

Fig. 14-13. A convenient "ship's ladder" with foot pads and handrails.

The moving parts of equipment, such as belt or gear drives, should always be protected by adequate metal guards (Fig. 14-12).

Ladders and portable platforms, widely used in laboratories, should be of proper size and height. Straight ladders should be provided with base safety grips, and even fastened in position or steadied by another man (Fig 14–13). A straight ladder is usually placed so that its base is at a distance of one-fourth the length of the ladder from the wall. For example, the base of a sixteen-foot ladder would be four feet from the supporting wall. All ladders should be inspected frequently for such defects as split rungs or rails and repaired or discarded when defective.

Halls, stairways, and passageways should be clear at all times and should not be used for storage purposes. Fire regulations require clear passageways in the event of an emergency requiring quick egress by personnel.

Heavy lifting or carrying should be avoided. The use of proper mechanical equipment, such as hand trucks and carrying cases, will avoid injury to personnel.

The use of special equipment requires special precautions. The use of ultraviolet light necessitates special protective glasses for operating personnel. Infrared heating lamps should be enclosed in protective devices to protect personnel from flying glass in the event of accidental breakage. X-ray equipment should be lead-shielded, and operators should wear film badges.

Apparatus setups on benches or in hoods should be well planned and organized so that the equipment is firmly installed. Good lighting and proper ventilation should be provided, especially in operations involving hazardous or toxic chemicals. The use of shatterproof glass in shields and hood doors is advisable.

The use of isolated areas for highly hazardous operations is advisable to minimize personnel exposure to incipient dangers.

It is incumbent upon research and laboratory personnel to report any malfunctions of mechanical equipment and to arrange for its repair.

References

1. *Identification of Materials,* Safety Guide SG-18, Manufacturing Chemists Association.
2. *Guide to Precautionary Labeling of Hazardous Chemicals* Manual L-1, Manufacturing Chemists Association.
3. *Case Histories of Accidents in the Chemical Industry,* Manufacturing Chemists Association, Vol. 1, 1962; Vol. 2, 1966; Vol. 3, 1970.
4. Nale, T. W., M.D., "The Federal Hazardous Substances Labeling Act," *Archives of Environmental Health* **4** (March 1962).
5. *Identification of the Fire Hazards of Materials,* NFPA No. 704 M—1969, National Fire Protection Association.

15

EVALUATING THE HAZARDS OF UNSTABLE SUBSTANCES

Continuing developments in the chemical industry lead to more new substances, many of which present undefined explosion hazards. Evaluation of the associated hazards must begin in the laboratory, and must accompany each stage of the development up to production. The larger the quantities of materials involved, the more catastrophic an accident can be and the more uncertain is the value of the conventional methods used to evaluate the potential hazards. In the following discussion, means of estimating the inherent hazards of individual chemicals or simple mixtures are considered, but no attempt is made to discuss in detail questions of compatability of mixtures. In general, however, the problem of compatability is accessible by many of the same techniques used to evaluate the stability and sensitivity of individual compounds.

Hazard evaluation frequently suffers from what may be called the "test syndrome," which develops from the compulsion to measure things, to seek relative numbers, and to ignore basic mechanisms. Admittedly the mechanisms of initiation and decomposition are complex and difficult to study. Nevertheless, a small but sustained attack by a few laboratories on the problem of basic mechanisms is beginning to bear fruit, and the blind dependence on comparative numbers that has characterized testing in the past is no longer necessary. This is an important gain, because numbers can, unfortunately, give a false impression of the relative hazards of two materials. This false impression may be due to idiosyncrasies of the test, so that apparatus-dependent variables may not apply in the same way to different materials or that size and mass effects may be completely ignored by the experimental configuration. Similarly, the gross variations associated with these tests tend to obscure small variations in the composition, or properties of the sample may have a significant but unsuspected effect on the test data. More important, a misconception as to the nature of the test result may confound its proper interpretation.

SENSITIVITY TESTS

Traditionally, methods for evaluating sensitivity have been divided into two categories: thermal and mechanical. Thermal methods comprise various procedures for heating samples, ranging from storage at elevated temperatures to extremely fast heating to high temperatures, with decomposition occurring in milliseconds or even microseconds. Mechanical tests are based on some type of friction, impact, or shock acting on the sample. However, there is now good evidence that all of these methods really involve thermal effects, that there are in fact no tribochemical or other exotic effects, although the physical processes of the mechanical methods may obscure the actual cause of heating and its localization into hot spots. The picture becomes clearer if the initial decomposition in the hot spot is regarded as simply a thermal process leading to a thermal explosion of the hot spot and subsequent ignition of the surrounding material. When homogeneous materials are subjected to intense shock waves, the bulk of the material is heated layer by layer as the shock wave passes through it. If the shock wave is sufficiently intense, a thermal explosion results after a time lag. These concepts will be discussed more fully later.

Thermal Tests

Despite the above statements on the commonality of basic heating mechanisms, it is convenient to discuss the thermal tests separately; these will be treated very briefly because most of them are in common use and are fairly well understood. Probably the earliest and simplest, and still useful, test involved dropping the material on a hot plate to see what happened. The first recorded "standard heat test" was that of Abel, who in 1867 employed potassium iodide indicator paper to observe the accelerated decomposition of nitrocellulose at elevated storage temperatures. Since that time, vacuum stability and Taliani (constant pressure) tests have been widely employed to evaluate thermal stability at elevated temperatures. Recently, gas chromatography has been applied to increase the amount of information produced by vacuum stability test.[1]

Employing still higher heating rates, differential thermal analysis and differential calorimetry are now in common use. These methods indicate the temperature at which an exothermic reaction begins, and give some measure of the quantity of heat released. Both methods involve rather complicated and specialized techniques that must be tailored to individual samples, so that no standard procedure is available. These methods too have been supplemented with gas chromatographic analysis of the decomposition products. For heating rates on the order of a few seconds to explosion, the technique of Henkin and McGill,[2] or the more recent version of Rogers,[3] is frequently employed. A small sample of the material confined in a metal capsule is dropped into a molten Woods metal bath, and the time to explosive decomposition is determined as a function of the bath temperature. If the investigator is ambitious and not overly concerned with scientific rigor, an activation energy can be computed.

The Wenograd[4] technique is near the fast end of the time scale of heating rates, exceeded only by the electric spark test. In this technique the sample is loaded into a hypodermic needle, which is heated very rapidly by a condenser discharge, the needle behaving somewhat like an exploding wire. Again the time to explosive decomposition is obtained as a function of temperature. Wenograd was able to find a close correlation between the temperature at which explosive decomposition occurs in 250 microseconds after the start of discharge and the drop weight impact sensitivity of a number of explosive materials (Fig. 8 of ref. 4). Gross and Amster[5] obtained a similar correlation employing an adiabatic heating method. They found that the impact sensitivity of a number of explosives correlated reasonably well with the critical temperature for a sphere with a radius of 0.1 cm. The theory of the role of thermal explosions in the drop weight test is discussed by Boddington,[6] Friedman, Macěk, and others.

Mechanical Tests

Turning now to impact sensitivity methods, we find many "standard" devices, such as those developed by the Bureau of Mines, the Explosives Research Laboratory (NDRC), and Picatinny Arsenal for solids, and the JANAF Test 4 apparatus (Olin Mathieson) for liquids. In all of these devices a weight is dropped from a given height onto the sample, which may be contained in a cup or spread on a flat anvil. The ERL (sometimes referred to as the NOL (Naval Ordinance Laboratory)) technique for solid samples differs from others in that a disc of sandpaper is inserted under the sample to provide a reproducible supply of hot spots for the initiation process. A British variation is the "Rotter technique,"[10] in which the firing pin, anvil, and sample are enclosed in a gas-tight box, and the amount of gas evolved by the sample upon impact is used to determine whether or not the sample reacted and to what extent.

The liquid impact of Test No. 4[9] is widely used in the United States. A number of problems were encountered in its early application, but these have been fairly well resolved.[11] Two instrumented versions have been described which sense the pressure produced upon impact.[12,13]

Recently Richardson[14] developed a thin-film impact device for evaluating the sensitivity of casting solvents to impact. In one version, the sample is surrounded by a gas-tight container, and the gases formed are analyzed by a nondispersive infrared analyzer; this allows detection of an extremely small amount of decomposition, which is important in investigating threshold conditions for initiation.

Common to nearly all impact sensitivity techniques is the use of the Bruceton up-and-down method.[15,16] This experimental design allows for a highly efficient determination of the stimulus that should cause initiation 50% of the time. This is usually expressed in terms of a height from which a constant weight is dropped, although it can also be a variable weight dropped from a constant height. This design, which concentrates the experimentation around the median value, gives a poor estimate of the distribution of the population, and thus of the standard deviation required to estimate the conditions for a low probability of initiation. Rejecting this technique, Richardson[4] has adopted a threshold initiation level (TIL) which he defines as the minimum condition for twenty consecutive negative results.

Numerous methods have been devised to evaluate the friction sensitivity of solids. In general, they take the form of frictional impact rather than pure friction; the sample is subjected to a glancing blow from a mallet swung by a practiced laboratory technician or from some mechanical device. Gurton[17] has reviewed a large mass of data obtained with a purer form of friction equipment which employed a loaded slider on a rotating plate. Although these empirical methods provide useful practical data for the choice of tools, utensils, and containers, the nature of the results and their significance in regard to the behavior of materials are somewhat more difficult to assess than for other tests. Again initiation occurs in a hot spot, the size of which is determined by the conditions of the experiment and the nature of the material tested.

Rifle bullet tests have been popular, deriving originally from concern over the initiation of munitions by small arms fire. In the more recent versions, the velocity of the projectile is varied to find a critical velocity for initiation. Following the lead of Eldh and others,[18] the U.S. Bureau of Mines has adopted a projectile test in which ½-in. by ½-in. brass cylinders are fired from a smooth-bore gun based on a Mauser 13-mm action chambered to accept M-2, 50-caliber ammunition. The powder load is adjusted to give the desired velocity, which is monitored by a chronograph. This procedure has proved useful in investigating initiation of both extremely cold (cryogenic) explosive systems when other sensitivity techniques posed experimental problems. The initiation process is either a combination of impact and frictional ignition which grows to detonaton or direct initiation by shock from the projectile.[19,20]

In yet another type of projectile test—the so-called Susa Test[21]—the material under study is contained in the projectile rather than struck by it. Crushing impact is provided by firing the projectile against armor plate at low velocity.

It is important to recognize that the reaction of the sample in the impact and friction tests involves ignition or initiation of a deflagration reaction and that the result typically gives little if any information as to the possible consequences of this ignition. The essential question is whether or not the deflagration, once initiated, will undergo transition to detonation (DDT). In impact tests, this transition takes place only with primary explosives, and it is indeed this facile transition to detonation that distinguishes primary explosives from secondary explosives.

DDT Methods

Some techniques have been developed to investigate the DDT process directly; although many of these are highly empirical, they can give results of practical value. Koenan and Ide[22] employ a vented bomb heated by a gas burner, and determine a critical vent diameter for the material. A critical vent size is that between a vent causing the bomb to explode and a larger one that does not. Typical results vary from diameters of 1 to 1.5 mm for dinitrotoluene to diameters of 20 to 24 mm for nitrocellulose, the latter being nearly equal to the diameter of the test bomb. Gipson and Macek[23] have investigated the DDT process in high explosives, and Richardson[24] has reported on a DDT test developed at Allegany Ballistics Laboratory in which the sample is contained in a pipe closed only at the bottom and is ignited with a bag igniter.

An instrumented DDT method has been developed by the U.S. Bureau of Mines. Combining the two concepts described above, it consists of a vented bomb in which the size of the vent can be varied, and employs a solid propellant igniter. Pressure transducers measure the rate of pressure developed as a function of time, and transition to detonation can be unmistakably identified by the expendable pressure transducer at the base of the charge (Fig. 15-1). Even such slow-burning materials as mixtures of ammonium nitrate and fuel can be made to undergo a transition to detonation in this equipment.

It is frequently but erroneously believed that the maximum pressure

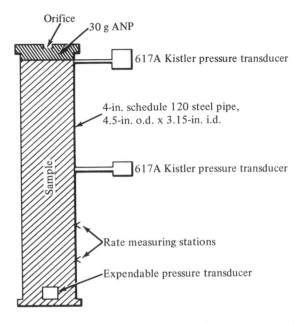

Fig. 15-1. Schematic diagram of the apparatus used to study transition to detonation of ammonium nitrate-fuel mixtures.

that can exist at the base of a pile of material is the hydrostatic pressure or, if confined, that pressure increases by the strength of the vessel. If an exothermic gas-forming reaction occurs at the bottom of a pile and gas production is too rapid for the gas to escape through the interstices of the pile, the pressure at the base of the pile is related to the acceleration given the pile by the equation

$$P = \rho ha$$

where ρ is the density, h is the height (or smallest dimension) of the pile, and a is the acceleration. In the case of a liquid in a pipe or tank, the maximum pressure is similarly determined by the inertial mass of the liquid to be accelerated or by the dynamic strength of the container.

Initiation of Detonation

Before considering methods of testing materials for their sensitivity to direct shock initiation, a brief review of the mechanisms involved in the initiation of detonation may be worthwhile. Although these mechanisms have been frequently discussed in the recent literature, they are still not clear to newcomers to the field. It is important to distinguish between be-

havior of homogeneous explosives, such as liquid explosives, and heterogeneous, or packed-bed solid explosives. The mechanisms are normally
quite different, although these differences tend to disappear in certain circumstances, as will be shown later. In homogeneous liquids, the entering
shock behaves as a normal, inert shock and moves through the liquid
with decaying amplitude and velocity. After a certain delay time, if the
shock has sufficient amplitude, a detonation begins at the interface first
shocked and proceeds through the compressed and heated medium at
abnormally high velocity (compared to that through unperturbed liquid)
until it overtakes the original shock. At this point the two fuse, and the
detonation proceeds at normal velocity.

On the other hand, in heterogeneous explosives the entering shock
wave immediately initiates a combustion reaction on the surface of the
grains if it is above some minimum value. Gaseous products resulting
from this reaction increase the amplitude and velocity of the shock wave.
The wave accelerates continuously up to stable detonation, the velocity
of which is defined by the density and charge dimensions for a specific
material. An understanding of the differences in the behavior of homogeneous and heterogeneous systems is important in evaluating the signifi-
cance of sensitivity test data. Ordinarily the critical shock strengths required to initiate detonation are much lower for solids than for liquids.

Gap Tests

Shock sensitivity of both solids and liquids is usually determined by one
or more of the numerous "standard" gap sensitivity tests. Among the more
popular versions are the NOL[25] and the JANAF tests.[27] In both versions,
an explosive donor, usually tetryl or pentolite, is separated from the test
charge by an attentuator consisting of a stack of cellulose acetate cards,
or a Plexiglas cyclinder, or a combination of both. The "up-and-down"
technique is employed to find the attenuator (gap) thickness at which
initiation occurs 50% of the time.

Solid explosive materials usually present little difficulty when evaluated
in the card gap test. The most important concern is the relationship of
the charge diameter to the critical diameter of the material under test.
The nearer the charge diameter is to the critical diameter, the more in-
sensitive the solid will appear, because of the greater difficulty of growth
to detonation from shocks of marginal strength. Again a "standard test"
may not reveal the true behavior of the material. Elevated temperatures
and small particle sizes reduce critical diameters. For example, one type
of fertilizer-grade ammonium nitrate may have a critical diameter as large
as 40 in. at ordinary temperatures and as small as 3 or 4 in. at 140°C

(melting point, 170°C). The same ammonium nitrate, if pulverized and packed at low density, has a critical diameter below 1 in. at ordinary temperatures.

Liquids, on the other hand, have presented disturbing anomalies from the earliest application of the gap test. The confusion that has persisted must be blamed largely on the reliance on a myopic witness plate as the indicator of the result. This steel plate was presumed to show unambiguously whether or not the sample fired. However, a number of years ago it was found that the plate was not necessarily simply perforated, indicating a positive result, or undamaged, indicating a negative result, but sometimes suffered all kinds of intermediate damage. In addition, it was found that the container sometimes had a profound influence on the test results. These effects differed from those which would be predicted on the basis of conventional concepts of confinement. Thus nitroglycerin gave a gap value of about 0.4 in. in a heavy-walled steel pipe, whereas in thin-walled aluminum tubing, gap values ranged from 2 to 4 in. Elucidation of these

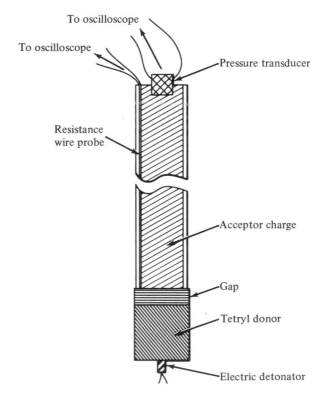

Fig. 15-2. Modified card gap arrangement.

ambiguous results required an extensive study using electronic and high-speed photographic techniques. With the development of the instrumented card gap test, it was established that the observed card gap values represent thresholds between high-velocity and low-velocity detonations rather than between initiation and failure.

The instrumented card gap test as developed by the Explosives Research Center employs two types of instrumentation and is shown in its latest version in Figure 15-2. One is a continuous velocity-measuring probe consisting of a resistance wire usually encased in an aluminum tube (0.020 in. inner diameter, 0.023 in. outer diameter) although sometimes it is simply wound around a second wire. The second employs an expendable pressure transducer inserted in the far end of the acceptor charge. This transducer measures pressures in the 1 to 75 kbar range and allows unambiguous interpretation of the result. Figure 15-3 illustrates the type of data that can now be obtained. These data were collected using a system of nitroglycerin-triacetin (a 50:50 mixture of nitroglycerin and ethylene glycol dinitrate was used in place of neat nitroglycerin because it was more readily available). With neat NG-EGDN in 16-in.-long, 1-in.-diameter aluminum tubes, the 50% point for the threshold between high-velocity

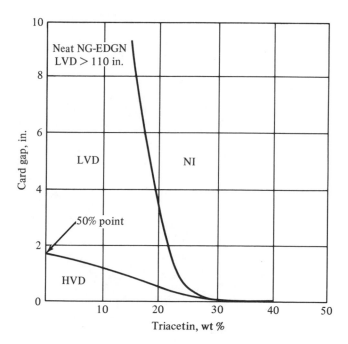

Fig. 15-3. Gap sensitivity of a 50-50 NG-EGDN triacetin system in 1-in. × 16-in. aluminum containers at 25°C (50-g donor).

and low-velocity detonation falls at about 2 in., whereas low-velocity detonation could still be initiated through 110 in. of Plexiglas. Clearly, this ease of initiation of low-velocity detonations (which propagate at 1.8 to 2.2 mm/μsec in most charge configurations) is the cause of greatest concern in the accidental initiation of nitroglycerin. The thresholds are diameter dependent. As the charge diameter increases, the threshold between low-velocity detonation and noninitiation moves to the right, and the threshold between high-velocity and low-velocity detonation moves to larger gap values. There is presumably some minimum charge diameter below which nitroglycerin will not propagate detonation but this has not yet been established. In ¼-in. aluminum tubing (0.035-in. wall), NG-EGDN still shows all three types of behavior—high-velocity detonations, and failures. This diameter effect again illustrates the problems that may be encountered in relying on a standard test.

A new type of gap test has recently been developed at the Explosives Research Center in which the behavior of liquid systems in thin films can be examined. In this version, a thin layer of the liquid is placed in an open tray consisting of ½-in.-thick Plexiglas plate with thin walls. A shock

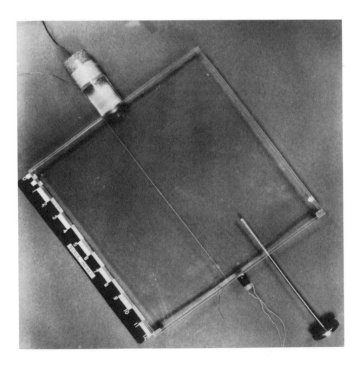

Fig. 15-4. A thin-film gap configuration.

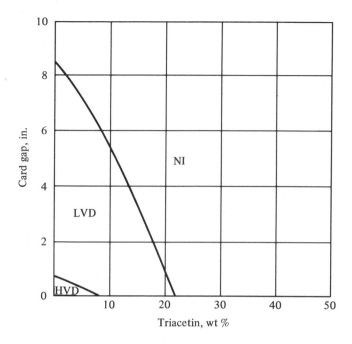

Fig. 15-5. Gap sensitivity of a 50-50 NG-EGDN triacetin system employing 1/16-in. film on 1/2-in. Plexiglas at 25°S (50-g donor).

source, consisting of a standard donor and Plexiglas attenuator, is employed to determine the limiting conditions for initiation. One test configuration is illustrated in Figure 15–4. Again, high- and low-velocity detonations have been observed in the NG-EGDN triacetin system. The highest % of triacetin which still yielded a high-velocity detonation under these conditions was 5% for a 1/16-in. layer, in contrast to the 25% value obtained in the 1/4-in. aluminum tube. These results are in reasonable agreement with those reported by Richardson,[14] who suggested that the region between partial propagation and no propagation lay between 80 and 85% nitroglycerin. The difference may well stem from the relative strength of the initiating stimulus and the fact that Richardson employed a polyethylene container with different sound velocity rather than a poly(methyl methacrylate) container used in this work. Figure 15-5 exhibits a threshold between a low-velocity detonation and failure at zero percent triacetin (neat NG-EGDN) in this system. This is in contrast to the results obtained in the usual card gap test where the threshold could not be determined for neat NG-EGDN. Thus it would appear that the new experimental approach may have considerable merit. Not only does it simulate the real hazard of accidental initiation of thin films of sensitive liquids by weak shocks, but it may allow better discrimination of systems as a result of the easily determined threshold.

Nature of the Low-Velocity Detonation

How does one explain the peculiar behavior of these nitroglycerin systems? Is this behavior typical of only these systems? The second question is easier to answer than the first. Many other systems do exhibit a low detonation velocity. It appears in many of the more energetic systems, but is not necessarily associated with all. For example, we have been unable to obtain a low-velocity detonation in the nitric acid-nitrobenzene system for reasons that are not yet apparent.

In an earlier report,[27] the role of precursor waves, moving through the container walls and setting up compression and rarefaction waves in the liquid, was discussed as being important in the initiation process. It has now been clearly demonstrated by the Explosives Research Center that these precursor waves cause cavitation and that the bubbles formed in the cavitation process become initiation sites. The surfaces in the bubbles which may be full of droplets undergo a deflagration reaction. The pressures developed by the rapid burning are transmitted ahead of the reaction zone, and support the precursor wall wave. The reaction assumes near steady-state conditions when the sound velocity of the container wall is greater than that of the liquid. When the sound velocity of the container wall is lower than that of the liquid, the reaction tends to become unstable, although it can still propagate for many charge diameters.

There is a clear analogy to the early stages of initiation of detonation observed in packed-bed, solid explosives, but the exact mechanism for initiation of reaction at or near the bubbles has not been satisfactorily worked out. Collapse of the bubbles with adiabatic compression has long been a favorite postulate, although there are many good arguments for a more complex process. Mader[28] has discussed the formation of "hydrodynamic hot spots" in the interaction of a bubble or other discontinuity with a shock wave. The breakup of the liquid and the possible role of micro-Munro jets have been discussed by Bowden.[29] Recently, Watson and Gibson[30] have demonstrated that micro-Munro jets from a bubble in nitroglycerin can initiate nitroglycerin.

In conclusion, we can return to the basic similarity of all of these "sensitivity" tests. All initiate a decomposition reaction by a stimulus which is converted more or less efficiently into heat. The consequences of the heat release vary considerably, depending on the nature of the material. Primary explosives undergo rapid transition to detonation. Secondary explosives may not detonate at all if the stimulus is weak, as in the impact test, or they may undergo a similar transition to detonation in the shock test. Homogeneous liquids subjected to extremely high amplitude shocks react in a thermal explosion process to give a detonation after a time delay that depends on the shock amplitude. Some energetic liquids when subjected to weak shocks exhibit a more complicated process, which again results in run-up to detonation, typically to a low-velocity detonation, which may or may not be stable. It appears to be stable if the liquid is in a container made of a material that has a sound velocity higher than that of the liquid itself; it may be unstable and decay or transit to a high-velocity detonation if the sound velocity of the container is lower than that of the liquid.

Thus, out of the confusion of heuristic tests which attempt to simulate real-life situations, there begins to emerge a basic understanding of the origin and growth of explosive reactions.

Bibliography

1. Frazer, J. W., and K. Ernst, "Chemical Reactivity Testing of Explosives," *Explosivstoffe* **12,** 4–9 (1964).
2. Henkin, H., and R. McGill, "Rates of Explosive Decomposition of Explosives," *Ind. Eng. Chem.* **44,** 1391–1395 (1952).
3. Rogers, R. N., "Incompatibility in Explosive Mixtures," *Ind. Eng. Chem., Product Res. Devel.* **1,** 169–172 (1962).
4. Wenograd, J., "The Behavior of Explosives at Very High Temperatures," *Trans. Faraday Soc.* **57,** 1612–1620 (1961).
5. Gross, D., and A. B. Amster, "Adiabatic Self-Heating of Explosives and Propellants," Eighth Symp. (Internat.) on Combustion, Williams & Wilkins Co., Baltimore, Md., 1962, pp. 728–733.
6. Boddington, T., "The Growth and Decay of Hot Spots and the Relation Between Structure and Stability," Ninth Symp. (Internat.) on Combustion, Academic Press, New York, N.Y., 1963, pp. 287–293.
7. Friedman, M. H., "A Correlation of Impact Sensitivities by Means of the Hot Spot Model," *Ibid.,* pp. 294–300.
8. Maček, A., "Sensitivity of Explosives," *Chem. Rev.* **62,** 41–63 (1962).
9. Chemical Propulsion Information Agency, Working Group on Liquid Propellant Test Methods, Test No. 4, Drop-Weight Test, May 1964.
10. Robertson R., "Some Properties of Explosives," *Trans. Chem. Soc. (London)* **1921,** 1–29.
11. Mason, C. M., "Drop Weight Impact Testing of Liquids," Proc. Intern. Conf. on Sensitivity and Hazards of Explosives, Explosives Research and Development Establishment, London, Oct. 1–3, 1963; also US Bureau of Mines Report of Investigations 6799 (1966).
12. Griffin, D. N., "The Initiation of Liquid Propellants and Explosives by Impact," Propellants, Combustion and Liquid Rockets Conference, Am. Rocket Soc., Palm Beach, Florida, April 26–28, 1961, Paper 1706–61.
13. Levine, D., and C. Boyars, "Measurement of Impact Sensitivity of Liquid Explo-

sives and Monopropellants," 149th Natl. Meet., Am. Chem. Soc., Div. of Fuel Chemistry, Detroit, Mich., April 4–9, 1965.

14. Richardson, R. H., "Sensitivity Aspects of Nitroglycerin Casting Solvents," ASESB, Sixth American Explosives Safety Seminar on High Energy Propellants, Barksdale Air Force Base, Shreveport, Louisiana, Aug. 18–20, 1964.

15. "Statistical Analysis for a New Procedure in Sensitivity Experiments," Applied Mathematics Panel Rept. No. 101.1R, July 1944. 58 pp.

16. Dixon, W. J., and F. J. Massey, Jr., *Introduction to Statistical Analysis,* 2nd ed., McGraw-Hill Book Co., New York, 1957, pp. 318–327.

17. Gurton, O., "Explosion of Nitroglycerin by Impact and Friction Between Metals and Nonmetals," Proc. Intern. Conf. on Sensitivity and Hazards of Explosives. Explosives Research and Development Establishment, London, Oct. 1–3, 1963.

18. Eldh, D., B. Persson, B. Ohlin, C. H. Johansson, S. Ljungberg, and T. Sjolin, "Shooting Test with Plane Impact Surface for Determining the Sensitivity of Explosives," *Explosivstoffe* **5,** 97–102 (1963).

19. Brown, S. M., and E. G. Whitbread, "The Initiation of Detonation by Shock Waves of Known Duration and Intensity," Les Ondes de Detonation, Editions du Centre National de la Recherche Scientifique, Gif S/Yvette, Sept. 1961.

20. Griffiths, N., and Laidler, R. McN. "The Explosive Initiation of Trinitrophenyl-methylnitramine by Projectile Impact," *J. Chem. Soc.* **1962,** 2304–2309.

21. Dorough, G. D., L. G. Green, and E. James, Jr., "Ignition of Explosives by Low Velocity Impact," Proc. Intern. Conf. on Sensitivity and Hazards of Explosives, Explosives Research and Development Establishment, London, Oct. 1–3, 1963.

22. Koenan, H., and K. H. Ide., "New Test Methods for Explosives," 31st Intern. Congr. of Ind. Chem., Liège, Sept. 1958.

23. Gipson, R. W., and A. Maček, "Flame Front Compression Waves During Transition from Deflagration to Detonation in Solids, "Eighth Symp. (Intern.) on Combustion, Williams & Wilkins Co., Baltimore, Md., 1962, pp. 847–854.

24. Richardson, R. H., "Hazards Evaluation of the Cast Double-Base Manufacturing Process," 16th JANAF Solid Propellant Group Meeting, Dallas, Texas, June 1960, Rept. No. ABL/X–47.

25. Amster, A. B., E. C. Noonan, and G. J. Bryan, "Solid Propellant Detonability," *Am. Rocket Soc.* **30,** 960–964 (1960).

26. Liquid Propellant Information Agency, JANAF Panel on Liquid Propellant Test Methods, Test No. 1, Card Gap Test for Shock Sensitivity of Liquid Mono-propellants, March 1960.

27. Gibson, F. C., C. R. Summers, C. M. Mason, and R. W. Van Dolah., "Initiation and Growth of Detonation in Liquid Explosives," Third Symp. on Detonation Phenomena, Princeton, N.J., Sept. 26–28, 1960.

28. Mader, C. L., "Shock and Hot Spot Initiation of Homogeneous Explosives," *Phys. Fluids* **6,** 375–381 (1963).

29. Bowden, F. P. "The Initiation and Growth of Explosion in the Condensed Phase," Ninth Symp. (Intern.) on Combustion, Academic Press, New York, 1963, pp. 499–516.

30. Watson, R. W., and F. C. Gibson, "Jets from Imploding Bubbles," *Nature,* **204,** 1296–1298 (1964).

16

EVALUATION OF NEW PRODUCTS AND PROCESSES

Prior to undertaking the preparation of a new product, a complete safety evaluation should be made. Such an evaluation is required to properly select the type of equipment needed, safety services and instrumentation required, and the secondary protection that must be built into the system to minimize the consequences should an accident occur. A complete safety evaluation is also necessary for the preparation of operating instructions, emergency procedures, and preventive maintenance schedules.

The safety evaluation is essentially a complete appraisal of the potential hazards that are associated with the materials to be handled, the production facilities planned, and the reaction mixtures to be processed. The data on fire, reactivity, health, and contact hazards must be critically reviewed as to validity and sufficiency. Where such information is incomplete, additional tests may be necessary if the data are not available from the literature, the raw-material suppliers, or other sources.

In appraising the fire hazard, the flash points, ignition temperature, flammable limits, vapor density, reactivity, and dust explosibility are the characteristics usually needed to determine such factors as the type of building facilities, processing equipment, electrical services, and ventilation required to minimize the danger of fire. Closely allied to the danger of fire is the reactivity of materials to be processed, including the intermediates. Here it is essential that foreseeable operating errors be carefully considered. What may happen if twice as much of one of the reactants is added to the batch or one of the reactants is omitted? Is the reaction mixture stable far above the operating temperature, or will it decompose violently within a few additional degrees that can be reached by an operational error or an instrument malfunction? Such information is needed to intelligently determine the operating technique, the barricading, isolation, special instrumentation, and other devices required to operate the process safely.

Health considerations are an important aspect in evaluating the safety of any new operation. The health hazards associated with the most common chemicals are well documented, but this is not true of intermediates. Unknown intermediates can, for example, have delayed effects on vital organs. Although the reactants and the end product may be relatively safe, transitory intermediates which are normally not isolated may sometimes be harmful if they contaminate the work area by leaking from the equipment. Nevertheless, the majority of health problems that are encountered are associated with solvent vapors or gases. Other potential hazards can also be encountered, e.g., corrosives which can burn body tissues on contact, materials which may be assimilated through the skin or by inhalation, and other materials that cause dermatitis or other allergic reactions. Some of the criteria for evaluating health hazards appear in Chapter 18.

All of these determinations are vital in properly planning a new process. Only when the hazardous characteristics of the materials and the intermediates are known can the hazards be adequately avoided. Occasionally a realistic appraisal of hazards will show that it is not economical to proceed with a process because of the risk involved. In such cases, usually another synthesis can be found that can be performed with reasonable safety.

One of the primary considerations is the location of the new process: outside or inside a building, and the type of construction desirable. Often some indoor processing is preferred, especially for the finishing steps. The steps that involve exceptional fire, health, or instability hazards can be located out of doors, if of course, geographic and other considerations are suitable.

If the process involves potentially hazardous reactions, such as some hydrogenations or nitrations, these should be isolated and barricaded. It is unwise to risk extensive damage to an entire operation and unnecessarily exposing workers to danger when a relatively small part of the process is critical from a safety viewpoint and can be adequately safeguarded.

Where highly toxic or flammable materials are handled, enclosed equipment is obviously preferred. For health reasons especially, it is often essential that the work area be as free as possible of the vapors, gases, and dusts which are usually produced as the reactions progress.

Adequate building ventilation, both general and specific, will be needed because there will be times when equipment may have to be opened for charging purposes or to remove samples, at which time some leaks invariably develop. The amount and types of ventilation will depend on the toxicity of the materials being produced. The number of air changes per hour should be sufficient to keep harmful vapors well below the level which may be injurious to personnel. When adequate ventilation is provided, it is nearly always sufficient to keep flammable vapors far below their lower explosive limit. Provisions should also be made to bring in fresh air, which is heated as necessary to compensate for the air exhausted from the building.

Other safeguards that may be required are discussed in more detail in other chapters, including electrical equipment to prevent ignition sources, protection against static sparks, methods of inerting, and secondary protection. Health, fire, and other hazards are discussed in other chapters. At this point, we are primarily interested in evaluating the stability of reaction mixtures.

TESTING FOR STABILITY

It is well known that some chemicals react explosively when mixed with other chemicals or may decompose or polymerize violently when subjected to heat or shock. In Chapter 18, a number of these potentially hazardous reactions are mentioned. The list is, however, far from complete, since it is almost impossible to determine all of the limits within which a chemical reaction can be perfectly safe. Further, there is no known method or test that will provide complete assurance that a reaction will not decompose violently.

In spite of these limitations, there are a number of measures available to the chemist which will give him reasonable assurance that the work he is undertaking is controllable. He can—and should—be familiar with the characteristics of the materials he is handling; he should refer to literature summarizing the hazardous characteristics of most common materials. He should know the classes of compounds which are highly reactive; then too, he can perform a few stability tests which will give a good indication as to whether the reactions he is planning are relatively safe.

Before starting a new synthesis, the chemist will make a literature search to learn all that he can about preparing the compound. Unfortunately, the literature is often lacking in safety information. The M.C.A.

Chemical Safety Data Sheets contain information on a limited number of common materials. Chemical Safety Reference No. 486, published by the National Safety Council, gives an indication of the data available on a wide range of compounds. Another useful reference is the Manual of Hazardous Chemical Reactions No. 491M, published by the National Fire Protection Association. *Dangerous Properties of Industrial Materials,* published by Van Nostrand Reinhold Company, New York, offers some guidance. The reactions listed in the appendix of this book may be helpful.

A chemist soon learns to recognize classes of compounds which can be troublesome. Some chemists theorize that any unsaturated compound with certain nitrogen bonds is to be regarded as being unstable. Others will make calculations to determine the available oxygen in the molecule or the oxygen balance of a reaction system. All of these criteria have merit as long as the chemist understands that the criteria are merely indicators. Although the literature search and a general knowledge of unstable chemicals are important, there are measures which the chemist should undertake to determine the stability of the reaction mixture he is preparing before making a significant quantity. These range from relatively simple tests to more sophisticated shock-sensitivity evaluations requiring specialized equipment and facilities. In all such testing, however, the quantities of reactants, temperatures, and/or pressures must be varied within reasonable limits. Occasionally a relatively slight variation in the composition of the mixture or the introduction of a contaminant will cause a normally stable reaction mixture to decompose explosively.

In addition to testing variations of the reaction mixture for stability, careful attention should be paid to cleaning solutions and other possible contaminants which may be accidentally introduced. Such materials as caustic soda solution, chromic acid, and nitric acid are often used to clean reaction systems and can often sensitize otherwise stable mixtures and cause explosive decomposition. Even an extremely small quantity of contaminant in a reaction mixture has caused a major explosion.

HOT PLATE TEST

A hot plate test should first be performed to ensure the safety of the investigator. The procedure is as follows: Heat the hot plate to at least 600°F. Check the surface temperature with the tip of a thermocouple. Drop about 0.1 g of the sample on the hot plate. Decomposition of the sample must occur for this test to have any validity. If the decomposition of the sample is accompanied by any pop or bang, proceed no further with the melting point test. This is an indication that the compound is sensitive. If the sample decomposes quietly or merely sizzles because of the evaporation of the liquids, proceed with the melting point apparatus test.

THERMAL STABILITY TEST

The purpose of the thermal stability test is to detect unexpected reactions that may occur in pure materials or mixtures on exposure to heat. For example, during the concentration of a solvent solution containing an organic reaction mixture, the bulk solution temperature gradually increases as the mole fraction of solvent becomes less (pressure is assumed constant). What will happen if the residue dries on the evaporator walls and is exposed to a high temperature?

Decomposition of organic chemicals is usually accompanied by one or more of the following effects: a spontaneous generation of heat, a change in physical appearance, or a liberation of gas. In open-tube and closed-tube thermal stability testing, we are trying to observe these symptoms while exposing samples to heat. By far the most important effect of a decomposition is the self-generation of heat. Once a decomposition starts at a minute point in a system, it can spread and consume the whole batch while entirely out of control. The final result may be the tearing of the head off a reaction vessel or worse.

Thermal stability tests can be made by using a differential thermal analyzer, such as the Du Pont 900 D.T.A. Such apparatus offers many advantages: small (3-mg) quantities can be used to minimize the hazards inherent in dealing with larger quantities, and the ΔT or differential temperature curve is plotted over a number of selected ranges.

OPEN-TUBE TESTING

About 0.5 g of a sample is placed in a glass tube behind a laboratory safety shield and heated at the rate of 5°C/min, using thermocouple readouts (Fig. 16.1). The temperature of the oil bath (T) and the differential temperature (ΔT) between the sample and the oil bath are recorded. In the temperature range of sample stability, the value of ΔT should always be negative (assuming a constant rate of heat-up) because the oil is hotter than the sample. If the ΔT becomes positive, it indicates that the sample is getting hotter than the oil, which can occur only if the sample is heating exothermally.

In addition to exotherms, one may observe gas evolution and other changes in the physical appearance of samples which may indicate difficult handling properties. The main disadvantage with open-tube testing is that the sample composition can change during the course of the test.

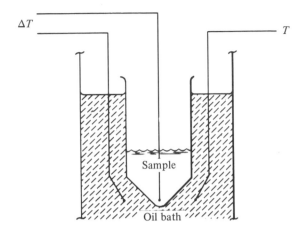

Fig. 16-1. Apparatus for open-tube testing.

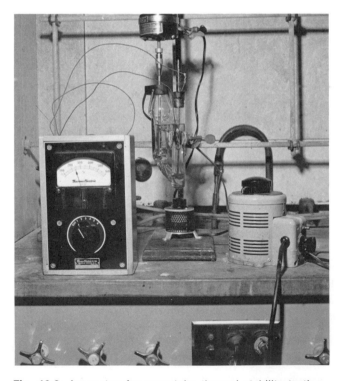

Fig. 16-2. Apparatus for open-tube thermal stability testing.

Closed-tube testing may be substituted if sample volatilization is a problem.

CLOSED-TUBE TESTING

The principle of operation and temperature read-outs for closed-tube testing are identical with open-tube testing. Instead of a glass tube to contain the sample, we now use a small sealed bomb. In this manner the sample composition is maintained constant over the range of temperature exposure. The inclusion of a miniature pressure transducer in the bomb permits a measure to be made of pressure buildup due to gas evolution.

EQUIPMENT

The equipment used for open- and closed-tube thermal stability tests can be extremely simple or highly sophisticated. For the heating bath, a modified melting point apparatus is adequate for open-tube testing (Fig. 16-2). Larger-scale open-tube tests may be run in a Dewar flask (Fig. 16-3).

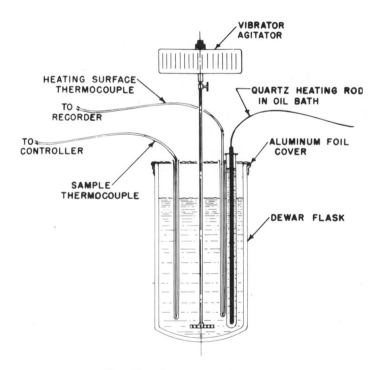

Fig. 16-3. Dewar flask test apparatus.

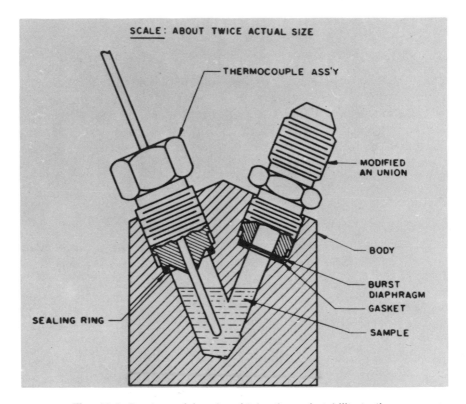

Fig. 16-4. Bomb used for closed-tube thermal stability testing.

Closed-tube testing requires a larger-size oil bath and barricaded remote-control facilities to protect the investigator (Fig. 16-4). Temperature read-outs should ideally be made on recording self-balancing potentiometers (Fig. 16-5) As a minimum, the temperature can be read using calibrated sensitive millivoltmeters. The heat-up rate should be controlled by a potentiometric program controller; however, a manually manipulated variac can be adequate.

If these tests indicate a possible extreme sensitivity, it may be desirable to have tests performed by a laboratory that specializes in determining explosive potential.

SHOCK SENSITIVITY TEST

Gap testing is a method of making stability determinations used by the U.S. Bureau of Mines and described in its publications (Fig. 16-6). This

Fig. 16-5. Instrumentation for thermal stability testing.

type of testing, which requires special barricaded and remote-firing facilities, is one of the more rigorous methods of testing reaction mixtures
and compounds. It consists of placing about 40 cm^3 of the sample in a
small cup which may be separated by a few plastic discs from a 50 g
tetryl booster in the lower portion of the cylinder. If the sample can be
made to detonate by the shock created by the tetryl charge, a hole will
appear in the ¼-in. steel plate resting at the top of the cylinder. The
energy release is so great and so sudden that the steel plate cannot
move out of the way. This test gets its name from the gap or distance
between the sample and the explosive charge which is measured by the
number of plastic discs used, zero gap being the more severe test.

It should be emphasized again that negative tests do not provide complete assurance that one is not dealing with a detonable compound. When
positive results are obtained, the reaction mixture can be varied to deter-

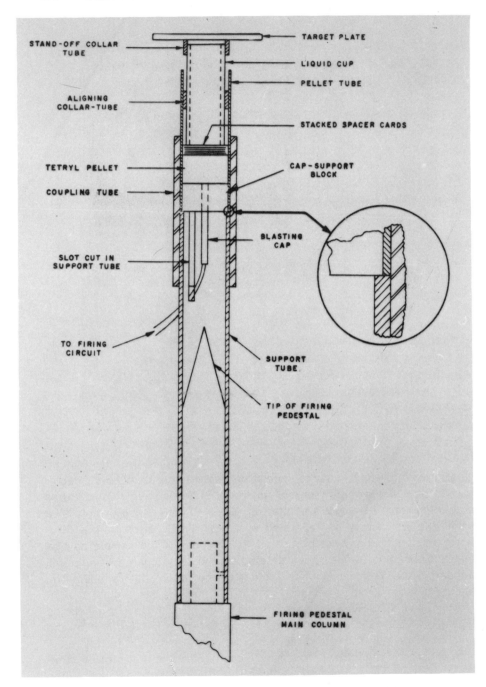

Fig. 16-6. Card gap test apparatus for shock sensitivity.

mine effects of desensitizers, and it is usually possible through design and instrumentation to carry out even the most unstable reactions with reasonable safety.

Only after every reasonable attempt has been made to learn of all the dangers in a chemical operation, and there is assurance that these dangers can be adequately safeguarded, should plans be completed for a new process.

Bibliography

Albisser R. H., and L. H. Silver, "Safety Evaluation of New Processes," *Ind. Eng. Chem.* **52** (11), 77A (Nov. 1960).

Fawcett, H. H., and W. S. Wood, *Safety and Accident Prevention in Chemical Operations,* John Wiley & Sons, New York, 1965.

Liquid Propellant Information Agency, Liquid Propellant Test Methods, Recommended by the Joint Army–Navy–Air Force Panel on Liquid Propellant Test Methods, Johns Hopkins, Silver Spring, Md.

Manual of Hazardous Chemical Reactions, National Fire Protection Association, 1966.

Safety in the Scale-up and Transfer of Chemical Process, Manufacturing Chemists Association, Safety Guide SG–14.

Steel, A. B., and J. J. Duggan, "Safe Handling of Reactive Chemicals," *Chem Eng.* **66,** (8), 157–168 (April 20, 1959).

Van Dolah, R. W. "Evaluating the Explosive Character of Chemicals," *Ind. Eng. Chem.* **53,** (7), 59A–62A (July 1961)

17

CORROSIVE IRRITANTS

This chapter discusses primary irritants, materials which, from the standpoint of chemistry, would be considered corrosive. These materials act by injuring the skin, eyes, and, if inhaled, the surface tissues of the respiratory tract. This later may lead to inflammation of the air passages and possibly the lungs themselves. The effects of the various irritants vary somewhat due to differences in solubility, physical state (solid, liquid, gas), and portion of the body exposed.

In any discussion of chemical irritants, some mention must be made of so-called secondary irritants. Some definitions would therefore be in order. A primary irritant is a material that causes little or no systemic toxic effect in concentrations that cause varying degrees of local injury. It is therefore apparent that the corrosive action is the primary hazard. An example of this type of substance is hydrochloric acid. A secondary irritant is a substance that, while being an irritant to exposed tissue surface, poses a greater hazard from systemic effects. An example of a secondary irritant is hydrogen sulfide.

Irritants can be classified rather loosely by their physical form, such as liquid, solid, or gas, and will be discussed in that order.

LIQUID IRRITANTS

A primary irritant may act on the skin rapidly or slowly depending on its concentration and length of contact. The more concentrated the irritant and the longer the contact, the more rapid and intense is the resulting inflammatory process. These irritants act directly on the skin either by chemically reacting with it, by dissolving or abstracting from it some of its essential components, by denaturing the proteins of the skin, or by disturbing the membrane equilibrium or osmotic pressure of the skin cells. These processes pertain, of course, to all irritants regardless of their physical form. However, concentrated liquid irritants are in most cases in the physical form most likely to cause immediate injury. It is therefore generally conceded that the liquid irritants are the most hazardous of all the irritants as far as external injury is concerned. This general class of materials causes the greatest number of skin reactions and chemical eye accidents in industry today.

Typical Examples of Liquid Irritants

Mineral acids

nitric acid	HNO_3
sulfuric acid	H_2SO_4
hydrochloric acid	HCl
hydrofluoric acid	HF
phosphoric acid	H_3PO_4

Organic acids

formic acid	$HCOOH$
acetic acid	CH_3COOH
chloroacetic acids	$CH_2ClCOOH$
cresylic acid	$C_6H_4OHCH_3$

Organic solvents
   petroleum solvents
   coal tar solvents
   chlorinated hydrocarbon solvents
   most liquid esters and ketones
   most alcohols
   carbon bisulfide
   turpentine and terpines

The above irritants are all liquids in their normal physical state. However, many cases of irritation occur from solutions of solid irritants, such as the caustics, the oxidizing and reducing salts, and many organic salts, such as organic acids, anhydrides, halogenated organics, and many others.

Precautionary Measures

When handling liquid irritants, a person should wear sufficient protective equipment to prevent accidental contact with these irritants. This may mean rubber gloves, chemical safety goggles, and a face shield. If there is the possibility of exposure to large quantities of these corrosives, it would be advisable to wear, additionally, rubber aprons and rubber boots.

Since many of the liquid irritants may also release irritant vapors, particularly at elevated temperatures, proper respiratory protection should be available. Self-contained breathing apparatus will provide adequate respiratory protection in all cases. Canister-type masks may not be suitable in some cases because of their limitations as to the maximum amount of contaminant and the minimum amount of oxygen in the atmosphere being entered. In emergency situations, it is often impossible to know vapor concentrations or oxygen levels, thus rendering canister-type masks hazardous.

First aid procedures should begin with immediate flushing of the affected skin or eyes with copious quantities of water. This flushing is recommended for at least fifteen minutes for eye contact; then the patient should be immediately referred to a physician.

SOLID IRRITANTS

The effects of solid irritants depend largely on their solubility in the moisture of the skin or other surfaces. These irritants can cause serious damage, both from their corrosive action and from their thermal heats of solution. This class of irritants is probably the least hazardous of all the materials handled, because a person has sufficient time to remove the material from exposed body surfaces before serious irritation occurs. However, in many cases the irritant is a liquid slurry and the potential for irritation is great.

Solid irritants can cause delayed irritation, as, for example, when solid material contacts the skin without one's knowledge or when it is incompletely removed following contact. Solid irritants, such as the caustic alkalies, are not immediately painful, as are the acids, and, if left in contact with the skin until pain is felt, serious injury has usually already occurred.

Typical Examples of Solid Irritants

Caustic alkalies
    alkaline sulfides   MS
    sodium hydroxide   NaOH
    sodium carbonate Na_2CO_3
    sodium silicate $Na_2O \cdot x\, SiO_2$
    potassium carbonate K_2CO_3
    ammonium carbonate $(NH_4)_2CO_3$

barium hydroxide	$Ba(OH)_2$
barium carbonate	$BaCO_3$
trisodium phosphate	Na_3PO_4
Lime (both hydrated	
and dehydrated)	$Ca(OH)_2$, CaO
calcium cyanamide	$N\equiv C-NCa$

Elements and salts

elemental sodium	Na
elemental potassium	K
elemental phosphorus	P
antimony and its salts	Sb salts
arsenic and its salts	As salts
chromium and the	
alkaline chromates	
copper sulfate	$CuSO_4$
copper cyanide	$Cu(CN)_2$
mercuric salts	Hg salts
zinc chloride	$ZnCl_2$
silver nitrate	$AgNO_3$

Probably the materials with the greatest potential for harm of all this group of solid irritants are the caustic alkalies because of their very wide application throughout the chemical as well as many other industries. This list does not by any means include all caustic alkalies, but includes the most commercially important ones and the most severely corrosive. A second group of solid irritants includes some of the elements and salts. This group contains many oxidizing agents, reducing agents, and elemental materials that produce irritation by thermal burns. Solutions of these materials probably pose the most serious problem. However, fine dusts of any of these solid materials will sometimes cause severe chronic irritation, particularly to persons who perspire freely.

Precautionary Measures

Solid materials in a finely divided state frequently cause chronic irritation. Dust control is a necessary adjunct to safe handling. In almost every case where finely divided solid irritants are handled, exhaust ventilation should be provided. Those handling large quantities of solid irritants should wear gloves, respirators, and protective clothing as the necessity arises. In some cases, where finely divided irritants are used, protective creams are recommended. When these creams are used, instructions for their application and removal are generally given with reasons why they should not be substituted for bathing and other personal hygiene practices.

As a first aid procedure, copious flushing of the skin or eyes with water is recommended. A physician should be summoned. It may be of interest to note that handling certain organic chemicals, such as phenol, which has a melting point of 41°C, or lactic acid, which has a melting point of 26°C, poses a problem characteristic of materials that melt in the general range of from 25 to 50°C. In ordinary chemical processing, these materials might be either liquids or solids. In the case of phenol, removal of the solid material from the skin is an absolute necessity. Copious flushing with water, followed by thorough washing with soap and water, is necessary to remove minute quantities of these irritants.

GASEOUS IRRITANTS

Although liquid and solid irritants afford considerable hazard from skin and eye contact, the most serious hazard associated with irritants in general is from materials in the gaseous state.

Widely different symptoms may result from the action of irritant gases and vapors. These differences are due primarily to the specific structures on which the irritant acts. Symptoms are therefore governed by the structures affected by the inflammation as well as by the mode of action of the irritant substance.

The site of action of gaseous irritants is influenced principally by the solubility of the irritant. For example, ammonia gas in high concentrations will cause intense congestion and swelling of the upper respiratory passages and possibly rapid death from spasm or edema of the larynx. If the immediate effects are survived, there may be little serious aftereffect since the deeper structures of the respiratory tract may not be seriously injured. On the other hand, phosgene, even in concentrations that cause little immediate irritation, may later be fatal owing to chemical pneumonitis or pulmonary edema through its action on the air cells (alveoli) of the lungs. Chlorine is intermediate (between ammonia and phosgene) in its action.

The selective action of the irritant gases in the respiratory tract determines the relative danger of death from them. The delicacy of the respiratory membranes, their susceptibility to injury, and the seriousness of the resulting damage are very different in the upper and lower respiratory tracts. The action of irritant gases and vapors in the nose and pharynx may produce an intense local reaction by inflammation without immediate danger or permanent damage. It is a more serious matter when the trachea and bronchi are injured. Not only are the local effects painful, but general systemic effects may develop. The most serious results of inhalation of an irritant gas or vapor occur when the lungs themselves are acted upon. The damage may give rise to an acute edema leading to suffocation or, if this danger is escaped, to pneumonia with prostration and circulatory impairment. The outcome of either process may be fatal.

The harmful effect of an irritant is not a straight line relationship with the product of duration of exposure and concentration as it is in the case of an asphyxiant like carbon monoxide. A single exposure to a high concentration of an irritant can have an intense effect and can terminate fatally. However, this acutely fatal exposure might be tolerated by the system if the concentration were halved as if it were inspired over a period twice as long. This fact leads to the conclusion that any reduction in the concentration of an irritant during its passage through the upper respiratory tract results in a more than proportionate sparing of the tissues of the lung.

Some Typical Examples of Gaseous Irritants

Gaseous irritants are generally grouped according to their locus of action. With the exception of the fourth group, these irritants are therefore grouped according to their relative solubilities, which is the physical property that dictates the site of absorption within the respiratory tract. This system of grouping is that presented by Henderson and Haggard in *Noxious Gases* (2nd ed., American Chemical Society Monograph No. 35, New York, Reinhold Publishing Corp., 1943).

Group I

Highly soluble. Affecting mainly the upper respiratory tract:

ammonia	NH_3
hydrochloric acid	HCl
sulfuric acid	H_2SO_4
hydrofluoric acid	HF
formaldehyde	$HCHO$
acetic acid	CH_3COOH
acetic anhydride	$(CH_3CO)_2O$
sulfur monochloride	S_2Cl_2
thionyl chloride	$SOCl_2$
sulfuryl chloride	SO_2Cl_2

Group II

Intermediately soluble. Affecting the upper respiratory tract and deeper structures such as the bronchi:

sulfur dioxide	SO_2
chlorine	Cl_2
bromine	Br_2
iodine	I_2
arsenic trichloride	$AsCl_3$
phosphorus trichloride	PCl_3
phosphorus pentachloride	PCl_5

Group III

Least soluble. Minimal primary irritation; cause delayed pneumonitis:

ozone	O_3
nitrogen dioxide	NO_2
phosgene	$COCl_2$

Group IV

There is no general rule as to the locus of action.

acrolein	CH_2CHCHO
dimethyl sulfate	$(CH_3)_2SO_4$
dichlorethyl sulfide (mustard gas)	$S(CH_2CH_2Cl)_2$
chloropicrin	CCl_3NO_2
ethyl chlorosulfonate	$C_2H_5OSO_2Cl$
dichloromethyl ether	$O(CH_2Cl)_2$
methyl chlorosulfonate	CH_3OSO_2Cl
xylyl bromide	C_8H_9Br

Group IV is by no means complete; there are literally thousands of such compounds. The biggest single class of Group IV compounds are the halogenated organic compounds. In this group, no simple solubility relationship exists. For example, acrolein (acrylic aldehyde) is only moderately soluble in water, yet it is very irritating to the eyes and the upper respiratory tract. Continued exposure to this compound may lead to lung damage manifested by edema and possible death. Dimethyl sulfate is a powerful irritant to the eyes and upper respiratory tract, and yet it is only slightly soluble in water; its action is apparently not entirely due to the decomposition of dimethyl sulfate to sulfuric acid.

Because of the anomalous properties of Group IV compounds, great care must be exercised in handling them. The best clue as to the possible seriousness of exposure to such materials would come from the volatility of the specific compound at the proposed operating temperature. A highly viscous material handled at room temperature would afford little hazard since the level would be extremely low. Many thousands of these compounds have been tested for toxicity; these data are frequently available in the literature or from the suppliers. It would be a wise course of action to obtain such information before handling unfamiliar chemical substances.

Precautionary Measures

The primary protection against accidental exposure to gaseous irritants is the use of respiratory protective equipment, keeping in mind, however, that additional protection may be necessary for protection of the skin and eyes.

If the irritant is such that dangerous concentrations are possible in the working atmosphere, exhaust ventilation is recommended as well as the necessary emergency protective equipment. Safety showers, eye baths, and other permanently installed safety devices should be available.

CONCLUSIONS

The control of irritant exposure is absolutely essential to the health of the staff. This control must include proper protective equipment, proper first aid training, properly designed ventilation equipment, and periodic exposure surveys, particularly where one of the insidious irritants, such as the Group II or III gaseous irritants, is concerned. When irritants are used, careful study must be made to evaluate the problem and to design control equipment.

In addition to control of process exposure, provisions must be made to protect all persons during such unusual conditions as spills or explosions. This requires thorough safety training in the proper use of gas masks and other protective equipment. Finally, it is imperative that all persons be trained in proper first aid practice and evacuation techniques.

The chemical industry has an excellent record in handling irritant materials. When one considers the tremendous tonnages of chlorine, ammonia, mineral acids, caustic alkalies, mineral salts, elemental substances, and even nitrogen dioxide that are produced and handled every day, this record is even more remarkable. Corrosive irritants can be handled safely if their hazardous characteristics are recognized and understood and proper protective measures are used.

18

TOXIC CHEMICALS

A gram of table salt will kill a rat. Kerosene, aspirin, and boric acid have proved fatal to children who swallowed them. An excess of any substance is harmful. Toxicity is the potential of a substance to cause injury by direct chemical action with the body tissues. Establishing the levels of safe and dangerous quantities of chemicals is the work of industrial hygienists and toxicologists, who also measure environmental exposures and examine workers and their body fluids for evidence of absorption.

MODE OF TOXICITY

The toxic or harmful effect of chemicals on body tissues is caused by interference with the function of the cells of the body tissues. Certain chemicals will pass through some cells, without interfering with their function, and then interfere with the function of other cells. This interference may be temporary or serious enough to cause permanent damage to the cell. The absorption of an extremely small amount of a chemical into a cell can interfere with the cell's ability to take in oxygen, which in turn will deprive it of its primary function. If the cell can dispose of the contaminant quickly enough, it will recover. Otherwise, the lack of oxygen will cause permanent damage or destruction.

Owing to the difficulty of observing the cell behavior, most toxicity studies are limited to reactions of the body to various dosages of the chemical under controlled conditions. Almost all such studies are performed on animals and then interpreted for application to the human body. The complexity of the combination of many factors, such as the inability to completely isolate the effects of the compound being studied without an infinite sampling, the variations of cell structures between animals and man, and variations of reactions of individuals in both man and animals, makes it impossible to establish exact toxicity standards. Therefore the published information must be used as a guide rather than an exact standard.

LOCAL AND SYSTEMIC TOXICITY

Toxic effects are usually described as local or systemic and in relation to the organs or tissues affected. Local injuries are those limited to the area of the body that has come in contact with the toxicant, the most common being the skin and the eyes. Local injury also may be encountered in the nose, throat, and lungs after inhalation of toxicants and in the mouth, throat, stomach, and intestine after swallowing.

Systemic injuries are those produced in any of the organs after the toxicant has been absorbed into the bloodstream. The chemical may have entered the body via the lungs, stomach, or the skin. Therefore the same chemical can cause both local and systemic effects. Distinctions are made between local and systemic effects because these effects do not show similar or parallel variations among different chemicals. Both local and systemic effects show considerable variation in type, duration, and seriousness to health and life.

ACUTE AND CHRONIC TOXICITY

Chemicals show a greater or lesser difference in toxicity, depending upon whether they act on the body for a short or a long time. Thus two substances may have nearly identical degrees of toxicity when inhaled or swallowed once. Their potency, however, may be greatly different when they are inhaled or swallowed every day for a long period of time. In addition, the nature of the toxic injury produced by a chemical can be different, depending on the duration of the exposure.

Acute toxicity is defined as that which is manifest on short exposure. "Short exposure" cannot be defined precisely; it is commonly thought of as a single oral intake, a single contact with the skin and eyes, or a single exposure to contaminated air lasting for any period up to about eight or possibly twenty-four hours. A typical acute intoxication might occur if a physicist spent all morning cleaning his optical equipment with carbon tetrachloride. In the evening he would probably become ill and vomit. The next morning he would be unable to urinate. Death might occur a few days later even though he had no subsequent exposure.

Chronic toxicity is defined as that which becomes manifest when the toxicants act on the body over a long period of time. The time period and the pattern of exposure cannot be stated precisely; they are usually thought of as regular daily exposures for periods measured in weeks, months, or even years. An example of a person who is chronically intoxicated might be one who has for several weeks been attempting the synthesis of a new organolead compound. His work habits are somewhat sloppy, and he contaminates his cigarettes and candy bars with his raw materials. A routine blood test, required by his laboratory for those working with lead, reveals that he has clinical evidence of lead poisoning. Each day he has taken in more lead than he can excrete, and it is now lodged in body systems where it can do harm. In this as in many other cases, the injured person will not normally realize the effects of chronic toxicity until it has reached advanced stages and has created considerable permanent damage.

UNITS OF TOXICITY

LD_{50} (lethal Dose, 50% kill). This represents the dose which, when administered to such laboratory animals as rats or mice, kills half of them. It is expressed in mg/kg (milligrams of toxicant administered per weight, in kilograms, of the animal). The route of administration, for example, oral, skin, intraperitoneal, is usually stated. If RNO_2 is exactly as toxic to man as to guinea pigs and the LD_{50} is 50 mg/kg, the LD_{50} for a 70 kg man is $70 \times 50 = 3.5$ g. Such extrapolation is always a guess, and may be inaccurate. However, in the absence of more accurate data, it can be helpful in comparing the relative toxicities of different chemicals.

LC_{50} (lethal concentration, 50% kill). This is always expressed in terms of the minutes or hours of exposure. The weights are not needed: small animals inhale less toxicant-laden air than do larger animals, but are physically less able to detoxify the agent. Units are ppm for time period (e.g., 100 ppm for 4 hr).

MLD (minimum lethal dose), MLC, LD_{100}, LC_{100}, LD_{90}. These are variations, usually found in the older literature, of LD_{50} and LC_{50}.

TLV (threshold limit values). These are estimates of the average safe airborne toxicant concentration, for all-day, every-day exposure. They are expressed in ppm or mg/m³ (milligrams of toxicant per cubic meter of air). Mineral dusts are expressed as mppcf (million particles per cubic foot). These values are published each year, including changes of pre-viously published levels based on additional information, by the American Conference of Governmental Industrial Hygienists.

MAC (maximum allowable concentration). These are estimates of con-centration of contaminants below which ill effects are unlikely to be experienced by any but hypersensitive individuals when exposed for eight hours every day. These units are also expressed in ppm and mg/m³.

TYPES OF EXPOSURE

There are only four types of exposure to chemicals: contact with the skin and the eyes, inhalation, swallowing, and injection. These are listed in order of importance in chemical work.

Contact with Skin and Eyes

Contact with the skin is the first in importance because of the frequency with which it occurs. The most common result of excessive contact on the skin is a localized irritation, but an appreciable number of materials are absorbed through the skin with sufficient rapidity to produce sys-temic poisoning. The main portals of entry for chemicals through the skin are the hair follicles, sebaceous glands, sweat glands, and cuts or abrasions of the outer layers of the skin. The follicles and glands are well supplied with blood vessels, which facilitate the absorption of chemicals into the body.

Minor breaks in the superficial layer of the epidermis will open lym-phatic channels through which chemicals may enter the body. Minor scratches which do not even penetrate the outer layer of the skin permit absorption of chemicals by the lymphatics below the abraded surface. The normal flow of lymph along the skin lymphatics is so rapid that even with the body at rest, dye introduced into the skin of the forearm reaches the lymphatics in the armpit in a few minutes.

The hair follicles and associated sebaceous glands are particularly vulnerable to fat solvents. Various detergents, such as hot or cold water, soap, alkali, and sulfonated fatty acids and alcohols affect the hair fol-licles. Overactivity of the sweat glands may assist in the absorption of chemicals in that the excess perspiration will pick up and dissolve chem-ical dusts. Contamination of shoes and clothing has particular signifi-cance, because the confinement of the toxicant materially increases the severity of exposure and of injury.

Contact of chemicals with the eyes is of particular concern because these organs are so sensitive and because impairment or loss of vision is tragic. Seemingly few substances are innocuous in contact with the eyes; most are painful and irritating, and a considerable number are capable of causing "burns" and loss of vision.

Inhalation

Inhalation of air contaminated with gas, vapor, dust, or fumes is of urgent concern in all large-scale operations. Even in small-scale laboratory operations, inhalation of such materials can be an important source of toxicants, since only a few grams of many substances can produce excessive contamination. Inhaled gas or vapor may be absorbed readily by the capillaries of the lungs and be carried into the general circulation. It is estimated that the internal surface of the human lung is about 100 m^2 compared with 1.5 m^2 of skin on the body. Thus the absorption of gas can be extremely rapid. The rate will vary with the individual, depth of respiration, and amout of blood circulation, which means that the absorption rate will be much higher when the person is active than it will when he is at rest. Inhaled dust is not carried to the lung unless the particles are less than five microns (μ) in size; those of about 1 μ in size are likely to gain entrance to the alveolar air sacs of the lung and be retained until they can be dissolved and disposed of by the body.

Swallowing

Swallowing of chemicals may occur from exposure to air contaminated with dust and fumes or by contamination of food and drink. The amount of such swallowing from the air is normally insignificant compared to the amount inhaled. The contamination of food and drink can be controlled by keeping food out of the laboratory and by prohibiting the use of laboratory beakers and equipment for drinking water, coffee, and other liquids.

Injection

Injection of chemicals would appear to be a type of exposure which could not occur in the chemical laboratory. The equivalent, however, can readily occur through mechanical injuries with glass or metal contaminated with chemicals. Injection may result from the careless handling of hypodermic needles, which are in common use in microchemistry. Injection may also result from work at high pressures, when a small leak will produce a "stream" of liquid material of sufficient force to penetrate the skin.

WARNING SENSES

The two sensory effects of smell and pain can be useful, but often their warnings come after some injury has occurred.

The sense of smell can be helpful in preventing excessive exposure to the gas or vapor of some materials if immediate action is taken to correct the situation when the material is smelled. Some substances are not detected by smell or only at concentrations considerably above safe levels. Some chemicals, such as hydrogen sulfide, will in lethal concentrations deaden the olfactory nerves before the odor can be detected.

Pain or discomfort in the eyes and breathing passages may be caused by gases, vapors, solids, or liquids dispersed in air. Some chemicals will produce so much pain that no one will voluntarily tolerate exposure to dangerous concentrations. However, pain or discomfort, like smell, should not be relied on as a primary defense but only as a backup when other devices fail.

THE HAZARD OF TOXICITY

The hazard of toxicity is not merely in the degree of toxicity but in the combination of the toxicity and the amount of exposure. A strong, well-sealed container of a highly toxic substance, such as hydrogen sulfide, is not hazardous as long as it is kept sealed; but a large container of a much less toxic material, such as a common solvent, used in a poorly ventilated room results in a hazardous exposure. The physical properties frequently play a major part in determining the amount of exposure. In a given situation, a material with a boiling point of 35°C will cause much more contamination of the air and resultant inhalation of vapor than one with a boiling point of 240°C. Liquid substances, upon contact with the skin, can spread much more than can large pieces of solid matter. Dust escaping into the air of the work space will produce more skin contact than a high-boiling liquid which cannot escape into the air.

Thus we see that not only the toxicity of chemicals but also chemical and physical properties and the nature and circumstances of handling and use determine whether or not chemicals will cause injury. Ultimately, the safe use of chemicals depends on the control of the hazard through proper handling techniques, good engineering design, and the use of protective equipment based on accurate knowledge of the characteristics of the materials.

HOW CAN TOXICITY BE DETERMINED?

A laboratory worker should try to learn the toxicity of every compound he works with. The TLV's, the MCA Data Sheets,[2] and books such as those by Spector,[3] Merck,[4] and Patty[5] are excellent sources. For more extensive data, indexes of medical industrial hygiene, and pharmacology journals should be consulted. Vendors of raw materials may have toxicity data; letters of inquiry should be directed to their safety or medical departments.

If a particular compound cannot be found in the literature, some reasoning by analogy is permissible. If some members of a class (e.g., carbamates) produce cancer, a new carbamate should be suspected. The $(RO)_2PS$ radical should be respected because of its analogy with toxic organic phosphates. If one metallic salt is toxic, probably all soluble compounds of the same metal will be toxic.

If a possibly toxic compound is going to be investigated and no data are available, it should be tested by a laboratory familiar with such work. Names of such laboratories can be found by examining specialty journals. Costs for the test of a single liquid or gas will usually be less than a thousand dollars; solid inhalation studies may be more expensive because of the intricacies of producing a uniform dust and keeping it in suspension in the exposure chamber. Chronic feeding studies, required before food, drug, or agricultural compounds are offered for sale, are extremely expensive.

PREVENTION OF INTOXICATION

The first rule of any laboratory should be: "Treat every substance as highly toxic unless you know definitely that it is not." Good personal hygiene is also essential. The wearing of laboratory coats, the use of personal protective equipment, and frequent washing of hands should all be required. If smoking is permitted, cigarettes should be carried in a polyethylene overpack; a pipe may act as a miniature furnace, and should not be smoked where there are dusts with toxic decomposition products. Contaminated clothing should be decontaminated or discarded at once. Unlabeled containers should not be permitted.

Antidotes, if known, should be on hand. Examples are: atropine for anticholinesterase agents, and amyl nitrate for cyanide poisoning. The medical department should be notified that work with certain highly toxic chemicals is in progress, so that they can be prepared for emergencies. They can also provide periodic physical examinations including special tests for known effects of toxicants, for example, stippled cells in lead exposure, urinary sulfate ratios in benzene exposures. These tests may show up problems before any clinical signs of illness.

TRAINING

Every incoming laboratory worker, whether an undergraduate or a Ph.D., must learn the dangers of his environment and how to avoid them. The undergraduate may not know the difference between benzene and benzine; the Ph.D. may be familiar with chemistry but not toxicology. In one widely read college text on organic chemistry, only polynuclear hydrocarbons are identified as carcinogenic. This is typical of much of the literature, which often details the preparation of extremely dangerous compounds but omits any warning.

Bibliography

1. Threshold Limit Values for 1971, available from American Conference of Governmental Industrial Hygienists, 1014 Broadway, Cincinnati, O.
2. Chemical Safety Data Sheets, Manufacturing Chemists Association, Washington, D.C.
3. Spector, *Handbook of Toxicology*, Vol. I, *Acute Toxicities*, W. B. Saunders Co., 1956.
4. *Merck Index*, 8th ed. Merck & Co. 1968.
5. Patty, F. A., *Industrial Hygiene and Toxicology*, 2nd ed., Vol. 2, John Wiley & Sons, New York, 1963.

Additional References

Grant, *Toxicology of the Eye,* Charles C. Thomas, 1963.
Sax, N. I., *Dangerous Properties of Industrial Materials*, Van Nostrand Reinhold, New York, 1968.
Gas Data Book, The Matheson Company.
Steere, N. V. *Handbook of Laboratory Safety*, Chemical Rubber Company, Cleveland, Ohio, 1970.
Kirk-Othmer, *Encyclopedia of Chemical Technology*, 2nd ed., John Wiley & Sons, New York.

19

FLAMMABLE LIQUIDS

Flammable liquids and gases play a highly important part in many laboratory operations. Flammable solvents are used in a great variety of analytical procedures and frequently also in cleaning apparatus in which water-insoluble materials have been used. Accidents with flammables can be avoided with the exercise of a reasonable amount of care. For example, billions of gallons of gasoline are safely dispensed each year into automobiles.

The two hazards inherent in flammable liquids and gases are toxicity and flammability. Since the toxicity of gases and vapors is treated in Chapter 18, the discussion in this chapter will deal exclusively with the flammability hazards of these materials. Flammability is a property that must be considered and evaluated in order to utilize such materials safely.

A fire hazard exists when a flammable liquid has sufficient volatility so that the vapors will mix with the air in ignitible proportions. The degree of hazard depends upon whether or not a material can burn, its volatility or vapor pressure, its ability to form flammable or explosive mixtures, the ease of ignition of these mixtures, and the relative densities of both the liquid with respect to water and also the vapor with respect to air.

The terms commonly used in describing the properties of flammable liquids are defined by the National Fire Protection Association[1,2] as follows:

The *flash point* of a liquid is the temperature at which it gives off vapors sufficient to form an ignitible mixture with the air near the surface of the liquid. Flash points at 175°F or lower are determined according to ASTM D 56-61, Tag Closed Cup Tester.

Flammable liquids are those having a flash point below 140°F and a vapor pressure not exceeding forty pounds per square inch absolute at 100°F.

Combustible liquids have a flash point at or above 140°F.

Liquefied compressed gases are *flammable liquids* with a vapor pressure above forty pounds per square inch absolute at 100°F.

The *ignition temperature* (autoignition temperature) of a substance is the minimum temperature required to initiate or cause self-sustained combustion without ignition from an external source of energy.

The *lower flammable limit*[3] (lower explosive limit, LEL) is the minimum concentration of vapor in air below which a flame is not propagated when an ignition source is present. Below this concentration, the mixture is too lean to burn; actually, the energy from combustion of one particle is dissipated before it can activate another particle to propagate the flame.

The *upper flammable limit* (upper explosive limit) (UEL) is the maximum concentration of vapor in air in which a flame can be propagated. Above this concentration, the mixture is too rich to burn; that is, the oxygen is used up in the combustion of one particle, and there is insufficient oxygen to burn the next adjacent particle of fuel. Products of combustion surrounding the first particle tend to quench the flame.

The *flammable range* consists of all concentrations between the lower flammable limit and the upper flammable limit (Fig. 19-1).

The *specific gravity* of a liquid is the ratio of its density to that of water under specified conditions. This term is important in that a material that does not mix with water will float if its specific gravity is less than 1 and will sink and be covered with water if its specific gravity is greater than 1.

The *vapor density* is expressed as the relative density of a vapor with respect to air at the same temperature. Thus, a vapor having a density less than 1 will tend to rise, and a vapor with a density greater than 1 will tend to sink. However, air turbulence can offset the settling effect of a vapor that is heavier than air.

Water solubility is sometimes important in determining whether water can be effectively used to flush away flammable liquids. It should be remembered, however, that a water solution of soluble solvents can give off sufficient vapors to burn. For example, a 5% solution of ethyl alcohol in water has a determinable flash point.

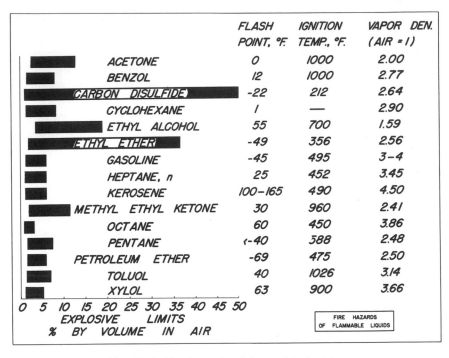

	FLASH POINT, °F.	IGNITION TEMP., °F.	VAPOR DEN. (AIR = 1)
ACETONE	0	1000	2.00
BENZOL	12	1000	2.77
CARBON DISULFIDE	-22	212	2.64
CYCLOHEXANE	1	—	2.90
ETHYL ALCOHOL	55	700	1.59
ETHYL ETHER	-49	356	2.56
GASOLINE	-45	495	3-4
HEPTANE, n	25	452	3.45
KEROSENE	100-165	490	4.50
METHYL ETHYL KETONE	30	960	2.41
OCTANE	60	450	3.86
PENTANE	<-40	588	2.48
PETROLEUM ETHER	-69	475	2.50
TOLUOL	40	1026	3.14
XYLOL	63	900	3.66

0 5 10 15 20 25 30 35 40 45 50

EXPLOSIVE LIMITS
% BY VOLUME IN AIR

FIRE HAZARDS OF FLAMMABLE LIQUIDS

Fig. 19-1. Fire hazards of flammable liquids.

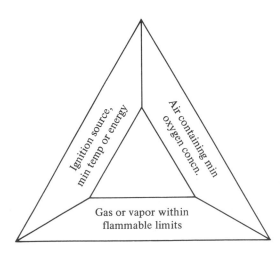

Ignition source, min temp or energy

Air containing min oxygen concn.

Gas or vapor within flammable limits

Fig. 19-2. Fire triangle.

For a fire to occur, all three components of the fire triangle (Fig. 19-2) must be present. These are: flammable gas or vapor within certain limits of concentration; oxidizing atmosphere, usually air containing a minimum concentration of an oxidizing gas, such as oxygen; a source of ignition incorporating sufficient temperature or energy to start combustion. Removal of any of the three components will prevent a fire or extinguish an existing fire. In most laboratories, air is present as an oxidizing atmosphere. The problem, therefore, usually resolves itself into preventing the coexistence in the same place of flammable vapors and an ignition source.

The safe handling of flammable liquids is based on the principle of preventing completion of the fire triangle. One of the most effective practices is to minimize the amount of flammable liquid present and exposed at any given location. When this is done, the chances of a flammable mixture occurring are minimized and the extent of a fire resulting from these materials will be limited. Maximum quantities of flammables in unprotected glass or plastic containers should be limited to one gallon per one hundred square feet of a laboratory area and should be in

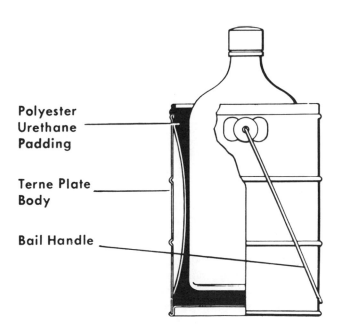

Polyester
Urethane
Padding

Terne Plate
Body

Bail Handle

Fig. 19-3. A padded metal bucket.

Fig. 19-4. Dip tank for storage of flammable materials (tray is removed).

containers no larger than one quart. The maximum quantity in ordinary metal cans should be five gallons per one hundred square feet of laboratory space. Where larger quantities are required, and where glass bottles are preferred to maintain purity, a padded metal bucket (Fig. 19-3) or a covered metal container (Fig. 19-4) will greatly reduce the possibilities of breakage and fire.

Fig. 19-5. Approved safety cans should be used to store flammable liquids in the laboratory.

Safety cans approved by Underwriters' Laboratories or the Factory Mutuals should be used whenever possible to store flammables in the laboratory (Fig. 19-5). These containers are of heavy welded construction, and have spring loaded caps and a flame arrestor (Fig. 19-6) to prevent backflash. Oval-shaped cans of one-gallon capacity (Fig. 19-7) are widely used in laboratories because they require less shelf space than round ones. They may be made of coated steel or stainless steel as required. Five-gallon safety cans are sometimes cradled in a tilt frame so that the user can operate the self-closing faucet with one hand. Safety disposal cans are also widely used to overcome the hazard of pouring solvents into drains or of having an unprotected can containing appreciable quantities of solvent. These can be obtained in stainless steel, coated steel, or even with glass linings. Double-wall steel storage cabinets are available to maximize the safety of storing flammables in the laboratory (Fig. 19-8).

All distillations, reactions, and the like involving flammables should be set up in a metal pan, preferably in a hood (Fig. 19-9). Ignition sources (flames, heaters, switches, relays, thermostats, open motors, and so on) should be remote from equipment containing flammable liquids. This pre-

Fig. 19-6. Cut-out showing flame arrester in place to prevent flashback.

caution does not preclude the proper heating of reactions or distillations with properly installed electric mantles. Glass bottles containing flammables should be protected from breakage and spill by the use of a metal bucket or acid carrier.

When flammables are being used in the laboratory, every effort should be made to eliminate sources of ignition. No smoking or open flames should be permitted while appreciable amounts of flammables are being poured or otherwise exposed to the air. Particular care should be taken to remove any sparking equipment from the laboratory bench on which the work is being done and from the floor area of the room. Transfers from one metal container to another should be preceded by electrical bonding in order to avoid a static spark that can result from pouring a solvent (Fig. 19-10).

Fig. 19-7. Oval-shaped safety cans require less shelf space.

When flammable materials are being handled in appreciable quantities on the laboratory bench, it is possible for significant concentrations of flammable or explosive vapor to accumulate. Ventilation is usually inadequate to handle such concentrations. A laboratory hood with a face velocity of not less than one hundred feet per minute should be utilized if possible for work involving solvents with flash point below room temperature, preferably for all solvents with flash points below 100°F. In pilot-plant areas, it may be necessary to utilize a portable blower and duct to effect local removal of flammable (or toxic) vapors or gases (see *Industrial Ventilation,* 10th ed., ACGIH).

Essential sparking electrical equipment should be explosion proof (Class I, Group D) or should be enclosed in such a way that flammable

Fig. 19-8. Storage cabinet for flammable liquids.

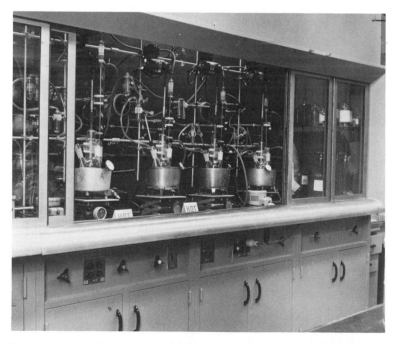

Fig. 19-9. Distillations involving flammable materials should be set up in a hood.

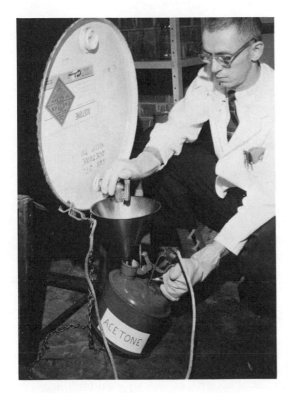

Fig. 19-10. Proper drum dispensing. The other end of the drum is grounded.

vapors cannot come in contact with the spark. The housing for such equipment may be purged with nitrogen, or the arcing contacts may be immersed in oil; the latter is a convenient arrangement for variable transformers.

Inerting techniques based on the exclusion of oxygen are used only infrequently, because of the difficulty in maintaining an oxygen-deficient atmosphere in most laboratory containers. An inerting atmosphere of nitrogen or carbon dioxide may occasionally be utilized to blanket a flammable liquid, but this is possible only under special conditions.

Properly arranged and segregated storage facilities are highly recommended if appreciable amounts of flammables are to be stored. The storeroom for flammables should not be used for general storage of other chemicals or equipment, and no wood or paper containers should be permitted. All construction, including shelving, should be fire resistant. The room should be equipped with an automatic fire extinguishing system, such as carbon dioxide in addition to sprinklers. The storage location should never be below grade unless dependable and adequate exhaust ventilation is arranged to take suction near the floor. Exhaust capacity should be such that air exchange takes place every two minutes in the entire room. The storage area is best located away from the main building. If, however, it is located in a large building, it should be above grade, preferably on the first floor. Construction of the storage room should be such that spread of a fire from this room to other parts of the building would be prevented. In dispensing from drums, either air-operated or hand drum pumps and approved spring-loaded spigots should be used. Sufficient bonding cables should be provided so that all transfers of solvents can be made without a static ignition hazard. (For other details on the transfer of flammable liquids, see Chapter 7.)

Disposal of accumulations of used flammables can be a difficult problem in a congested laboratory. Where large amounts of a single solvent are recovered, it may be economical to redistill in order to recover the solvent for reuse. For the disposal of unknown mixtures, there are generally two major possibilities. One is to find a contractor who will pick up and dispose of chemical wastes; the other is to feed the materials to a waste solvent burner of proper design.

In case of emergency, such as spillage of a large amount of flammable material, prompt action must be taken in order to prevent fire or explosion:

Immediately shut down all flames and sparking equipment. Some laboratories are equipped with master switches to shut off all power.

Stop the source of the flammable material as quickly as possible.

Vacate personnel from the affected area.

Alert the fire department for standby service.

Ventilate the area as effectively as possible. Forced ventilation from near the floor is best.

Use a flammable gas detector for monitoring the area before readmitting personnel or starting up equipment.

Review the procedure to prevent recurrence.

Bibliography

1. *Fire Hazard Properties of Certain Flammable Liquids, Gases and Volatile Solids,* NFPA No. 325M, 1969.
2. Flammable Liquids Code, NFPA No. 30, 1966.
3. Zabetakis, M. G., *Flammability Characteristics of Combustible Gases and Vapors, U.S. Bureau of Mines,* Bulletin 627, 1967.
4. *Fire Protection Guide on Hazardous Materials,* NFPA, 3rd ed., 1969.
5. *Sax, N. I., Dangerous Properties of Industrial Materials,* 3rd ed. Van Nostrand Reinhold, 1968.
6. Mellan, I. *Industrial Solvents Handbook,* Noyes Data Corp., Park Ridge, N.J., 1970.

7. Steere, N. V., *Handbook of Laboratory Safety,* Chemical Rubber Co., Cleveland, 1970.
8. Marsden, C., *Solvents Manual,* 2nd ed., Interscience Publishers, New York, 1963.
9. Birchall, J. D., *Classification of Fire Hazards and Extinction Methods,* Ernest Benn Limited, London, 1954.
10. *Classification of the Hazards of Liquids,* Underwriters' Laboratories, Research Bulletin No. 29, 207, 1943.
11. Dunlap, A.P., "Fire Safety," *Ind. Eng. Chem.* **48,** (2), 7A–10A (1956).
12. *Fire Protection Handbook,* 13th ed., National Fire Protection Association, 1969.

20

SAFE HANDLING OF COMPRESSED GASES IN CYLINDERS

A compressed gas is defined by the Department of Transportation as "any material or mixture having in the container either an absolute pressure exceeding 40 pounds per square inch at 70°F, or an absolute pressure exceeding 104 pounds per square inch at 130°F, or both; or any liquid flammable material having a Reid vapor pressure exceeding 40 pounds per square inch at 100°F."[1,4]

For the purposes of safety, all volatile materials and mixtures packaged in cylinders should be considered compressed gases.

HAZARDS

The handling of compressed gases must be considered more hazardous than the handling of liquid and solid materials because of the following properties unique to compressed gases: pressure, diffusivity, low flash points for flammable gases, low boiling points, and no visual and/or odor detection of many hazardous gases.

These unique properties give rise to the following hazards:

With pressure, hazards may arise as a result of equipment failure and leakage from systems that are not pressure tight. Also, improper pressure control may cause unsafe reaction rates due to poor flow control.

Diffusion of leaking gases may cause rapid contamination of the atmosphere, giving rise to toxicity, anesthetic effects, asphyxiation, and rapid formation of explosive concentrations of flammable gases.

The flash point of a flammable gas under pressure is always lower than ambient or room temperature. Leaking gas can therefore rapidly form an explosive mixture with air.

Low-boiling-point materials can cause frostbite on contact with living tissue. This is common among the cryogenic liquids, such as nitrogen and oxygen, but it also can result from contact of the liquid phase of liquefied gases, such as carbon dioxide, fluorocarbons, and propylene.

Other effects of some compressed gases that are similar to hazards found with other chemicals are corrosion, irritancy, and high reactivity.

The procedures adopted for the safe handling of compressed gases are mainly centered on containment of the material, to prevent its escape to the atmosphere, and proper control of pressure and flow. All rules and regulations are directed toward these ends. Knowledge of emergency procedures is important to limit property damage or injury, but is usually necessary only because a basic rule of handling has been broken. It is far better to observe the rules and avoid the need for emergency measures. A listing of the hazards of some common compressed gases appears at the end of this chapter.

CYLINDER INFORMATION

The following information (referring to Fig. 20-1) will serve to familiarize laboratory personnel with cylinder parts and terminology as well as the meaning of important cylinder markings:

1. Valve handwheel: used to open and close cylinder valve. Valves are occasionally not equipped with handwheels, and require special wrenches to effect operation.
2. Valve pack nut: contains packing gland and packing around stem. Adjusted only occasionally; usually tightened if leakage is observed around valve stem. Should not be tampered with for diaphragm-type valves.
3. Valve outlet connection: for connection to pressure- and/or flow-regulating equipment. Various types of connections are provided to prevent interchange of equipment for incompatible gases. Usually identified by a CGA (Compressed Gas Association) number; for example, No. 350 for hydrogen service.
4. Safety device: to permit gas to escape if the temperature gets high enough to endanger the cylinder by increased unsafe pressures.

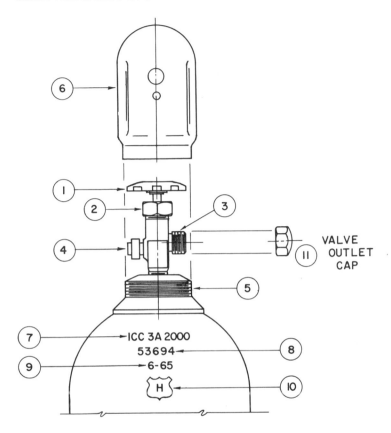

VALVE
OUTLET
CAP

ICC 3A 2000
53694
6-65
H

Fig. 20-1. Diagram of a cylinder head. (Numbers are explained in the text.)

5. Cylinder collar: holds cylinder cap (6) at all times, except when regulating equipment is attached to cylinder valve.
6. Cylinder cap: to protect cylinder valve.
7. This number signifies that the cylinder conforms to Department of Transportation specification DOT-3A governing materials of construction, capacities, and test procedures, and that the service pressure for which the cylinder is designed is 2000 pounds per square inch gage at 70°F.
8. This number is the cylinder serial number.
9. This number indicates the date (month and year; in this case, June 1965) of initial hydrostatic testing. Thereafter, hydrostatic pressure tests are performed on cylinders, for most gases, every five years to determine their fitness for further use. At this time new test dates are stamped into the shoulder of the cylinder. Present regulations permit

visual tests in lieu of hydrostatic tests for low-pressure cylinders in certain gases free of corrosive agents; special permits allow for hydrostatic pressure tests at ten-year intervals for cylinders in high-pressure service for certain gases.

10. Original inspector's insignia for conducting hydrostatic and other required tests to approve the cylinder under DOT specifications.

11. Valve outlet cap: protects valve threads from damage and keeps outlet clean; not used universally.

CYLINDER HANDLING UPON RECEIPT

When a cylinder is delivered to the user, it should have attached: an identification label and (or) markings indicating contents: a DOT label (may not be required in certain intrastate shipments); and a valve-protection cap (Fig. 20-2). Under no circumstances should the identification markings be removed from the cylinder. The valve-protection cap should

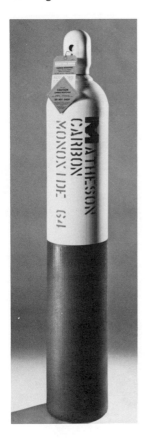

Fig. 20-2. Standard high-pressure cylinder showing identification, warning label (and DOT label), and valve-protection cap.

also remain in place until the cylinder has been transported to the point of use and secured in place. DOT labels and commodity names (chemical names or commonly accepted names) are required for cylinders in interstate (and some intrastate) transportation. These labels have a minimum of precautionary handling information, and will classify the cylinder contents as either flammable, nonflammable, poison, or acid. Some suppliers provide adequate labels with as much information on them as possible, warning against possible hazards associated with the cylinder contents. A standard for the method of identification marking on compressed gas cylinders and a guide for the preparation of labels to identify contents are available.[2,3] Cylinders may be received with no identification other than a color code. Such cylinders should not be accepted. Color codes are of value in helping the supplier to segregate large numbers of cylinders into various gas services or as a secondary means of confirming cylinder contents. If a conflict exists between the color coding and the written identification, do not attempt to use the material. Return it to the supplier.

Fig. 20-3. Example of a good compressed gas storage facility.

PROPER CYLINDER STORAGE

After cylinders are received, they should be stored in a detached and well-ventilated or open-sided building. Storage buildings or areas should be fire resistant, well ventilated, located away from sources of ignition or excessive heat, and dry. Such areas should be prominently posted with the names of the gases being stored (Fig. 20-3). Indoor storage areas should not be located near boilers, steam or hot water pipes, or any sources of ignition. Outdoor storage areas should have proper drainage and should be protected from the direct rays of the sun in localities where high temperatures prevail. Subsurface storage areas should be avoided. Cylinders should be protected against tampering by unauthorized personnel.

Cylinders should be chained in place or put in partitioned cells to prevent them from falling over. Where gases of different types are stored at the same location, cylinders should be grouped by types of gas, and the groups arranged to take into account the gases contained; for example, flammable gases should not be stored near oxidizing gases. Storage in a laboratory should be confined to only those cylinders in use. In all cases, storage areas should comply with local, state, and municipal requirements as well as with the standards of the Compressed Gas Association and the National Fire Protection Association.[6,8]

CYLINDER TRANSPORTATION

When cylinders are being moved from a storage area into the laboratory, the valve-protection cap should be left in place. The cylinder should then be transported by means of a suitable hand truck (Fig. 20-4). Such a hand truck should be provided with a chain or belt for securing the cylinder on the truck. If a large number of cylinders must be moved from one area to another, a power device, such as a fork truck equipped with a special container and provided with some means of securing them, can be used. Do not lift cylinders by the cap. Avoid dragging or sliding cylinders. Use hand trucks even for short distances.

SECURING CYLINDER PRIOR TO USE

When the cylinder has reached its place of use in the laboratory, it should be secured to a wall, a bench, or some other firm support, or placed in a cylinder stand (Fig. 20-5, 20-6). An ordinary chain or belt of the type commonly available from a laboratory supply house can be used. Once the

Fig. 20-4. Cylinder hand trucks properly designed for in-plant transportation of cylinders.

cylinder has been secured, the cap may be removed, exposing the valve. The number of cylinders in a laboratory should be limited, to minimize the fire and toxicity hazards.

CYLINDER SAFETY RELIEF DEVICES

Safety devices are incorporated in all DOT compressed gas cylinders, except those in poison or toxic gas service, where the risk of exposure to fumes is considered more hazardous than that of a potential cylinder failure. Gases for which safety devices are not permitted usually require cylinders having a higher safety factor than do other compressed gases. Safety devices are incorporated in the cylinder valve, in plugs in the cylinder itself, or both (Fig. 20-7). In certain types of gas service, and in many cylinders over a particular length, two safety devices may be required, one at each end of the cylinder. Where safety devices are required to meet DOT regulations, such devices must be approved by the Bureau of Explosives.[4] These safety devices are of four basic types: spring-loaded safety relief, used mostly for low-pressure, liquefied, flammable

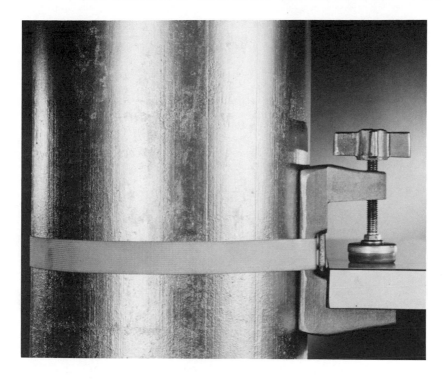

Fig. 20-5. Bench-clamp-type cylinder support.

Fig. 20-6. Floor-stand-type cylinder support.

Fig. 20-7. Examples of types of cylinder valves.

gases; frangible disc, used mostly for high-pressure cylinders; frangible disc backed up by a fusible metal; and fusible metal. The spring-loaded safety relief type consists of a spring-loaded seat which opens to relieve excessively high pressures and then closes when the pressure returns to a safe value. The frangible disc will burst out considerably above the service pressure but not below or at the hydrostatic test pressure of a cylinder, and will release the entire cylinder contents. The frangible disc backed up by a fusible metal will function only if the temperature is high enough to melt the fusible metal, after which excessive pressures will burst the disc, resulting in release of the entire contents of the cylinder. The fusible metal devices melt at excessive temperatures (approximately 160-212°F), allowing the entire contents of the cylinder to escape. Any of these devices will prevent a cylinder from bursting due to excessively high temperatures. However, the latter two devices may not prevent a cylinder from bursting in cases where an overfilled cylinder is exposed to a temperature which is excessive but not high enough to melt the fusible metal safety devices. Since the proper functioning of cylinder safety devices depends to a large extent on the proper filling of a cylinder, such filling should never be attempted by the user unless express permission has been obtained from the gas supplier. Safety devices may also fail to function properly if an intense flame impinging on the side wall of a cylinder weakens the metal to the point of failure before heat or pressure can cause the safety device to function properly. Finally, it must be emphasized that tampering with cylinder safety devices is extremely hazardous.

GENERAL RULES[5,6,8]

It is of the utmost importance that the hazardous properties of a compressed gas (such as flammability, toxicity, chemical activity, and corrosive effects) be well known. Every effort should be made to learn these various properties before the gas is put to use.

In addition to the properties of various gases, it is important to know what materials of construction must be used with many of them to prevent equipment failure due to corrosion. Another important factor in the choice of materials of construction concerns the possible formation of hazardous compounds, such as acetylides formed by the reaction of copper with acetylene or gases containing acetylene as an impurity, or the possible formation of an explosive compound when mercury is used in the presence of ammonia.

The hazards of toxic, flammable, and corrosive gases can be minimized by working in well-ventilated areas. Where possible, work should be done in a hood, employing cylinder sizes that will ensure use of all the gas within a reasonable amount of time. Leaks should not be allowed to go unchecked. Advise the supplier immediately of cylinder leaks that cannot be stopped by simple adjustments, such as tightening a packing nut.

In addition, the user of compressed gases should familiarize himself with the first aid methods to be used in cases of overexposure or burns caused by a gas. A plant doctor should be familiar with whatever further treatments may be necessary. Unnecessary delay in the treatment of a patient overcome by a toxic gas or burned by a corrosive gas could cause permanent damage to the patient and might even result in death. Authorized personnel should administer first aid, but medical treatment should be administered only by a physician. One should be notified immediately.

LEAK DETECTION

When using toxic gases, it is advisable that some device or indicator be used to give warning of toxic concentrations. For example, strips of lead acetate paper can be hung in an area where hydrogen sulfide is being used. Although this gas has a disagreeable odor, it soon deadens the sense of small, resulting in the user's being incapable of detecting increasingly dangerous concentrations by odor.

There are numerous monitoring devices available for detection of dangerous concentrations of gases in the atmosphere (Fig. 20-8). There are also available appropriate chemical procedures for the detection of leaks in lines and equipment and for determining dangerous concentrations of

Fig. 20-8. Thermal conductivity-type leak detector.

gases in the atmosphere. The user of gases should familiarize himself with suitable control procedures for the determination of such dangerous concentrations. Instructions are usually supplied in the data sheets associated with the particular gas being used.[5,6]

Cylinders that develop leaks should be treated as follows. Cylinder valve packing leaks of acidic gases, such as chlorine, hydrogen chloride, hydrogen sulfide, and sulfur dioxide, can usually be corrected by tightening the valve packing nut (turn clockwise as viewed from above). If valve leaks persist or if leaks appear at any portion of the cylinder, advise the supplier immediately. Remove the cylinder to a hood or location where the leakage cannot cause damage until the contents can safely be disposed of by venting or absorption in water or caustic solution or until information on other means of disposal as recommended by the supplier is obtained. Cylinder valve packing leaks of basic gases (ammonia and amines) can be stopped by tightening the packing nut (turn counterclockwise as viewed from above). Leaks of these basic gases through the valve outlet that cannot be controlled by turning the handwheel to the closed position should be stopped by plugging the outlet with ⅜-in. pipe plugs. If leaks persist, remove the cylinder to a hood and ask the supplier for information on how to remedy the problem.

Check cylinders and all connections under pressure for leaks prior to using the contents.

On rare occasions, emergency action may be necessary in order to move a leaking cylinder to a location where it can vent safely, or it may have to be removed from a building and brought outdoors. In such instances, an emergency plan should be put into effect to: (1) properly warn all personnel required to evacuate a building or section of a building; (2) shut off electrical power to prevent ignition of a leaking flammable gas; (3) determine shortest route to point of gas disposal; (4) obtain satisfactory conveyance, such as hand truck, to move cylinder swiftly; (5) post area where cylinder is venting to prevent tampering by unauthorized personnel.

Such an emergency plan can function efficiently only if a trained safety crew is educated in the proper handling of gas cylinders, with training in the procedures to be followed in cases of emergency with all the gases handled by the facility. Equipment such as self-contained gas masks must be available for handling toxic gases or for handling asphyxiating gases in close confines. Emergencies involving flammable gases must be managed with the utmost care in order to prevent ignition. The aftermath of gross leakage is extremely important. All areas must be adequately vented before the restoration of power in cases of flammable gas leakage. Areas contaminated by corrosive gases must be adequately vented and completely washed down to prevent subsequent degradation of delicate instruments, electrical contacts, and so on.

HANDLING OF CORROSIVES

Corrosive gases should be stored for the shortest possible periods before use, preferably less than three months. Storage areas should be as dry as possible. A good supply of water should be available in case of emergency leaks. Most corrosive gases can be absorbed in water.

In cases of leaks, corrosive gases should not be stored in areas containing instruments or other devices sensitive to corrosion. These gases should be segregated as to type, and rotated so that the oldest stock is used first.

The smallest cylinders possible should be used to ensure reasonable turnover of cylinders. Cylinders used and then put back in storage should have all appurtenances removed from the valve outlet and preferably flushed with dry nitrogen or air to keep them in good working order.

When corrosive gases are being used, the cylinder valve stem should be worked frequently to prevent freezing. The valve should be closed when the cylinder is not in use. Regulators and valves should be closed when the cylinder is not in use and flushed with dry air or nitrogen after use. Such control devices should not be left on a cylinder, except when it is in frequent use. When corrosive gases are to be discharged into a liquid, a trap, check valve, or vacuum break device should always be employed to prevent dangerous suckback.

Corrosive gas cylinder valve stems occasionally become stuck ("frozen") and cannot be turned. This condition may be corrected by first plugging the cylinder valve outlet with a solid plug, gasketed cap, or a closed needle valve, loosening the valve packing nut, and then trying to turn the valve stem manually with a correctly fitted wrench (no longer than eight inches) or by slight tapping of the wrench with a light tool. Once the stem is turning freely, shut the valve, tighten the packing nut, and safely vent the gas trapped in the valve port. This operation should be done in a hood or out of doors. If the cylinder valve cannot be unfrozen by this procedure, ask the supplier for information or return the cylinder to him.[7]

GENERAL PRECAUTIONS[5,6,8]

Some general precautions for handling, storing, and using compressed gases follow.

Never drop cylinders or permit them to strike each other violently.

Cylinders may be stored in the open, but should be protected from the ground beneath to prevent rusting. Cylinders may be stored in the sun, except in localities where extreme temperatures prevail; in the case of certain gases, the supplier's recommendation for shading should be observed. If ice or snow accumulates on a cylinder, thaw at room temperature or with water at a temperature not exceeding 125°F.

The valve protection cap should be left on each cylinder until it has been secured against a wall or bench or placed in a cylinder stand and is ready to be used.

Avoid dragging, rolling, or sliding cylinders, even for a short distance. They should be moved by means of a suitable hand truck.

Never tamper with safety devices in valves or cylinders.

Do not store full and empty cylinders together. Serious suckback can occur when an empty cylinder is mistakenly attached to a pressurized system.

No part of a cylinder should be subjected to a temperature higher than 125°F. A flame should never be permitted to come in contact with any part of a compressed gas cylinder.

Cylinders should not be subjected to artificially created low temperatures (−20°F or lower), since many types of steel will lose their ductility and impact strength at low temperatures. Special stainless steel cylinders are available for low-temperature use.

Do not place cylinders where they may become part of an electric circuit. When electric arc welding, precautions must be taken to prevent striking an arc against a cylinder.

Bond and ground all cylinders, lines, and equipment used with flammable compressed gases.

Use compressed gases only in well-ventilated area. Toxic, flammable, and corrosive gases should be handled in a hood. Only small cylinders of toxic gases should be used.

Cylinders should be used in rotation as received from the supplier. Storage areas should be set up to permit proper inventory rotation.

When discharging gas into liquid, a trap or suitable check valve should be used to prevent liquid from getting back into the cylinder or regulator.

When using compressed gases, wear appropriate protective equipment, such as safety goggles or face shield, rubber gloves, and safety shoes. Well-ventilated barricades should be used in extremely hazardous operations, such as in the handling of fluorine. Gas masks should be kept available for immediate use when working with toxic gases. These masks should be placed in convenient locations in areas not likely to become contaminated, and should be approved by the U.S. Bureau of Mines for the service intended.

When returning empty cylinders, close the valve before shipment, leaving some positive pressure in the cylinder. Replace any valve outlet and protective caps originally shipped with the cylinder. Mark or label the cylinder EMPTY (or utilize standard DOT "EMPTY" labels) and store in a designated area for return to the supplier.

Before using cylinders, read all label information and data sheets associated with the gas being used. Observe all applicable safe practices.

Eye baths, safety showers, gas masks, respirators, and/or resuscitators should be located nearby but out of the immediate area, which is likely to become contaminated in the event of a large release of gas.

Fire extinguishers, preferably of the dry chemical type, should be kept close at hand and should be checked periodically to ensure their proper operation.

PROPER DISCHARGE OF CONTENTS

Liquefied Gases

For controlled removal of the liquid phase of a liquefied gas, a manual valve is used. It must be remembered that withdrawal of liquid must necessarily be done at the vapor pressure of the material. Any reduction of pressure will result in flashing of all or part of the liquid to the gas phase.

Liquid may be removed from a cylinder by inverting the cylinder. If the cylinder is fitted with a gooseneck eductor tube, the cylinder should be placed on its side with the valve outlet pointing up. If the cylinder is equipped with a full-length eductor tube, liquid is withdrawn with the cylinder in a normal vertical position. Request information from the vendor if there is any question as to whether a particular cylinder is equipped with an eductor tube and the type.

Rapid removal of the gas phase from a liquefied gas may cause the liquid to cool too rapidly, causing the pressure and flow to drop below the required level. In such cases, cylinders may be placed in a water bath heated to not more than 125°F. Rapid gas removal can also be effected by transferring the liquid to a heat exchanger, where the liquid is vaporized to a gas. This method imposes no temperature limitations on the material. However, care should be taken to prevent blockage of the gas line downstream from the heat exchanger, which may cause excessive pressure to build up in both the heat exchanger and the cylinder. Liquid should not be permitted to be trapped in any part of the piping between valves unless a safety device has been placed in that section, since dangerous hydrostatic pressures can build up with rises in temperature.

Nonliquefied Gases

An automatic pressure regulator is the device most commonly used to reduce pressure to a safe value for gas removal of nonliquefied gas or for gas phase control of a liquefied gas (Fig. 20-9). This device contains a spring-loaded diaphragm, which controls the throttling of an orifice. Delivery pressure will exactly balance the delivery pressure spring to give relatively constant delivery pressure.

Before attaching a regulator to a cylinder valve outlet, wipe the outlet with a clean rag to remove any dirt. When using oxygen, blow any dirt from the outlet by opening the cylinder valve momentarily and then closing it. Do not wipe or touch the valve outlet of an oxygen cylinder valve so as not to leave any organic residues which might be subsequently ignited by exposure to high oxygen pressure.

A regulator should be attached to a cylinder without forcing the threads. If the inlet of the regulator does not properly fit the cylinder outlet, no effort should be made to try to force the fitting. A poor fit or mismatch of fittings probably indicates that the regulator is not intended for use with the particular compressed gas.

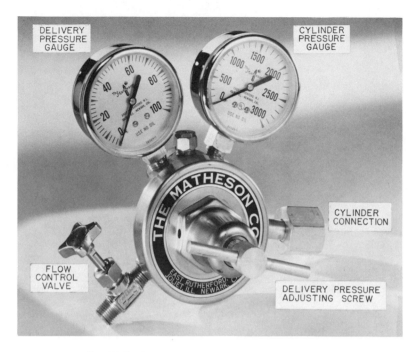

Fig. 20-9. Common automatic pressure regulator.

REGULATOR OPERATION

The following procedure should be used to obtain the required delivery pressure.

After the regulator has been attached to the cylinder valve outlet, turn the delivery pressure adjusting screw counterclockwise until it turns freely.

Open the cylinder valve slowly until the tank gauge on the regulator registers the cylinder pressure. At this point, the cylinder pressure should be checked to see if it is at the expected value. A difference may indicate that the cylinder valve is leaking, possibly at the valve, the safety device, or plugs that may be in the cylinder.

With a flow-control valve downstream from the regulator in closed position, turn the delivery pressure adjusting screw clockwise until the required delivery pressure is reached. Control of flow can be regulated by means of a valve supplied in the regulator outlet or by a supplementary valve placed in a line downstream from the regulator. The regulator itself should not be used as a flow control by adjusting pressure to obtain different flow rates. This defeats the purpose of the pressure regulator, and in cases where higher flows are obtained by this manner, the pressure setting may be in excess of the design pressure of the system. This will be evident when pressure goes higher as flow is restricted again.

Fig. 20-10. Low-pressure-type pressure regulator.

Fig. 20-11. Gas-dome-type regulator for high-pressure control up to 6000 psig.

REGULATOR TYPES

The proper choice of a regulator depends on the delivery pressure range required, the degree of accuracy of delivery pressure to be maintained, and the flow rate required (Figs. 20-10 and 20-11). There are two basic types of automatic pressure regulators: the single-stage, and the double- or two-stage. The single-stage type will show a slight variation in delivery pressure as the cylinder pressure drops. It will also show a drop in delivery pressure greater than a two-stage regulator does as the flow rate is increased as well as a higher "lock-up pressure" (difference between pressure under flowing conditions and pressure at a zero flow condition) than does the two-stage regulator. In general, the two-stage regulator will deliver a more constant pressure under more stringent operating conditions than will the single-stage regulator.

Where intermittent flow control is needed and an operator is to be present at all times, a manual needle valve may be used. This type of control is simply a valve which is operated manually to deliver the proper amount of gas. Fine flow control can be obtained, but dangerous pressure can build up in a closed system or in one that becomes plugged, since no means is provided for automatic prevention of excessive pressures.

REGULATOR MAINTENANCE

Regulators designed for gas service must be dependable as pressure controls. Regulators should be checked periodically to ensure proper and safe operation. This periodic check will vary depending on gas service and usage. Regulators used with noncorrosive gases, such as nitrogen, hydrogen, and helium, require relatively little maintenance, and a quick check on a monthly basis is usually adequate. Regulators used with corrosive gases, such as hydrogen chloride, chlorine, and hydrogen sulfide, require considerably more checking—once a week is recommended.

The procedure for checking out any regulator is as follows:

Drain all pressure from the system; gages should read zero.

Open cylinder valve and turn adjusting screw counterclockwise until it turns freely. The high-pressure gage should register the cylinder pressure, and the delivery pressure gage should not indicate any pressure.

With the regulator outlet needle valve closed and after waiting from five to ten minutes, the delivery pressure gage should not indicate a pressure increase, which would indicate leakage across the internal valve system.

Turn the adjusting screw clockwise until a nominal delivery pressure is indicated. The inability to attain a proper delivery pressure setting or abnormal adjustment of the screw indicates improper operation. Continued wear on a regulator valve and seat assembly will cause pressure to rise above a set delivery pressure, termed "crawl." Regulators showing crawl should not be used.

Close the cylinder valve and observe the pressure on both inlet and delivery sides of the regulator. A drop in the pressure reading after five or ten minutes may indicate a leak in the system, possibly at the inlet or through the needle valve, safety devices, or diaphragm.

An excessive fall in delivery pressure under operating conditions and normal flows indicates that an internal blockage exists or that the cylinder valve has not been sufficiently opened.

Any deviation from the normal in the above checkout will require repair. Regulators should be repaired only by the supplier or his authorized agent or by thoroughly qualified personnel.

STANDARD OUTLETS[9]

The Compressed Gas Association has developed various cylinder valve outlet connection types for different families of gases, to prevent the interchange of regulating and control equipment between gases which are not compatible. These connections have also been adopted as standards by the American National Standards Institute.

Identification of outlet connection styles has been accomplished with a numbering system developed by the Compressed Gas Association. Although the separate parts of a connection have subassembly numbers, the complete connection is usually referred to when identifying a particular outlet style. All numbers defining cylinder valve outlet connections utilize a three-digit number (except some air connections utilizing four digits). The number representing the complete connection ends in zero. The next to last digit is an even number for right-hand threads and odd for left-hand threads. The mating nut for a left-handed thread connection is usually notched.

The use of adapters (Fig. 20-12) defeats the intent of varying outlet designs, and is prohibited in many laboratories. Adapters should be used with care only on gases definitely known to be compatible and always with special authorization.

Equipment for certain gases, such as oxygen, should never be interchanged for use on other compressed gases. Gases which are pumped through an oil lubricated compressor may cause an oil film to coat the internal parts of regulators and associated equipment, which, if subsequently used with oxygen, can cause a fire or explosion.

Unfortunately, not all companies conform to standard valve outlets,

Fig. 20-12. Adapter fitting used to permit utilization of non-mating components.

making it doubly important to prevent interchange of equipment on cylinders that may have a common valve outlet but contain different gases.

HAZARDS OF SOME COMMON COMPRESSED GASES

The threshold limit values, flammability limits in air, and major hazards associated with some of the more common gases are listed in Table 20-1.

TABLE 20-1. DATA FOR COMMON GASES

Gas	Threshold Limit Values, ppm[10]	Flammability Limits in Air, % by Vol[11]	Major Hazards
Acetylene	Not established (nontoxic, produces anesthetic effects)	2.5–81.0	Flammable; asphyxiant
Ammonia	50	15–28	Toxic
Argon	Not established (nontoxic)	None	Asphyxiant
Boron trifluoride	1	None	Toxic; causes burns
1,3-Butadiene	1000	2–11.5	Flammable; skin irritant
Butane	Not established (nontoxic, produces anesthetic effects)	1.9–8.5	Flammable
Carbon dioxide	5000	None	Asphyxiant
Carbon monoxide	50	12.5–74.0	Flammable; toxic
Chlorine	1	None	Toxic; severe irritant; causes burns; corrosive
Ethane	Not established	3.0–12.5	Flammable;

TABLE 20-1.   (Continued)

Gas	Threshold Limit Values, ppm[10]	Flammability Limits in Air, % by Vol[11]	Major Hazards
	(nontoxic, produces anesthetic effects)		asphyxiant
Ethylene	Not established (nontoxic, produces anesthetic effects)	3.1–32.0	Flammable; asphyxiant
Ethylene oxide	50	3.0–100.0	Flammable; toxic; can cause burns when trapped by clothing or shoes
Helium	Not established (nontoxic)	None	Asphyxiant
Hydrogen	Not established (nontoxic)	4.0–75.0	Flammable; asphyxiant
Hydrogen bromide	3	None	Toxic; causes burns; corrosive
Hydrogen chloride	5	None	Toxic; causes burns; corrosive
Hydrogen fluoride	3	None	Toxic; causes severe slow healing burns; corrosive
Hydrogen sulfide	10	4.3–45.0	Toxic; flammable; irritant
Methane	Not established (nontoxic)	5.3–14.0	Flammable; asphyxiant
Methyl bromide	20	13.5–14.5	Toxic; causes burns
Methyl chloride	100	10.7–17.4	Toxic; flammable
Methyl mercaptan	0.5	Unknown	Toxic; flammable
Nitrogen	Not established (nontoxic)	None	Asphyxiant
Nitrogen dioxide	5	None	Toxic; corrosive
Oxygen	Nontoxic	None	Highly reactive
Phosgene	0.1	None	Toxic
Propane	Not established (nontoxic, produces anesthetic effects)	2.2–9.5	Flammable; asphyxiant
Sulfur dioxide	5	None	Toxic; causes burns
Vinyl chloride	500	4.0–22.0	Flammable; causes burns

SOME COMMON LABORATORY APPLICATIONS OF COMPRESSED GASES

The past decade has seen a tremendous increase in the use of compressed gases in the chemical laboratory. Table 20-2 indicates some of the common applications for certain gases in laboratory use.

TABLE 20-2. USES FOR COMPRESSED GASES

Gas	Applications
Acetylene	Fuel for atomic absorption spectroscopy; intermediate in organic syntheses
Air	Oxidizer used in flame ionization detectors associated with gas chromatography
Ammonia, anhydrous	Chemical synthesis; neutralizing agent; solvent (in liquid form) for chemical reactions
Argon	Inerting atmosphere; carrier gas for gas chromatography
Boron trifluoride	Catalyst in such diverse reactions as isomerization, alkylation, polymerization, esterification, and condensation
Carbon dioxide	Inerting atmosphere; organic nitrogen analysis; refrigerant for freezing microtomes
Chlorine	Chlorinating agent; chemical reactant
Ethylene oxide	Gas sterilization; organic syntheses
Helium	Inerting atmosphere; carrier gas in gas chromatography
Hydrogen	For catalytic hydrogenations; reducing atmosphere; carrier gas in gas chromatography
Hydrogen bromide	Chemical reactant; catalyst
Hydrogen chloride	Chemical reactant; catalyst and condensation agent
Hydrogen fluoride	Catalyst; fluorinating agent
Hydrogen sulfide	Qualitative and quantitative analysis of heavy metals; phosphor development
Nitrogen	Inerting atmosphere; flushing gas
Oxygen	Carbon and hydrogen determination by combustion; calorimetric determinations
Phosgene	Organic syntheses
Vinyl chloride	Organic syntheses; production of plastics

Bibliography

1. *Hazardous Materials Regulations of the Department of Transportation,* Agent T. C. George's Tariff No. 23, issued by T. C. George, Agent, 1969.
2. *American National Standard Method of Marking Portable Compressed Gas Containers to Identify the Material Contained,* American National Standard Z 48.1–1954, Pamphlet C–4, Compressed Gas Association, Inc., New York.
3. *A Guide for the Preparation of Labels for Compressed Gas Containers,* Pamphlet C–7, Compressed Gas Association, Inc., New York.
4. *Safety Relief Device Standards,* Part I, *Cylinders for Compressed Gases,* Pamphlet S–1, Part 1, Compressed Gas Association, Inc., New York.
5. *Gas Data Book,* 4th ed., The Matheson Company, Inc., East Rutherford, N.J., 1966.
6. Handbook of Compressed Gases, Reinhold Publishing Corp., 1966.
7. *How to Open Cylinders of Hydrogen Chloride, Anhydrous,* The Matheson Company, Inc., East Rutherford, N.J.
8. *Compressed Gases,* Safe Practices Pamphlet No. 95, National Safety Council, Chicago, Ill.
9. *American National Standard Compressed Gas Cylinder Valve Outlet and Inlet*

Connections, ANSI B57.1–1965 and Canadian Standard B96–1965, CGA Pamphlet V–1, Compressed Gas Association, Inc., New York.

10. *Threshold Limit Values,* American Conference of Governmental Industrial Hygienists, Cincinnati, Ohio.

11. Zabetakis, M. G., *Flammability Characteristics of Combustible Gases and Vapors,* Bulletin 627, U.S. Bureau of Mines, U.S. Government Printing Office, Washington, D.C.

12. Sax, N. I., *Dangerous Properties of Industrial Materials,* Van Nostrand Reinhold, 1968.

13. Steere, N. V. *Handbook of Laboratory Safety,* Chemical Rubber Co., Cleveland, Ohio.

21

HAZARDOUS REACTIONS

Chemical reactions in general use have been classified by numerous authors and several major works on the subject are available as listed in the bibliography at the end of this chapter. This section will, therefore, not attempt to cover the many classifications considered in the reference works, but only those which include reactions of a hazardous nature. Likewise, no attempt will be made to list the general precautions that should be observed in all chemical laboratory work to guard against such hazards as flammability, toxicity, and extreme conditions, since these are covered in detail in other sections of this book.

A list of hazardous reaction experiences, with many literature references, prepared by W. F. Elmendorf of Merck and Company, is presented in the appendix.

ALKYLATION REACTIONS

Friedel-Crafts Reaction

Explosions of Friedel Crafts reactions have been attributed to two causes: the solvent used in the reaction, and the aluminum chloride. The use of carbon disulfide as a solvent is extremely dangerous because of its low ignition temperature and explosive limits. Therefore nitrobenzene has been recommended as a solvent for these reactions. Explosions are due to the spontaneous decomposition of aluminum chloride in closed containers. An explosion often occurs when the container is opened. It has also been reported that ethylene, especially in the presence of aluminum chloride as a catalyst, may react violently.

Precautions. If carbon disulfide must be used as a solvent, check for any possible leakage, since carbon disulfide vapor is ignited by steam, hot plates, or motor sparks.

High-Pressure Alkylation

Papers by Kharasch, Jensen, and Urry, as well as many new industrial processes, have made high-pressure alkylation (initiated by free radicals) an important means of organic synthesis. As a result of this method, there have appeared several reports of uncontrolled free-radical reactions resulting in dangerous explosions. The reported explosions have been encountered in attempting to react carbon tetrachloride with ethylene. The unpredictable nature of this reaction requires extreme caution when dealing with this and similar reactions. The major factors in controlling this reaction appear to be: the effective dissipation of the heat of reaction, and the rate of the reaction. In view of this the following recommendations have been made:

Water appears to be an efficient moderator in the ethylene–carbon tetrachloride reaction. No accidents have occurred using 100 g CCl_4, 100 g H_2O, and 0.23 g benzoyl peroxide as a charge treated with ethylene. The presence of water assists in the dissipation of the heat of reaction. It is efficient because it is inert and has a high specific heat.

In a free-radical reaction of this type, the quantity of the reaction initiator should be kept at a minimum.

DIAZONIUM COMPOUNDS

Diazonium Salts

In the standard texts, the preparation of mercaptans by the replacement of the diazonium group of a diazonium salt is given as a straightforward method of synthesis. This reaction, however, continually results in explosions and complete loss of product. Nawiasky, Ebersole, and Werner report that the slow addition of the diazonium chloride of 4-chloro-*o*-toluidine to solutions of sodium bisulfide, sodium sulfide, and sodium polysulfide always results in explosions when the reactions are carried out at 25°C, and 5°C. Reversal of the addition process did not change the results. Hodgson reports similar experiences with the diazonium salts of *o*-

nitroaniline and *m*-chloroaniline. This reaction is not a normal diazo synthesis but a potentially dangerous reaction.

HALOGENATION

Chlorination of Alkylthioureas

An excellent method for the preparation of many alkylsulfonyl chlorides consists of the reaction of chlorine upon *S*-alkylisothiourea salts, in aqueous solution.

$$RNHCH_2-S-C-NH_2 \cdot HCl + Cl_2 \xrightarrow{\quad H_2O \quad} RCH_2SO_2Cl + NH_2CN \cdot HCl$$

Two serious explosions have been reported with this reaction in preparing ethylsulfonyl chloride. In the first case, the reaction mixture was allowed to stand in a refrigerator for a few days; in the second, the chlorination was run over a period of ten hours. The explosions were attributed to the formation of NCl_3, brought about by standing or excess chlorine. If the proper precautions are taken to avoid over-chlorination, this reaction may be used.

Precautions. The chlorination should be stopped after the first signs of excess chlorine appear in the solution. Do not allow the reaction mixture to stand for lengthy periods of time. Until clear differentiation can be made as to when the formation of the sulfonyl chloride is complete, all safety precautions should be taken.

The preparation of sulfonyl chlorides by the use of thionyl chloride, instead of direct chlorination, also resulted in a violent explosion.

Chlorination of Ethane

Hexachloroethane can be obtained by direct chlorination of ethane in the presence of activated carbon in 70–75% yields, at a temperature of 350°C. Explosions have been reported with this reaction.

Precautions. Diluting the reacting gases with carbon dioxide reduces the danger of explosion. The most favorable gas ratios are $C_2H_6:Cl_2:CO_2$ 1:6:12.

Monochloroacetone

An explosion, apparently of a spontaneous nature, of monochloroacetone during storage has been reported. A similar explosion occurred while trying to prepare bromoform from acetone by the haloform reaction.

NITRATION

Acetyl Nitrate

Acetyl nitrate is an important nitrating agent, used to prepare o-nitro com- pounds, but it will detonate spontaneously. An explosion has been reported during vacuum distillation of crude acetyl nitrate. Picket has exploded acetyl nitrate by touching it with a glass rod.

Precautions. Do not attempt to isolate acetyl nitrate as a pure product. A safe procedure in using acetyl nitrate as a nitrating agent is to dissolve the compound to be nitrated in acetic anhydride and then add a cold mix- ture of nitric acid and acetic acid through a dropping funnel. During the addition the reaction mixture should be kept at 5 to 10°C.

o-Nitro Aromatic Compounds

Explosions have been experienced when attempting to distill o-nitroben- zoyl chloride. Small quantities of the compound do, however, distill satis- factorily. The pure compound can be prepared by the use of thionyl chloride without having to resort to a distillation. When prepared by this method, the gaseous reaction products are removed by aspiration with dry illumination gas. o-Nitrophenacetyl chloride is also likely to explode.

Precautions. It is considered dangerous to heat any quantity of o-nitro- benzoyl chloride above 100°C. It is also suggested that this and similar compounds be prepared and utilized in solution.

Nitromethane

Although no specific accidents caused by nitromethane have been re- ported, there are certain precautions which should be taken when using this reagent. Though it is not easily exploded on impact, its sodium and ammonium salts are extremely sensitive and may explode without apparent provocation. Experiments show that nitromethane is only moderately sen- sitive to explosion by detonation or heat and pressure. When it does ex- plode, however, it is a powerful explosive.

Precautions. Do not subject nitromethane to severe shock, high tem- perature and pressure, or to reaction with metallic sodium or potassium.

OXIDIZING AGENTS AND PEROXIDES

It has been mentioned that the formation of peroxides is responsible for many ether explosions. Peroxide formation is not limited to ethers. Organic liquids, which tend to form peroxides in the presence of air, can be protected from oxidation by keeping them in contact with activated charcoal. This precaution minimizes the risk of explosion in stored liquids which contain alcohols, ethers, ketones, esters, and unsaturated compounds.

Butadiene

When heated under pressure, butadiene may undergo violent thermal decomposition. In contact with air, it may form violently explosive peroxides, which may be detonated by mild heating or mechanical shock. Solid butadiene absorbs enough oxygen at subatmospheric pressures to make it detonate violently when heated slightly above its melting point.

Precautions. Addition of an inhibitor is recommended to prevent peroxide formation. Treatment with a strong sodium hydroxide solution (47%) will destroy peroxides safely and effectively.

Peracetic Acid

Attempts to prepare peracetic acid by autoxidation of acetaldehyde, using cobalt acetate as a catalyst, gave fair yields of peracetic acid but resulted in violent explosions.

Precautions. It is recommended that peracetic acid be prepared from hydrogen peroxide and acetic acid or acetic anhydride and used in solution without isolation.

Benzoyl Peroxide

Benzoyl peroxide has been reported to explode for no apparent reason. Other explosions, presumably due to friction, have occurred while opening bottles containing benzoyl peroxide.

Precautions. Do not leave bottles of benzoyl peroxide open or allow contamination with organic matter. Store in containers having tops which minimize friction on opening.

In the standard directions for the preparation of perbenzoic acid, it is stated that impure benzoyl peroxide should be recrystallized from a small amount of hot chloroform. Several attempts at such crystallizations have resulted in explosions.

Precautions. Benzoyl peroxide can be recrystallized without danger by the addition of methanol to a solution of benzoyl peroxide in chloroform at room temperature.

Acetyl Peroxide

Acetyl peroxide is more sensitive and unpredictable than benzoyl peroxide. It is reported that 5 g of acetyl peroxide, after being removed from an ice chest, detonated with sufficient violence to tear off a worker's hands.

Precautions. The preparation of acetyl peroxide should be carried out without interruption. The material should be used immediately, and should be handled with long tongs a sufficient distance from the body, using a protective screen. When working with peroxides, be especially careful to avoid rapid changes in temperature.

Monopersulfuric Acid

Caro's acid in contact with primary or secondary alcohols often results in explosions. An abnormally high concentration of Caro's acid in any organic medium may prove dangerous.

Selenium Dioxide Oxidation

Selenium precipitated during a selenium dioxide oxidation usually reacts more vigorously with nitric acid than does ordinary selenium in the initial step of the recovery of selenium dioxide. This is probably the result of traces of organic matter present. The direct reoxidation of selenium, by burning in oxygen, has resulted in a vigorous explosion, probably due to organic matter present.

Precautions. When reoxidizing selenium by direct oxidation with oxygen, it is advisable to reheat the selenium slowly in air almost to its boiling point until all organic matter is expelled before any attempt is made to burn it. The nitric acid method for the recovery of selenium dioxide is, however, a more convenient laboratory procedure.

REDUCTION

Catalytic Reduction

Carswell has reported the explosion of 400 g of o-nitroanisole during catalytic reduction using nickel as a catalyst. Adkins attributes this explosion to ill-advised procedures, and recommends the following precautions when attempting to carry out high-pressure catalytic hydrogenations.

Precautions. The bomb to be used in the hydrogenation should be of sufficient strength. Adkins recommends a bomb 6 cm inside diameter and 2.5 cm thick, bored out of rolls of Ni-Cr steel. The bomb should be provided with a thermocouple which has only a slight lag in temperature between the reading device and the actual temperature in the bomb. The temperature of the reaction and the amount of catalyst used should not be excessive, since the reactions are exothermic.

Explosions have been experienced when dioxane has been used as a solvent in hydrogenations employing Raney nickel as a catalyst.

Precautions. Dioxane is suitable as a solvent only when the temperature to be used is low. At high temperatures and pressures, dioxane is likely to decompose explosively.

Lithium Aluminum Hydride, Aluminum Hydride, and Sodium Aluminum Hydride

In reductions using these reagents and dimethyl ether, ethyl ether, or dimethyl cellosolve as a solvent, explosions have occasionally occurred. The details of the explosions are described by Schlessinger, who attributes them to an impurity of carbon dioxide in some of the samples of the solvents. Reactions involving these reagents can be used with perfect safety if the precautions suggested are observed.

Precautions. When reducing organic compounds, it is recommended that the initial reaction product be hydrolyzed before evaporating the solvent. If it is desired to evaporate the solvent before hydrolysis of the initially formed organic salts, the safety of the procedure should first be determined on a small sample.

Activated Metallic Catalysts

Activated zinc and Raney nickel are pyrophoric catalysts, and will tend to glow or burn if they are allowed to dry. The following general precautions are recommended when working with any activated metallic catalyst.

Precautions. Store activated metallic catalysts in tightly closed containers. Raney nickel should be stored in a glass or metal container under a solvent. Clean up all spills immediately and flush with copious amounts of water. When filtering, do not allow the catalyst to become air dried. The disposal of the catalyst should be carefully controlled. Small amounts may be deactivated by water containing 1% sodium sulfide. If the catalyst starts to glow, flush the area with liberal quantities of water. If the catalyst is part of a solvent fire, use carbon dioxide to extinguish the fire prior to flushing with water.

Hydrogen

Hydrogen has an unusual and relatively unknown characteristic which under certain conditions presents a great hazard. Unlike other gases, hydrogen heats up if it is expanded at a temperature above its "inversion point" of $-80°C$. This is known as the "inverse Joule-Thomson effect." It is well known that a cylinder of hydrogen will occasionally emit a flash of fire when the cylinder valve is opened suddenly, permitting a rapid escape of gas. It is thought by some that the "Inverse Joule-Thomson effect" plus the static charge generated by the escaping gas may cause its ignition.

INORGANIC REAGENTS

Perchloric Acid

The most recent perchloric acid disaster caused the death of seventeen persons and the destruction of 116 buildings in Los Angeles. The explosion was caused by the contamination of a mixture of 70-72% perchloric acid and acetic anhydride with "easily oxidizable" organic material. Tests run by the U.S. Bureau of Mines indicate that many mixtures of perchloric acid with oxidizable materials are more hazardous, as regards sensitivity to impact and heat, than many common explosives. Anhydrous perchloric acid is even more dangerous and is stable only at very low temperatures.

Perchloric acid is often used in analysis as a method of destroying organic matter, and has often resulted in violent reactions. In destroying organic matter, a preliminary treatment with nitric acid is recommended to destroy first all easily oxidizable material.

Precautions. Avoid formation of the anhydrous acid, and contamination with organic material. Store perchloric acid under fireproof conditions. Clean up spillage immediately; the spill may not result in fire or explosion until some later occurrence, such as friction or impact, sets it off.

Hydrogen Sulfide

Two deaths and three almost fatal injuries caused by hydrogen sulfide have recently been reported. Hydrogen sulfide is as toxic as hydrogen cyanide, 1 part of the gas in 200 parts of air (5000 ppm) being rapidly fatal. The well-known "rotten egg" odor cannot be depended upon as a warning that hydrogen sulfide is present, since at high concentrations it has a sweet odor. A concentration of 600 ppm (0.06%) is fatal in 30 min, but only a few breaths of the pure gas will cause instant death by respiratory paralysis. Hydrogen sulfide also forms explosive mixtures with air. The lower flammability limit is 4.4% H_2S, and the upper limit

45% H_2S. When working with hydrogen sulfide, a fume hood should be used at all times.

Ammoniacal Silver Nitrate Solutions

Explosions of ammoniacal silver nitrate solutions often occur when the solution is allowed to stand. These explosions are caused by the presence of silver hydroxide and ammonia, which form fulminating silver and water.

Precautions. Do not attempt to dissolve precipitated silver hydroxide in ammonium hydroxide. In preparing the solution, dissolve silver nitrate in ammonium hydroxide and then add sodium hydroxide. Use the solution as soon as possible, and immediately discard any excess.

Magnesium Perchlorate

When magnesium perchlorate was used as a drying agent for unsaturated hydrocarbons, an explosion occurred from unknown causes when the mixture was heated to 220°C.

Precautions. Magnesium perchlorate should be kept out of contact with acids, and should not be used as a drying agent.

MISCELLANEOUS REACTIONS

Preparation of Acetyl Chloride

In the preparation of acetyl chloride from acetic acid and phosphorus trichloride, several laboratory explosions have occurred. The cause of these explosions has been attributed to the generation of phosphine by local overheating of the phosphorous acid formed.

Precautions. The distillation should be carried out on a water bath and not over an open flame.

Ball's Reaction

Ball's reaction is a method of determining sulfur compounds in petroleum distillates. The method involves the removal, and then the estimation, of hydrogen sulfide, mercaptans, disulfides, and free sulfur. The aliphatic sulfides are removed by the formation of complexes with mercurous nitrate, and the amount calculated from the reduction of sulfur content. Aromatic sulfides and thiophenes are determined in a somewhat similar manner. It has been reported that a violent reaction occurred when a naphthenic type gas oil was treated with mercuric nitrate. It is believed that for some reason a nitration reaction was initiated, with the mercuric nitrate acting as a catalyst and nitrating agent. The reaction was probably accelerated by a lack of cooling and by the exothermic decomposition of mercuric nitrate. A similar explosion occurred when attempting to analyze cracked naptha by Ball's method.

Furfuryl Alcohol and Formic Acid

It is known that furfuryl alcohol will polymerize rapidly, and sometimes with explosive violence, in the presence of strong mineral acids. It now appears that a similar explosive reaction may occur in the presence of concentrated formic acid. While attempting to prepare furfuryl formate from furfuryl alcohol and concentrated formic acid, an explosion occurred, producing a hard brown resin.

N-Nitro-N'-2,4-Dinitrophenylurea

This reagent has been suggested as a means of characterizing primary and secondary amines, since it gives easily purified derivatives. It should be recognized as an explosive of power lower than picric acid but with a slightly greater sensitivity to friction and impact.

Precautions. Do not store this reagent in bottles with ground-glass stoppers.

Ethyl Sulfate

An explosion of ethyl sulfate in a metal container was attributed to moisture in the container, causing the formation of sulfuric acid and alcohol. The presence of sulfuric acid resulted in corrosion of the container and the formation of hydrogen, causing excessive pressure.

Precaution. Store ethyl sulfate and similar reagents in glass containers and dry places.

Diazomethane

Besides being extremely toxic, diazomethane will undergo violent thermal decomposition. If the vapor is heated above 200°C, it may explode violently. Explosions also occur at low temperatures, owing to traces of organic matter. It is hardly possible to control the behavior of the substance, even at −80°C.

Precautions. When working with diazomethane in any form, be it gas, liquid, or solution, use all possible precautions against explosion.

Ethylene Oxide

Considerable danger may be present when working with ethylene oxide in high pressure reactions. Two explosions of this nature have been reported: one with a mercaptan, and the other with an alcohol. These, like many other reactions, are especially hazardous because they were performed many times before any accident occurred.

Preparation of Triacetin

The preparation of triacetin from glycerol and acetic anhydride, using phosphorus oxychloride as a catalyst, leads to a violent reaction. There appears to be no safe means of controlling this reaction. Other safe methods for the preparation of triacetin are given in the literature.

Bibliography

Albisser, R.H., and L.H. Silver, "Safety Evaluation of New Processes," *Ind. Eng. Chem.* **52** (11) 77A–79A (Nov. 1960).

Case Histories of Accidents In The Chemical Industry, Vol. 1, 1962; Vol. 2, 1966; Vol. 3, 1970, Manufacturing Chemists Association.

Fawcett, H. W., and W. S. Wood, *Safety and Accident Prevention in Chemical Operations,* John Wiley & Sons, Inc., New York, 1965, Chaps 7 and 8, pp. 80–107.

Groggins, P. H., *Unit Processes in Organic Synthesis,* 5th ed., McGraw-Hill Book Co., New York, 1958.

Kent, J. A., *Riegel's Industrial Chemistry,* Reinhold Publishing Corp., New York, 1962.

Manual of Hazardous Chemical Reactions, NFPA #491–M, National Fire Protection Association, 1968.

Shabica, A. C., "Evaluating the Hazard in Chemical Processing" *Chem. Eng. Prog.* **59** (9), 57–66 (Sept. 1963).

Shreve, R. N., *The Chemical Process Industries,* 3rd ed., McGraw-Hill Book Co., New York, 1967.

Steele, A. B., and J. J. Duggan, "Safe Handling of 'Reactive' Chemicals," **66** (8), 157–168 (April 20, 1959).

Appendix 3 lists 256 cases of explosions and literature references and cross references.

22

FIRE FIGHTING

BEFORE THE FIRE

The design and construction of laboratory facilities should include a careful evaluation of fire hazard. Construction and protection should be selected so that injury to persons will be prevented and damage to equipment and structures reduced to a minimum.

Automatic sprinklers are recommended for most laboratory situations. Although water may not be the best extinguishing agent for the material involved in the original fire, it will limit damage and prevent spread. Even in instances where such active metals as sodium are used, sprinklers are usually recommended. Sodium will of course continue to burn under sprinklers, but the fire is controlled by being confined to the metal itself.

In a few instances, for example, in physical chemistry and microanalytical work, where the quantities of combustible chemicals are extremely small, and where there is a great deal of complicated instrumentation and electronic equipment, sprinklers may not be desirable. In this case, the provision of portable fire extinguishers is essential.

In petroleum laboratories or other operations where large quantities of flammable liquids are involved, especially those which are not miscible with water, fire protection in the form of automatic high expansion foam systems in addition to automatic sprinklers and/or portable hand extinguishers may be desirable.

Whether sprinklers are used or not, adequate floor drainage is necessary to take care of water used in fire fighting as well as any liquids that may be spilled.

MANUAL FIRE FIGHTING

There are two phases to fire fighting operations in the laboratory: the first is the use of first aid measures by laboratory personnel; the second, the measures used by fire departments, whether these are industrial or public departments.

EMERGENCY OR "FIRST AID" FIRE FIGHTING

Equipment

The most commonly used first aid fire extinguisher in laboratories is the carbon dioxide type, which relies principally on a smothering effect.

For this reason, CO_2 may be used on small fires involving flammable liquids, often putting out the fire without damaging instruments or electronic equipment. CO_2 extinguishers require skill in their effective use, but proper and routine training of all personnel in their use can be arranged through local fire departments or the extinguisher manufacturer if necessary. The CO_2 extinguisher has no continued blanketing effect; the entire fire must be smothered in a few seconds, or it will reflash.

Except where the quantities of flammable liquids are extremely small, it is advisable to back up the CO_2 extinguisher with the dry chemical type (Fig. 22–1). The dry chemical (such as sodium bicarbonate) creates a cleanup problem and may damage equipment, but is a more effective extinguisher and does have a more lasting effect. If ordinary combustibles (wood, cloth, paper) may be involved, the "ABC" or all-purpose type of dry chemical extinguisher may be used. If the laboratory operations involve appreciable quantities of flammable liquids, the potassium bicarbonate-type dry chemical extinguisher should be used.

It is usual practice to provide a water-type extinguisher or small hose for fire in ordinary combustibles. However, in some cases the "ABC" dry chemical has been preferred, to avoid the possibility of using water on a fire where it is not suited.

Where combustible metals may be involved, a special extinguisher for metal fires may be advisable. These extinguishers must be suited to the type of metal involved, and are effective only when the amount of metal involved is small in relation to the amount of extinguishing powder. A practical method of dealing with a metal fire is to provide pails and

Fig. 22-1. A 200-lb dry chemical extinguisher located adjacent to a high-hazard area.

shovels, so that the burning material may be scraped up, dumped into a pail, and carried outside. The pail must be kept dry and should contain some material which is inert to the particular metal involved, and more of the same dry material should be on hand.

It should be noted that some reactive metals will continue to react with ordinary dry chemical extinguishing materials and with sodium bicarbonate, potassium carbonate, soda ash, and even with sand and limestone.

It is often satisfactory to provide extinguishers for the flammable liquids and simply let the metal fire burn itself out if there is no serious fume or containment problem.

Extinguishers should be located close to exit doors and not close to the actual hazard. This permits laboratory personnel to leave safely and then, if it appears reasonable to attempt fire fighting, to return, always keeping themselves between the fire and a safe exit.

A few positive rules should be enforced in regard to laboratory fire fighting, as follows:

Turn in an alarm or see that someone else does so before fire fighting.

Be sure that you are safely between the fire and the exit.

Avoid fighting a fire alone, if possible; at least two persons should always be present so that the fire fighter can be assisted in case of difficulty. As with most positive rules, there are exceptions, such as the small fire which cannot spread or the circumstance where rescue is involved. A considerable judgment factor is always involved in any fire situation.

Self-contained breathing equipment should be provided in the laboratory and worn if there is any possibility of involvement with toxic gases or smoke. Again, this rule requires judgment in interpretation.

FIRE DEPARTMENT OPERATIONS

The laboratory should consult with the fire department which will respond to alarms, review hazards and, as far as possible, preplan the fire fighting operations.

In particular, the fire department should be made aware of toxic materials and reactive chemicals, especially those which will react with water.

In general, the fire department should always operate with self-contained breathing equipment when fighting laboratory fires.

Bibliography

National Fire Protection Association
  Fire Protection Handbook, 13th ed., 1969
    pp.   1–56   causes of fires in laboratories
    pp. 16–80   sprinklers
  NFPA Standard No. 10   *Portable Fire Extinguishers,* 1969
  NFPA Standard No. 13   *Sprinkler System,* 1969
National Safety Council
  Accident Prevention Manual, 6th ed., 1969
    pp. 1232–33   safety showers
    pp. 1179–82   self-contained breathing apparatus
    pp.   244   first report forms
Fawcett, H. F., and Wood, W. S., *Safety and Accident Prevention in Chemical Operations,* John Wiley & Sons, 1965, pp. 485–526.
Handbook of Industrial Loss Prevention, 2nd ed., Factory Mutual Engineering Division, McGraw-Hill Book Co., New York, 1967.
Sax, N. I., *Dangerous Properties of Industrial Materials,* 3rd ed., Van Nostrand Reinhold Co., New York, 1968, pp. 194–207.

23

RADIATION

Radiation is as old as the universe and Man has always been exposed to it from his environment; but, beginning with the discovery of x-rays by Roentgen in 1895, he has been additionally exposed also to artificial ionizing radiation.

Natural radioactivity was discovered by Becquerel in 1896 and by 1911, nearly forty different radioisotopes were known.

Some deleterious effects of radiation on people began to be observed shortly after the discovery of x-rays. Recognizing a need for radiation-exposure control, the International Commission of Radiation Protection was formed in 1928, with representatives from Great Britain, Germany, Sweden, and the United States. Since then the States, the Federal Government, and the AEC in America have set up controls and guidelines to protect not only the workers with radiation but the general population.

Artificial radioisotopes have become plentiful and less expensive with the advent of nuclear reactors. Radioisotopes with varying degrees of radioactivity and half-lives may be produced at will by the use of nuclear reactors. Products resulting from the splitting of fissionable material may also be used. Other sources of radiation include neutron beams from nuclear research reactors and polonium or plutonium-beryllium neutron sources, x-ray machines, electron beams, and particle accelerators.

Experience has shown that radiation-producing machines and radioisotopes are useful tools and, like other hazardous material, can be safely stored, handled, and used if the hazardous properties are fully understood and the necessary precautions are observed.

RADIATION UNITS

The basic units for measuring radiation are the curie and roentgen. A curie is equivalent to 3.7×10^{10} dps (disintegrations per second) or 2.2×10^{12} dpm (disintegrations per minute) and describes the activity level of a source. A roentgen is defined as the quantity of x or gamma radiation for which the associated corpuscular emission per cubic centimeter of air produces ions carrying 1 electrostatic unit (esu) of electricity of either sign and describes the dosage level of a given situation. Radiation dose units are roentgen (r) roentgen absorbed dose (rad), and roentgen equivalent man (rem) which result from radiated energy released in air and tissue.

The dose rate from any gamma-radiation source, such as cobalt-60, can be approximated by utilizing the following formula:

$$r/hr \text{ at } 1 \text{ ft} \cong 6 \ CEn$$

where $C =$ number of curies; $E =$ gamma ray energy in millions of electron volts (MEV); and $n = \gamma$ quanta/disintegrations.

Radiation dose rates can also be obtained by direct radiation measurements with radiological monitoring instruments.

RADIATION EXPOSURE

Acute radiation exposures exceeding 450 rem are expected to be fatal to 50% of humans exposed, but many known accidental radiation exposures above this value have not resulted in fatalities. However, strict controls are necessary to prevent radiation exposures above recommended guide values, in order to insure protection of laboratory personnel. Guide values established by the Federal Radiation Council take into consideration both acute and chronic radiation exposure problems. Chronic radiation exposures above the Radiation Protection Guides (RPG) may result in somatic and genetic damage. The recommended yearly Radiation Protection Guide maximum dose in roentgens is $5(N\text{-}18)$, where N is the person's age. Most laboratories using radioactive sources limit weekly radiation exposure of personnel to 0.3 rem. Also, accumulated radiation exposure for laboratory personnel should not exceed 3 rem per quarter. Additional information concerning these guides can be found in National Bureau of Standards Handbook 69 and the Federal Radiation Council Bulletins.

Man may also be exposed to ionizing radiation from internally deposited radioisotopes. Radioactivity Concentration Guides are utilized in relating the concentration of radioisotopes in the environment to levels of assimilation by laboratory personnel. Assimilation of radioisotopes may result from inhalation, ingestion, or injection. Once inside the body, radioisotopes expose the entire body or specific organs to ionizing radiation. The total accumulative radiation exposure resulting from this depends on the type of radiation (alpha, beta, gamma), energy, biological half-life, and the critical organ exposed.

Radiation Exposure Control

Exposure of laboratory personnel to radiation can be effectively controlled by designing laboratories to handle radioactive materials safely; preparing written procedures for handling, monitoring, and inspecting radioactive materials; instituting emergency procedures; instructing personnel who handle radioactive materials in radiological hazards and safe handling techniques; providing radiation dosimeters and monitoring instruments to determine personnel exposure and contamination hazards; and maintaining permanent records of radiological surveys, inspections, and dosimeter results.

LABORATORY DESIGN

Radioisotope laboratories should be specifically designed to handle certain types of radioactive materials. Laboratories handling encapsulated radioactive materials would not require the same protective measures as those handling liquid or loose radioisotopes. Alpha-emitting radioisotope laboratories usually require more stringent controls because of the hazard from body assimilation. Where highly radioactive sources are handled, provision should be made to shield and store the source effectively. Remote handling facilities may also be necessary. Personnel radiation exposure can be effectively reduced by installing shielding, increasing the distance from the source, or reducing the exposure time.

PROCEDURES

Written, detailed procedures are necessary to ensure adequate radiation and contamination control in radioisotope laboratories. Frequent changes in personnel and the material handled may result in unnecessary exposures unless written procedures are followed. Written emergency procedures are also necessary to keep minor incidents from developing into exposure and contamination involving many persons and entire buildings.

PERSONNEL TRAINING

Effective contamination control cannot be achieved unless laboratory personnel are adequately trained. Specific, formalized training should be given to each person who handles radioactive materials to ensure that the necessary safety precautions are taken. Follow-up is essential.

MONITORING

Because the detection of radiation requires special detection instruments, personnel must have personal as well as laboratory monitoring instruments. Personnel routinely receiving radiation exposure should always wear film badges (Fig. 23-1), and, where high exposures are possible, self-reading dosimeters (Fig. 23-2). Permanent records of personnel exposure should be maintained to ensure that personnel do not exceed the Radiation Protection Guides.

Airborne radioactive contamination concentrations can be easily determined by establishing routine filter paper air sampling programs. The filter paper can then be analyzed by means of standard laboratory scalers and counters.

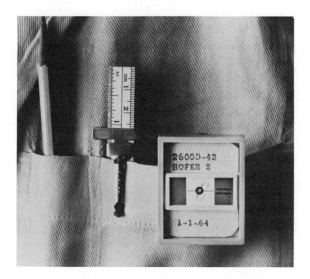

Fig. 23-1. Personnel film badge dosimeter.

Fig. 23-2. (left to right) Indirect- and direct-reading pocket dosimeters (ionization chambers) and readers.

Radiological laboratory surveys should be conducted frequently to ensure safe handling of radioisotopes. Each survey should involve radiation and contamination audits. Ionization-type instruments such as "Cutie Pies" (Fig. 23-3) and Geiger counters (Fig. 23-4) should be utilized for beta-gamma surveys. Contamination surveys can be effectively conducted by rubbing a paper towel over the surface of the area or item being surveyed and then monitoring the towel with a counter. Where repeat surveys are conducted, standard areas (one square foot per towel smear) should be checked to improve the accuracy of the results. Routine survey results should be maintained as permanent records. Figure 23-5 illustrates an alpha survey meter.

RADIATION SAFETY ENGINEER

Because of the unusual nature of the hazards and the special monitoring techniques, the safety considerations associated with radioisotope han-

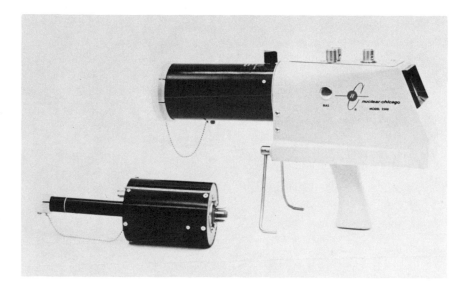

Fig. 23-3. "Cutie pie," ionization-type survey meter.

dling should be placed under the control of a trained radiation safety engineer, who should be responsible for: training laboratory personnel in radiological hazards and monitoring techniques; conducting radiological survey audits to determine exposure rates and degree of con-

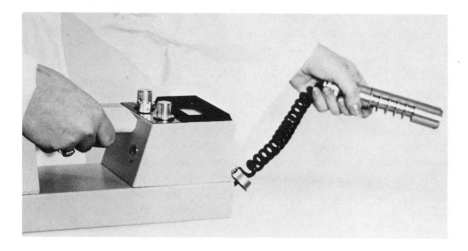

Fig. 23-4. Beta-gamma survey meter.

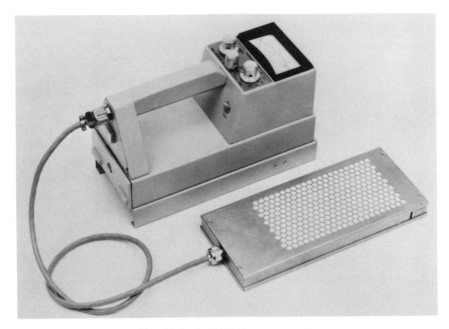

Fig. 23-5. An ALPHA survey meter.

tamination; maintaining records of personnel training, radiation surveys, monitoring instrument calibration and inventory, and radioisotope and radiation source inventories; issuing reports to laboratory management concerning personnel radiation exposures and unusual incidents; and ensuring that AEC and health department regulations are followed.

EMERGENCY RADIOLOGICAL CONTROLS

Procedures should be established to provide necessary protective action when emergencies arise. Some of the protective measures that should be considered are: evacuation of personnel from the incident site; isolation of the laboratory; monitoring and decontamination of personnel; reentry into the contaminated laboratory under controlled conditions with adequate protective clothing, respiratory devices, and monitoring instruments; obtaining expert advice concerning the decontamination operations from the AEC or public health departments; and devising personnel protective measures that must be taken during decontamination operations to insure that Radiation Protection Guides are not exceeded.

Bibliography

Chase, G. D., and J. L. Rabinowitz, *Principles of Radioisotope Methodology*, 3rd ed., Burgess Publishing Co., Minneapolis, Minn.

Malsky, et al., "Measurement of Radiation Dosage" *JAMA* 187 (11), 839–841 (March 14, 1964).

Nuclear Terms, A Brief Glossary, U.S.A.E.C., Oak Ridge, Tenn., 1964.

The Radiochemical Manual, 2nd ed., The Radiochemical Centre, Amersham, England, 1966.

Radioisotopes, a reprint with revisions from Atomic Energy Facts, item No. 220, U.S. Atomic Energy Commission, Div. Isotopes Devel., Washington, D.C.

Safe Handling of Cadavers Containing Radioactive Isotopes, National Bureau of Standards Handbook 56, Oct. 26, 1953.

Sax, N.I. *Dangerous Properties of Industrial Materials*, 3rd ed. Van Nostrand Reinhold Co., New York, 1968, Sect. 5, 8, 12.

24

STORAGE

Many potential hazards are associated with the storage and handling of materials used in a chemical laboratory. These potential hazards will always exist, but accidents can be eliminated by: acquiring a thorough knowledge of the properties of the materials to be stored and handled and planning a safe procedure by which they can be stored and handled; and informing all personnel who will come in contact with these materials of the hazards involved and the safety precautions which must be taken.

Numerous problems arise in laboratory chemical storage because of the thousands of different chemicals which may have to be stored. Careless storage practices, which lack planning and control, invite injury to personnel and damage to facilities. If, on the other hand, the storage area is carefully planned and supervised, most accidents can be avoided.

The chemical requiring storage may be a solid, a liquid, or a gas. It may be contained in a paper container, a metal container, a glass bottle, a carboy, a drum, or a cylinder. The hazardous nature of each chemical must then be considered individually and in relation to other chemicals which may be stored in the same area. To facilitate these considerations, chemicals can be grouped into the following general categories: flammable, toxic, explosive, oxidizing agents, corrosive, compressed gases, and water-sensitive chemicals.

FLAMMABLE CHEMICALS

Flammable liquids are stored in most chemical laboratories. To select or design the proper storage facilities, the properties of each chemical must be known. Such information may be obtained from the supplier of the chemical, from the literature, or by laboratory testing. A reference file should be maintained. It should list information such as boiling point, flash point, explosive limits, autoignition temperature, products of combustion, and fire extinguishing agents. For further discussion of the flammable nature of chemicals, see Chapter 19.

The type of storage container for flammable liquids will depend in part on the volume and the rate of use. The quantity of flammable liquid stored should always be kept to a realistic minimum.

Whenever possible, large quantities of flammable liquids should be stored outside. It is recommended that drum storage be limited to 100 drums per lot, with these large quantity lots located not closer than sixty feet from important buildings (NFPA 30) Smaller lots can be located closer to buildings; however, it is wise to maintain a minimum distance of approximately 10 feet. Drum lots of flammable liquids of high vapor pressure should be protected from the sun or cooled by water spray. High vapor pressure may be defined as above 150 mm at 40°C (2.9 psia at 104°F). Drains or drainage ditches should be located so that spilled flammable liquids will not flow under or into adjacent drum lots. All drains should discharge to a safe location. Yard fire hydrants should be located so that all drums can be covered by hose streams.

Inside storage of flammable liquids in drums or a large quantity of smaller containers should be limited as much as possible. Where inside storage of a large quantity of flammables is necessary, a fire-resistant flammable liquid storage vault with automatic water sprinkler system should be provided. If possible, a separate building should be used for indoor flammable liquid storage. Adequate ventilation must be provided for the removal of normal or accidental vapors, and all drains should be trapped and discharged to a safe location.

Whenever practical, glass containers should be avoided for storing flammable liquids. Small quantities of flammable liquids (less than 5 gal) should be kept in properly labeled, approved metal safety cans. Stainless steel safety cans are available if purity of materials is a consideration. Smoking should be prohibited in or near any area where flammables are stored. Electrical equipment should conform to the National Electric Code. Safe storage practices for compressed and liquefied gases, many of which are flammable, will be discussed later in this chapter.

TOXIC CHEMICALS

Most chemicals are considered toxic. In order to adequately evaluate the danger involved on exposure to them the relationship between frequency, duration, and concentration of exposure and the toxic hazard must be known. The probable mode of entry of a toxic substance into the body should also be known. Toxic substances can enter the body by inhalation, ingestion, absorption through the skin, or by any combination of these routes. Some chemicals will decompose to form toxic materials when in contact with heat, moisture, or acids. These chemicals, even though they are not toxic in their normal form, must be carefully considered in light of their potential hazard. Information concerning toxicity and potential toxic hazards may be obtained from the supplier of the chemical, from the literature, or by laboratory testing with animals. It is also important that information be obtained concerning personnel protective equipment, to guard against exposure, and medical treatment, to be used if exposure should occur. For a more detailed discussion of the toxic nature of chemicals, see Chapter 18.

The quantity of toxic chemicals stored should always be kept to a realistic minimum. Whenever possible, large quantities of toxic chemicals should be stored outside the building. Drums of toxic liquids of high vapor pressure should be protected from the direct rays of the sun. Toxic materials should not be stored near flammables.

When inside storage of toxic chemicals is necessary, the area should be well ventilated and cool. Flammables should not be stored in the same area. Chemicals which decompose on contact with moisture to form toxic materials should be protected from contact with water. Chemicals which decompose on contact with acid to form toxic materials should not be stored in the same area with acids.

All personnel in an area where toxic chemicals are stored must be instructed concerning the potential hazard of these chemicals. The eating of food and the drinking of coffee and other liquids should not be permitted in areas within the laboratory where toxic materials are stored, dispensed, or used. Access to toxic materials by unauthorized personnel must be prevented by adequate security measures. Protective equipment, if required, must be available, and personnel must be periodically instructed in the use of this equipment. The symptoms of exposure and necessary first aid and medical treatment must be known. A warning should be posted cautioning fire fighters to use self-contained breathing equipment.

EXPLOSIVE CHEMICALS

Some chemicals used in the laboratory are sensitive to shock or impact. Explosives are in this category. Many peroxides are shock and impact sensitive. These materials, on exposure to shock, impact, or heat, may release sudden energy in the form of heat or an explosion. Close control of the storage of these materials and stringent security measures are required. The quantity stored must always be kept to a minimum.

Storage facilities for explosive chemicals should be well identified and isolated from other areas. The type of storage area required will depend upon the particular chemical and the quantity stored. Frequently, shielded storage facilities are required. The best source of assistance in the selection and design of proper storage facilities for these materials is the supplier of the chemical.

Quantity-distance tables should be followed for the storage of chemicals which are classified as high explosives. See the *American Table of Distances for Storage of Explosives.*

OXIDIZING AGENTS

Oxidizing agents are chemicals which can supply oxygen to a reaction. Some examples of oxidizing agents are oxides, peroxides, nitrates, nitrites, bromates, chromates, chlorates, dichromates, perchlorates, and permanganates. Since oxidizing agents can initiate the combustion reaction, these materials present a definite fire hazard when stored with combustibles. Some oxidizable materials will react with oxidizing agents at room temperature to produce a fire or an explosion.

Oxidizing agents should not be stored in the same area with any fuel, such as flammables, organic chemicals, dehydrating agents, or reducing agents. Any spills in the storage area should be cleaned up immediately. Good housekeeping practices are essential. The storage area for oxidizing agents should be fire resistant (shelving included), cool, well ventilated, and preferably remote from other operations. The floor of the storage room should be fire resistant, water tight, and without cracks in which these materials can lodge. Sprinklers are recommended for the storage area.

CORROSIVE CHEMICALS

Many acids and alkalies are corrosive to their containers, other materials in the storage area, and body tissue. Acids react with many metals to form hydrogen gas. Alkalies may form hydrogen gas on contact with aluminum. Since hydrogen forms an explosive mixture with air, accumulation of hydrogen in storage areas must be prevented.

Corrosive liquids should be stored in an area which is cool but maintained above the freezing point of the chemical. This area should be dry and well ventilated, with provisions for good corrosion-resistant drainage and hoses for clean-up of spills. With some corrosive liquids, such as sulfuric acid, periodic venting of drums may be necessary to relieve the accumulated internal pressure of hydrogen formed by the reaction of the corrosive with the metal drum. Safety showers, eye wash fountains, and other required protective equipment should always be operable and available for personnel handling corrosive chemicals or working in the storage area. For more specific information on Corrosives, see Chapter 17.

COMPRESSED GASES

Compressed gases may be classified as liquefied gases, nonliquefied gases, or gases in solution. All are a potential hazard in the laboratory because of the pressure within the cylinders and their flammability and/or toxicity (see Chapter 20).

Compressed gases are supplied to laboratories in cylinders of varying sizes. These cylinders should be handled carefully to insure that they are not dropped or permitted to strike other objects. Valve caps should be kept on all cylinders that are not being used. When small lecture cylinders are furnished without valve caps, storage racks and holders should be provided to keep the cylinders upright and also to protect the valve from mechanical damage. Should the valve on a gas cylinder be accidentally knocked off or the cylinder ruptured in any way, the escaping gas could propel the cylinder with tremendous force.

Cylinders should be identified and stored in a well-ventilated area away from flammable materials. Cylinders stored outside should be protected from excessive variations in temperature and from direct contact with the ground. Possible external corrosion of cylinders by corrosive liquids or vapors must be prevented.

Compressed gas cylinders should be stored in the vertical position and secured so they will not fall. Full cylinders should be stored in a separate location away from empty cylinders. If storage space requirements make it necessary to store cylinders containing different types of gases in the same area, the cylinders should be grouped separately by type of gas. Groups of flammable gases should be separated from groups of oxidizing gases by noncombustible partitions. Wherever possible, flammable gases and oxygen cylinders should be kept outside, with proper manifolds and distribution piping to points of use.

WATER-SENSITIVE CHEMICALS

Some chemicals react with water to evolve heat and flammable or explosive gases. Potassium and sodium metals and metal hydrides react on contact with water. Hydrogen is produced with sufficient heat to ignite with explosive violence. Certain polymerization catalysts, such as alkyl aluminum, react and burn violently when in contact with water.

Separate storage facilities for water-sensitive chemicals must be designed to prevent their accidental contact with water. This is best accomplished by keeping all sources of water out of the storage area. Sprinklers should be eliminated over water-sensitive materials where large quantities of materials are involved and where the reaction will definitely spread or magnify a fire or cause an explosion. Sprinklers, however, have been proved effective over such materials as magnesium in controlling a fire and protecting the building from burning or collapsing. Storage construction should be fire-resistant, and other combustible materials should not be stored in the same area.

RADIOACTIVE CHEMICALS

Radioactive chemicals are discussed in Chapter 23.

INCOMPATIBLE CHEMICALS

Separate storage areas should be provided for "incompatible chemicals," chemicals which may react together and create a hazardous condition because of this reaction. Some examples of these incompatible chemicals are listed in Table 24-1.

TABLE 24-1. EXAMPLES OF INCOMPATIBLE CHEMICALS[a]

Chemical	Keep Out of Contact With:
Acetic acid	Chromic acid, nitric acid, hydroxyl compounds, ethylene glycol, perchloric acid, peroxides, permanganates
Acetylene	Chlorine, bromine, copper, fluorine, silver, mercury
Alkaline metals, such as powdered aluminum or magnesium, sodium, potassium	Water, carbon tetrachloride or other chlorinated hydrocarbon, carbon dioxide, the halogens
Ammonia, anhydrous	Mercury (in manometers, for instance), chlorine, calcium hypochlorite, iodine, bromine, hydrofluoric acid (anhydrous)

TABLE 24-1.   (Continued)

Chemical	Keep Out of Contact With:
Ammonium nitrate	Acids, metals powders, flammable liquids, chlorates, nitrites, sulfur, finely divided organic or combustible materials
Aniline	Nitric acid, hydrogen peroxide
Bromine	Same as for chlorine
Carbon, activated	Calcium hypochlorite, all oxidizing agents
Chlorates	Ammonium salts, acids, metals powders, sulfur, finely divided organic or combustible materials
Chromic acid	Acetic acid, naphthaline, camphor, glycerin, turpentine, alcohol, flammable liquids in general
Chlorine	Ammonia, acetylene, butadiene, butane, methane, propane (or other petroleum gases), hydrogen, sodium carbide, turpentine, benzene, finely divided metals
Chlorine dioxide	Ammonia, methane, phosphine, hydrogen sulfide
Copper	Acetylene, hydrogen peroxide
Cumene hydro- peroxide	Acids, organic or inorganic
Flammable liquids	Ammonium nitrate, chromic acid, hydrogen peroxide, nitric acid, sodium peroxide, the halogens
Fluorine	Isolate from everything
Hydrocarbons (butane, propane, benzene, gasoline, turpentine, etc.)	Fluorine, chlorine, bromine, chromic acid, sodium peroxide
Hydrocyanic acid	Nitric acid, alkali
Hydrofluoric acid, anhydrous	Ammonia, aqueous or anhydrous
Hydrogen peroxide	Copper, chromium, iron, most metals or their salts, alcohols, acetone, organic materials, aniline, nitromethane, flammable liquids, combustible materials
Hydrogen sulfide	Fuming nitric acid, oxidizing gases
Iodine	Acetylene, ammonia (aqueous or anhydrous), hydrogen
Mercury	Acetylene, fulminic acid, ammonia
Nitric acid (concentrated)	Acetic acid, aniline, chromic acid, hydrocyanic acid, hydrogen sulfide, flammable liquids, flammable gases
Oxalic acid	Silver, mercury
Perchloric acid	Acetic anhydride, bismuth and its alloys, alcohol, paper, wood
Potassium	Carbon tetrachloride, carbon dioxide, water
Potassium chlorate	Sulfuric and other acids
Potassium per- chlorate (see also Chlorates)	Sulfuric and other acids
Potassium permanganate	Glycerin, ethylene glycol, benzaldehyde, sulfuric acid
Silver	Acetylene, oxalic acid, tartaric acid, ammonium compounds

TABLE 24-1. (Continued)

Chemical	Keep Out of Contact With:
Sodium	Carbon tetrachloride, carbon dioxide, water
Sodium peroxide	Ethyl or methyl alcohol, glacial acetic acid, acetic anhydride, benzaldehyde, carbon disulfide, glycerin, ethylene glycol, ethyl acetate, methyl acetate, furfural
Sulfuric acid	Potassium chlorate, potassium perchlorate, potassium permanganate (or compounds with similar light metals, such as sodium, lithium)

ᵃAdapted from the *Dangerous Chemicals Code,* 1951, Bureau of Fire Prevention, City of Los Angeles Fire Department. This list is not complete, nor are all incompatible substances shown.

CHEMICAL STORAGE FACILITIES

The number and total quantity of chemicals which require storage will vary with each laboratory. The availability of adequate storage space is often given little consideration in the design of laboratory buildings. Insufficient storage space often creates a hazard due to overcrowding, storage of incompatible chemicals together, and poor housekeeping. Adequate, properly designed storage facilities must be provided to insure personnel safety and property protection.

In small laboratories, with only a few rooms, most chemicals may be safely and economically stored in the individual rooms. As the number of rooms and personnel increases, it becomes far more economical to reduce each room's storage space to the minimum required for adequate storage of the materials in current use. Excess chemicals and equipment should then be stored in a central storage area.

CENTRAL CHEMICAL STORAGE FACILITY

The size and complexity of a central chemical storage area will depend upon the variety and quantity of materials which must be stored. Since the quantity of chemicals stored in the individual laboratory rooms should be kept to a minimum, careful consideration of storage-space requirements, based upon current and anticipated chemical usage, is required before a central storage area can be designed. Proper organization of supplies is a necessity. A competent individual must be placed in charge of the storage area.

Small containers of chemicals should be stored in a cool, well-lighted, well-ventilated room isolated from the rest of the laboratory by fire walls. The containers must be well labeled and stored in an orderly manner. Good housekeeping is extremely important in a chemical storage area.

In determining the arrangement of chemicals in a storage area, the hazardous nature of each chemical, as discussed earlier in this chapter, must be considered. It is not sufficient to store chemicals on the shelves in alphabetical order by name. Incompatible chemicals should not be stored together. A technically trained person familiar with the hazardous nature of the chemicals being stored should assist in the arrangement of the chemicals in the storage area.

Larger containers present additional storage hazards because of their size and weight. Boxed acid carboys should be stored out of the direct rays of the sun and stacked not more than two tiers high. The carboy boxes should be inspected carefully for corrosion of the nails or weakening of the wood due to acid leakage from the carboy. The carboy should be inspected to insure that the stopper is securely in place and wired. If multiple-level drum storage is required, the drums should be placed on racks, which must be easily accessible to facilitate placement, removal, and inspection of the drums.

Fire protection and necessary personnel protective equipment must be considered in the arrangement and operation of a chemical storage area. Adequate electrical grounding must be provided for all equipment, including pipes and ducts entering the room. The importance of good housekeeping practices in any storage area cannot be stressed too often. Easy access to all areas is necessary. The type of fire protection equipment required will depend upon the quantity and type of chemicals being stored. Automatic sprinkler systems, standpipes, portable fire extinguishers, and the like should be provided as required. Sprinkler systems provide one of the best and frequently most inexpensive means available to adequately protect chemical storage facilities. Protective equipment, such as safety glasses, goggles, face shields, rubber or cloth gloves, aprons, boots, safety showers, eye wash fountains, and air-supplied masks, must be provided as required by the nature of the chemicals being stored. Personnel must be thoroughly instructed in the use of all protective equipment.

INDIVIDUAL LABORATORY STORAGE

Storage of small quantities of some chemicals will be necessary in individual laboratory rooms even when a central storage area is provided. The storage area in the individual rooms should, however, be kept to a realistic minimum, dependent upon the nature of the work being performed. Excessive storage in individual laboratories wastes space and presents unnecessary hazards due to the flammable, toxic, or corrosive nature of the chemicals being stored.

In the individual laboratory rooms, stock reagents should be stored within easy reach. Large containers should be stored in well-ventilated areas on low shelves or beneath benches. Corrosive chemicals should be stored in lead or plastic trays as near the floor level as possible. The trays should be large enough to hold the contents of the bottles in case they leak. Flammable liquids in quantities greater than one liter should be stored in metal safety cans. If these cans cannot be used, glass containers should be stored in trays. Quantities requiring containers larger than five-gallon safety cans should not be stored in the individual laboratory rooms. Chemicals which must be maintained at temperatures cooler than room temperature may be stored in a refrigerator or freezer. If a home type refrigerator or freezer is used for chemical storage, all electrical or spark-producing equipment, such as lights and switches, must be removed from the interior. Explosion-proof refrigerators and freezers are available for chemical storage. Toxic chemicals should always be stored in a well-ventilated area, preferably in a hood. The quantity and size of compressed gas cylinders should be kept to a minimum. Compressed gas cylinders and low-boiling liquids should not be stored near radiators or other sources of heat.

CHEMICAL STORAGE TIME LIMIT

Another factor which deserves consideration is the length of time chemicals are to be stored. A first-in, first-out system of stockkeeping is vital. Ethers, liquid paraffins, and olefins form peroxides on exposure to air and light. Since these chemicals are packaged in an air atmosphere, peroxides can form even though the containers have not been opened. The longer the storage period for these chemicals, the greater the amount of dangerous peroxides that may form. Ethers, such as isopropyl ether, diethyl ether, dioxane, tetrahydrofuran, glyme, and diglyme, are some of the worst offenders. When possible, these ethers should be purchased containing an inhibitor. Ethers should not be kept in storage for more than a year unless they contain an inhibitor known to prevent the formation of peroxides. Opened containers of ethers should be discarded within six months after they are first opened. All ethers should be dated when received and when they are first opened, so that the storage time can be determined and controlled.

Chemicals which are no longer needed should be discarded. Few laboratories have sufficient chemical storage space. The storage of chemicals which are not needed results in the loss of valuable storage space, and in time will lead to crowded conditions as well as serious housekeeping problems and fire hazards.

The condition of the containers in the chemical storage area should be checked frequently. Damaged containers present a serious hazard, and should be removed or repaired immediately.

Bibliography

Sax, N. I., *Dangerous Properties of Industrial Materials* 3rd ed. Van Nostrand Reinhold Co., New York, 1968, Sect. 7.

Steere, N. V. *Handbook of Laboratory Safety,* Chemical Rubber Company, Cleveland, Ohio, 1970.

25

SCALE-UP FROM RESEARCH TO MANUFACTURING

SAFETY IN DANGEROUS ENVIRONMENTS

In the chemical manufacturing industry, the scale-up of manufacturing operations is of great importance, and enlarging the scale of reactions brings its share of hazards and safety requirements (Figs 25-1 and 25-2). A review of procedures, investigation, and communication of knowledge involved in scale-up and transmittal of a process from research to manufacturing is therefore in order. A safety engineer must cover the scale-up process all the way from laboratory bench to pilot plant and from pilot plant to manufacturing unit.

REQUIREMENTS FOR SAFETY

This chapter is devoted to one important facet of the accident-prevention problem in the chemical industry—learning all the important properties of the raw materials, the nature of the reactions, the intermediates, and the finished products. Once this information is obtained, definite procedures must be established for communicating all the essential knowledge to those concerned.

Many of the severe accidents in the industry involve overpressuring of equipment, explosions, or fires. It is a tragic fact that ignorance is the cause of many of these occurrences. It may be lack of knowledge of the chemistry of a process; of the safe parameters of time, temperature, pH,

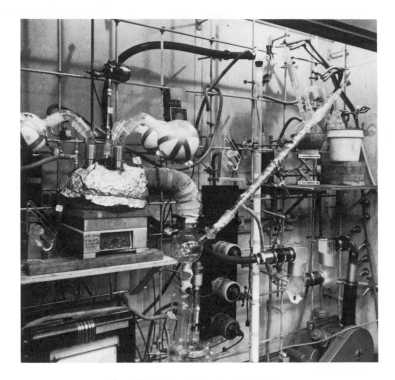

Fig. 25-1. Bench-scale apparatus.

or other variables; or of the consequences of varying the specified quantities of reactants. It may be lack of knowledge on the part of an operator because he has not received detailed instruction on the job or—just as important—because he has not been given a full understanding of the *consequences* which can result from not doing the job in the prescribed way. Ignorance of a process and failure to communicate sufficient knowledge to ensure a safe operation are inexcusable.

The following typical examples illustrate the types of accidents that could occur:

1. During the course of stripping methylisobutyl ketone from a batch of crude product, the temperature of the batch rose to 185°F, 10°F above normal operating temperature. The operator put full cooling water on the kettle, but the temperature of the batch continued to rise. When the temperature reached 195°F, the operator heard a "loud hissing" and evacuated the immediate area around the kettle. Seconds later the batch decomposed violently. There were no injuries, but damage to

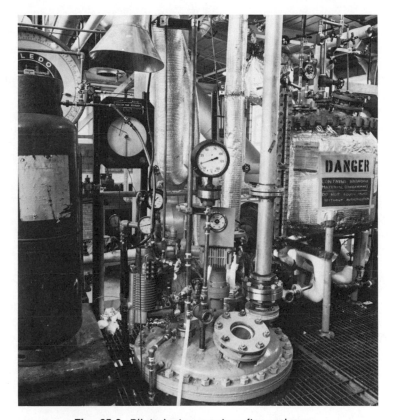

Fig. 25-2. Pilot-plant apparatus after scale-up.

the building and equipment were extensive. On examination it was found that there was not sufficient heat transfer in the mix and the material decomposed through uncontrolled exothermic reactions.

2. In the course of the manufacture of an organic phosphate an inadequately trained operator's helper added 16.6 gallons of a heat-sensitive reactant instead of 166 gallons as the operating instructions dictated. The reason for the mistake was simple: the meter through which the reactant was added showed the last number not as a full unit but as one-tenth unit. All of the numbers on the meter, including the last digit, were marked in black; there was no differentiation on the meter. Three men who were working in the area of the explosive reaction received second degree burns about the face, hands, and legs. The roof and a large portion of the siding of the building were blown off; other structural damage was done inside the building; three ASME coded vessels

were extensively damaged; and there was other property damage and loss of product.

3. An exothermic reaction was being conducted with concentrated sulfuric acid, water, and a reactive flammable liquid of low flash point. Inadvertently, the agitator was not started until after the total charge of reactants had been put into the kettle. When the agitator was started, pressure from the resulting uncontrollable reaction ruptured the vessel and released flammable vapors into the building with an ensuing explosion and fire. Again, a simple oversight produced extremely severe results.

These illustrations serve to emphasize the need for proper procedures in scale-up of laboratory reactions.

Attention is now directed to the safety requirements of large-scale chemical processing.

GENERAL PROCEDURES FOR SAFETY IN SCALE-UP

The procedure covering scale-up and transfer for a chemical company should cover two situations: scale-up within a given laboratory or pilot plant; scale-up and transfer of a process from pilot plant to manufacturing plant.

RESPONSIBILITY

The primary responsibility for determining or obtaining all necessary process information, including full information on safety aspects, lies with laboratory and pilot-plant group leaders and their technical associates. This responsibility is an integral part of their jobs and cannot be delegated elsewhere. The responsibility begins as soon as work is started on a process, whether it be on a laboratory, prepilot, or pilot-plant scale. Much of the responsibility can be met by the preparation and distribution of a check list carefully designed for this purpose. A guide of this type is desirable in all stages of process development, but it becomes most urgent from the safety aspect when a scale-up from one stage to another is planned. This is because certain of the items that will appear on a large-scale check list may not be critical in small-scale operation but are critical upon scale-up.

In preparing information for the check list and guide, the chemist at the bench should obtain all necessary information on thermal stability, shock sensitivity, heats of reaction, and other characteristics which might apply, and should transmit this information to the pilot plant. The pilot-plant engineer, on the other hand, has the responsibility of appraising the product and the process and satisfying himself that all of the laboratory investigative work necessary or desirable has been accomplished. If it has not been accomplished, he must have the blank areas explored by the chemist.

Technical and engineering persons in manufacturing should become involved with the new process while it is still in the process development stage. In this way there can be early recognition, anticipation, and correction of potential hazard areas.

When transfer of a product from pilot plant to a production unit or directly from the laboratory to production is contemplated, review of process data, using the check list as a guide, must be started with the plant personnel as early as possible before transfer. Again, it becomes the pilot-plant group leader's duty to supply to the manufacturing establishment all of the data necessary for the safe manufacture of the product, and it is the responsibility of the manufacturing department to make sure that it has all of the data necessary for safe production.

SAFETY REVIEWS

Certain important considerations are involved in the scale-up to a manufacturing operation. There must be a review with the safety and fire protection engineers and others concerned in the manufacturing department. This review should include a complete beginning-to-end narrative of the process, with a flow sheet showing the proposed equipment. The material balance on the equipment flow diagram should be set up, and there should be a clear definition of potential pollution and health problems as well as physical and chemical hazards.

In the review for scale-up to the manufacturing of the product, the process should be analyzed critically for the possible consequences of operator error and malfunction of equipment. Process procedures and equipment should be thoroughly examined and attempts made to visualize effects of such variations as temperature, pH, pressure, sequence of and rate of addition of materials and under- or over-charges of materials that may result through operating error, mechanical failure, or loss of services.

Before operating instructions for the manufacture are completed, it is most desirable that a job safety analysis or a study of the operation, element by element, be made to identify and anticipate hazards and to remove them or neutralize them by clearly defined means.

PREPARATION OF A CHECK LIST

Safety across a scale-up of processing depends on many interrelated factors, including types of equipment, process variables, properties of materials in process, manning, as well as many other considerations. Our attention is directed particularly to safety and health. With this emphasis the check list which is designed by a given research and development unit or a manufacturing unit might well include the following health and safety items:

Chemistry of Process

Safe parameters of such variables as temperature, pressure, pH, rates of addition of reactants, under- and over-charge of materials.

Process kinetics and thermochemistry.

Known side-reactions.

Possible side-reactions.

Stability of raw materials, reaction system, intermediates, and final product to heat, light, air, water, metals, oils, pH, and storage time.

Process Flow Sheets

Description of the process.

Materials flow streams.

Heat duties and other services.

Equipment and instrumentation.

Necessity of dual instrumentation.

Necessity of interlocking.

Fail-safe requirements.

Materials of construction.

Effect of improper control or side reactions on materials.

Chemical Hazards

Possible induction effects, hang-fire reactions.

Estimated decomposition energies of reactants and products.

Flammability characteristics of materials, for example, flash point, explosive range, autoignition temperatures.

Hazards of drying and grinding.

Equipment Hazards

Effect of failure of services, vacuum failure, air or water leakage, and metal exposure.

Fouling of heat transfer surfaces and instrument-sensing units.

Health Hazards

Acute and chronic toxicity data on reactants, products, and by-products, including oral, dermal, vapor, and eye data.

First aid treatment and antidotes for various exposures.

Notification of medical department of work on toxic materials.

Procedures for safe handling of materials.

Procedures for decontamination of toxic or obnoxious materials.
Personal protective equipment.

Operating Procedures

Detailed description of recommendations.
Possible effects of deviations from recommended range of operating variables on safety of operation.
Preparation of step-operating chart.
Procedure for discarding unsatisfactory product or intermediates.
Procedure for waste disposal.
Emergency shutdown procedures or action to be taken in the event of having to "kill" a reaction.

Analysis and Process Controls

Analysis of reactants and products.
Controls, physical and chemical, during processing.

Final Product

Labeling requirements.
Container size and type.
Pertinent DOT regulations.
Shelf life or storage stability.
Special warehousing requirements.
Sensitivity to contamination.

Joint planning by all concerned is most important to ensure a manufacturing facility which will not only produce the required yield of the necessary quality, but will be designed fully with a view to safe maintainability. During construction, therefore, the new production facility (or the old one, modified as may be necessary to produce the given product) must be continually reviewed by competent personnel. It is well known from experience that even the best of planning and preparation may sometimes omit important details; that field changes may be made; and that engineering drawings and models cannot be so precise and detailed as to show distribution, location, and arrangement of instruments and controls or to produce an optimum situation for safe and practical operation and necessary maintenance. It is certainly much better to keep new construction or existing equipment modifications under constant surveillance and have corrections made as the work develops than to wait until completion and then prepare a large-scale "punch list." This procedure, however, definitely does not exclude the necessity of a final examination and preparation of a punch list by a cadre composed of those competent to do so, and this would include the technical, engineering, maintenance, production, and safety and loss-prevention departments.

Often in the period between the transfer to those doing the scale-up and the final transfer to manufacturing, many inadequacies in the original experimental data show up. It is well to document, with explanation, all the principal precautions which have been found necessary for the safe operation of the process. Feedback of this information to research and development will result in a continuing improvement of process transfers.

26

EXTREME
CONDITIONS

HIGH-PRESSURE EQUIPMENT

The need for high-pressure equipment in the average laboratory is limited but its use is inherently very hazardous if not properly operated by qualified personnel. Such equipment, when required, should be certified by the manufacturer to meet all the requirements for pressure vessel design and fabrication as set forth in the ASME Code for Pressure Vessels and applicable state or local codes. If it becomes necessary to fabricate equipment locally, the design should be carefully checked by a qualified mechanical engineer. Regardless of the source of equipment, the entire completed assembly, including all piping, valves, and fittings, should be hydrostatically tested at 150% of the maximum allowable working pressure. For equipment that is to be used above 3000 lb per square inch working pressure, special design criteria and tests are needed. X-rays of welds and stress relieving of welds and vessels sections may be advisable.

All pressure equipment should be protected against over-pressure by relief valves or rupture discs, with the relief discharging to a safe location. The relief device should satisfy the ASME Code, so that at maximum calculated discharge rate it will prevent the pressure from rising more than 10% above the maximum allowable working pressure. Calculations of the relief capacity should include all internal pressure possibilities as well as external heat from fire exposure. The largest probable combination of heat input or pressure should dictate the size of the relief.

Chemicals subject to reaction should not be left in pipelines or equipment isolated by block valves. There is not only danger of pressure developing from temperature change, but thermal decomposition and violent reaction could result in equipment rupture or even explosion. Similarly, liquids with appreciable coefficients of expansion should not be left in pipelines or tubing systems between closed valves, since they may be ruptured. If necessary to use valves in such systems as liquid chlorine, expansion chambers or relief valves should be included between all valves.

All pressure vessels should be visually inspected internally and externally at least biennially and should be hydrostatically tested at 150% operating pressure at regular intervals. The frequency of pressure tests is usually determined by the rate of corrosion. If there is no evidence of corrosion and usage is unchanged, tests at five-year or longer intervals may be sufficient. Where corrosion is a significant factor, tests may be in order as frequently as every three months. Relief valves should be tested at a frequency determined by inspection of valve parts. If corrosion is a problem, it may be necessary to install a rupture disc upstream of the valve to protect it from corrosive chemicals. When this is done, a small weep hole or a pressure gauge should be installed in the line between the rupture disc and the relief valve, in order to detect a small leak in the disc which might build up pressure between the devices and result in as much as 100% over-pressure on the system before the relief valve would operate.

High-pressure equipment should be guarded by surrounding barricades or isolation if there is any danger of rupture. When there is a minor corrosion factor and the reaction is known to be predictable and non-exothermic, shielding and isolation may not be justified since the vessel design includes a safety factor of four to one or more.

HIGH TEMPERATURE

Most laboratory work is conducted at relatively low temperatures, such as those obtained with direct gas flame, electric and steam hot plates, and electric mantels. No special protection is required for personnel in this work other than normal body, eye, and hand protection. When electric furnaces and other means of reaching extremely high temperatures are employed, special protection is not only required for personnel but additional knowledge of the effect of temperature on the equipment becomes important (Fig. 26-1). Additional protection for the eyes is required, but

Fig. 26-1. High-temperature furnace.

this is easily provided by available shades of glass to absorb the infrared or glare rays. Body and hand protection may involve asbestos or treated, insulating gloves and aprons. Tongs for direct handling of heated objects are standard equipment.

All hot surfaces should be insulated and warning signs supplied where necessary to warn employees against touching hot equipment and material. Warning color schemes may be used in addition.

If there is any question of the stability of equipment or its ability to withstand internal pressure or vacuum under high temperature, the effect of the high temperature on the equipment should be thoroughly investigated and understood. Many metals and alloys undergo drastic changes at certain critical temperatures, especially as they approach the softening range. Tensile strength decreases and the ability to withstand compression is drastically reduced. In addition, there is frequently a decided increase in chemical reaction between the containing vessel or piping and the chemicals contained at higher temperatures.

All equipment used in high-temperature work should be checked frequently for any deterioration which might affect personnel safety or result in a fire hazard. Electrical controls and wiring should also be checked for condition of insulation or any signs of wear which might result in shock or fire hazard.

CRYOGENICS

Safe handling of the more common cryogenic fluids, oxygen and nitrogen, depends primarily upon knowing their properties and using common sense to avoid hazards. However, anyone handling them for the first time should carefully review the subject and obtain up-to-date information, which is available from the suppliers of the fluids and reliable reference sources.

Liquefied hydrogen and helium have become important cryogenic fluids in the missile and electronics industries. The properties of these two fluids present unusual hazards, requiring special equipment and handling techniques. Among potential hazards are the following:

1. Hydrogen is easily released in the gaseous form, and is extremely flammable over a wide range (4% to 75% by volume in air).
2. Both liquid hydrogen and helium are extremely cold. Helium is the coldest of all liquids. Both are capable of solidifying air that may come in contact with them.
3. The extremely low temperatures of uninsulated cryogenic containers or piping systems may condense air, which in turn could fractionate to yield high concentrations of oxygen. If hydrocarbons or other combustibles are present, the resulting mixture can be very dangerous.

The following paragraphs will summarize some of the problems connected with the handling of liquefied hydrogen and helium, and suggest some of the more important precautions that should be observed. Many of the items will apply also to the less exotic cryogenic fluids. For specific and more detailed information, see the reference sources listed at the end of this chapter and similar more up-to-date material that may be available.

The extremely low temperatures of liquefied helium, hydrogen, nitrogen, and oxygen can cause skin injuries similar to high-temperature burns. Contact with cold gas evaporating from the liquid or piping containing the cold liquid may have the same effect. Since the eyes could be seriously damaged by exposure to the cold liquid or gas, the eyes and face should be well protected by a face shield at all times. Clean, loose-fitting nonporous gloves should be worn together with a long-sleeved lab coat or jacket. Open pockets, trouser cuffs, and similar places where a liquid spill might lodge should be avoided. Never touch uninsulated cold equipment with any unprotected part of the body (Fig. 26-2).

At liquid helium temperature, all other gases will solidify; at liquid

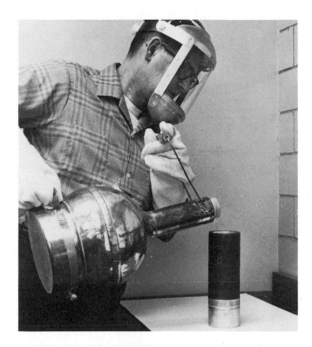

Fig. 26-2. Proper method for handling cryogenic materials.

hydrogen temperature, all gases except helium and hydrogen will solidify, Including, of course, air. This can cause serious trouble if the vent paths to containers become plugged with solidified air and overpressure results. Also, oxygen can be either frozen solid or concentrated in liquid form from the air, according to the temperature, both on direct contact with cold vapor or on the outside of containers. This presents a most serious hazard because of the danger of forming explosive mixtures with hydrogen or other liquid-state flammables. For this reason, liquefied helium and hydrogen must be handled in closed systems, and air must never be allowed to enter the container. The technique of handling containers, precooling systems, venting, and the like requires special knowledge and training, and should never be attempted by technicians who have not been thoroughly instructed in the proper procedures.

Liquefied hydrogen presents the additional hazard of vaporizing into a highly flammable gas. Hydrogen has a high diffusion rate in air and a wide flammable limit. Apparatus and containers should therefore be vented to the outdoors or set up in a hood with a high rate of ventilation. A continuous analyzer and alarm system is recommended to monitor the area and all sources of ignition should be carefully guarded against. Electric motors, switches, and fixtures cannot be obtained with explosion-proof rating for hydrogen-air atmospheres (Class I, Group A). Usually, however, satisfactory protection can be provided by air or inert gas pressure purging of equipment and cabinets and otherwise complying with the requirements of Section 500 of the National Electrical Code.

Careful attention should be given to the selection of materials for handling cryogenic fluids, since serious embrittlement of some metals, alloys, and many other materials can result from the extremely low temperatures involved. Specific information can be obtained from the leading producers of liquefied gases. In general, vessels and fittings should be made of aluminum, austenitic stainless steels, nickel alloy steels properly heat treated, high-nickel alloys, copper, or suitable copper alloys. All plastics should be avoided unless specifically recommended for some particular service.

LIGHT

The use of high-frequency collimated beams of intense, coherent, electromagnetic radiation produced by masers and lasers is increasing rapidly in laboratories and industry. When the frequency of radiation is in the visible portion of the spectrum, the device producing the radiation is known as a laser. Because of the high degree of collimation of the laser beam, the intensity diminishes only slightly with increasing distance. Thus eye damage can result from looking directly into the beam from a distance of several miles. Also, the beam is capable of being reflected in appreciable intensity from a smooth surface.

Recent developments include the direct laser emission of blue-green frequencies, which opens the whole visible spectrum to laser action, previously concentrated at the red end. Proposed uses include underwater probing, communications, color displays, and laboratory instruments. Mixtures of the inert gases, argon, krypton, xenon, and neon, may be used to simultaneously emit different wavelengths that can be individually detected and utilized. The laser is being effectively used in many metallurgical applications as an excitation source for a commercial Raman spectrophotometer and other instruments requiring a precise light source.

The precautions that should be taken are not too different from the precautions required with direct-beam x-ray. Accidents have been relatively few, but as the uses of high-intensity beams increase, certain precautions should be observed in laboratories. The following general safety guides for laser users are the result of considerable study, and should prove useful to those who are using or planning to use lasers. They are reproduced here with the permission of the Argonne National Laboratory.

Laser Operation Safety Guidelines: The area in which a laser is used will be regarded as a controlled area, limited to those whose work requires access to the room. Control shall be indicated by the use of signs on each door to the room.

An audible signal shall be associated with the charging of the laser condenser and shall be sufficiently intense and distinctive to alert all in the area of imminent firing of the laser.

All lasers will be equipped with key-lock switches which will permit removal of the key only when the circuit is open. The key or keys will be assigned to the authorized users of the equipment who shall be responsible for its operation at all times.

The potential for specular reflections shall be minimized by the removal of all unnecessary shiny surfaces in the general direction of the beam. Such reflections may arise not only from mirrors and the front surfaces of lenses, but even objects which we ordinarily regard as innocuous from this point of view, such as doorknobs, polished table tops or walls, metal or glass containers, etc. Laser safety glasses are available which are effective for reducing the intensity of laser reflections. These safety glasses must be tested periodically to assure continued protection since some have been found to deteriorate with time and exposure.

No one should look directly into a laser beam even while wearing optical absorbing lenses or at a distance calculated to be adequate. Off-axis viewing may be equally as dangerous as on-axis viewing since reflections are difficult to predict.

All lasers should be contained within some suitable shield so that uninformed personnel who may gain access to the area cannot be injured by accidental exposure. The light source (pumping light) of the solid-state laser should also be shielded to protect operating personnel from the brilliant flashes of light and the fragments from exploding lamps.

Electric shock and burns resulting from input power or capacitor discharge may cause serious injury or death. Proper maintenance of cables, connectors, cabinets, and switches is essential. Capacitors should be discharged before cleaning or before repairs are made. Since capacitors can retain a charge even when the power is disconnected, bleeder resistors should be provided and left in the circuit until the laser is to be employed again. Units should be flashed once after the switch is turned off to remove most of the residual charge from the capacitors. The operator should not leave the equipment until all voltage is removed from the capacitors as indicated by a zero voltage reading on the meter. Covers over high-voltage circuits should be interlocked to prevent access to energized components. A tagout or lockout system should be used to provide assurance that all connections are made with power supplies disconnected. Components not carrying current, including instrumentation, chassis, etc., should be grounded.

The laser should be operated, where possible, in a well lighted room to avoid enlarging the pupils and thereby minimizing access of the beam to the retina of the eye.

Various liquefied gases used as coolants for the crystals of pulsed lasers can cause severe burns if they contact the skin or eyes. Minimum protection for personnel who handle liquefied gases should be gloves (impervious, quick removal type), face shields, and safety glasses.

The area in which a laser is employed should be free of all unshielded combustible material because of the demonstrated ability of a laser beam to ignite paper at distances up to twenty feet. Shielding drapes should be tested by exposure to the beam before use.

Any incident involving persisting after-images of any light source should be reported to an eye specialist immediately.

PLASMA

At extreme temperatures, atoms are stripped of their electrons. When this happens the atoms become electrically charged particles called ions. A collection of ions, freely moving electrons, and any "unstripped" atoms or molecules make up a plasma.

When a gas is ionized, it becomes an excellent conductor of electricity; hence a plasma will respond to electrical and magnetic forces. The interaction of magnetic fields and electrically conducting fluids (gases in this case) has given rise to a field of study known as magnetohydrodynamics, or MHD.

The neon light is a form of relatively "cold" plasma, where only a few of the neon gas atoms are ionized, but enough to make the gas a good electrical conductor. Lightning, the aurora borealis, and the H-bomb at the moment of explosion are forms of plasma. Much of the present study of plasma phenomena is motivated by an attempt to harness or develop power from controlled fusion of light atoms in a reactor.

Magnetic fields can be used to confine a plasma; in effect it seems possible to erect a controlling magnetic fence around it. This principle is the basis for the plasma-arc spray process. The plasma cutting torch provides an extremely high-temperature, constricted arc. The hazards involved include high voltages (up to 400 volts) and the possibility of toxic vapors or gases emitted at the high temperature of metal cutting. Special training should be given to anyone handling this equipment.

Plasma-arc spraying or plasma casting is accomplished by spraying molten particles on a mandrel of the proper shape and later removing the mandrel. A variety of intricate shapes can be obtained by this method, which is really precision casting of fine particles of molten material. It is especially useful for fabrication of refractory metals and compounds as well as of other difficult-to-handle materials and components. The process is especially advantageous for parts requiring close tolerances, intricate shapes, or thin walls. It has been used in the production of refractory metal components for aerospace and electronics applications, and will undoubtedly find many uses in technical laboratories. The hazards of the process are essentially the same as those with the plasma-arc cutting torch, described in the preceding paragraph.

Bibliography

Bahun, C. J., and R. D. Engquist, "Metallurgical Applications of Lasers," *Metals Eng. Quart.* Feb. 1964.

Mash, D. R., and I. MacP. Brown, "Structure and Properties of Plasma-Cast Materials," *Metals Eng. Quart.,* Feb. 1964.

Neary, R. M., "The Air Condensing Cryogenic Fluids," Safety Maintenance, June 1964.

Weintraub, A. A., "Control of Liquid Hydrogen Hazards at Experimental Facilities: A Review," Health and Laboratory, U.S. Atomic Energy Commission, New York, May 1965.

27

PLANNING FOR EMERGENCY OPERATION

Every laboratory, no matter how small, should make definite plans in advance for the proper handling of emergency situations. In large institutions or industrial laboratories, these plans may become complex and must be integrated with a general emergency plan of the organization.

Even in the larger laboratories, however, it is possible to develop plans which have the essential elements of simplicity and flexibility. The very nature of emergencies indicates that the occurrences cannot be fully anticipated; there must be dependence upon judgment and initiative.

The principal advantage of preplanning is that those handling an emergency situation can forget the details already provided in the plan and put their full abilities to solving the problem immediately at hand.

ORGANIZATION

Someone must be selected to take charge in an emergency. In a small laboratory, this may be a single individual. He must of course have alternates in case of his illness, vacations, and the like. Usually there should be at least two alternates for each key person in any emergency plan.

In a large organization, the general emergency plan will include the laboratory, so that laboratory personnel will not be involved in much of the detail. There should, however, be at least one person in each local laboratory operation who is prepared and can take the essential first steps.

Because of the wide variation in types of organizations, it is not practical to suggest an emergency plan that would be suitable for all types of laboratories. The sections of this chapter are therefore arranged to indicate methods of taking care of certain types of emergencies and related matters.

Clothing Fires

The principal concern in fire is injury to persons. Clothing fire resulting from a spill or flash fire is the most serious fire emergency involving real danger to an individual.

Safety showers should be provided at locations on the normal exit path and well away from the points of major fire hazard. A good location is in a hallway just outside a laboratory door. A person with clothing on fire usually tries to escape from the area by the exits he uses every day. He will not usually pull the safety shower on himself (although experienced and well-trained men will). This means that someone must be prepared to get the involved person under the shower. Considerable force may be needed, and there is also danger to the rescuer. Emergency showers must be tested periodically (lat least annually), and, from time to time, practice in actually getting under the shower is very much worthwhile. The emergency man for the laboratory must think out his actions in advance for any type of clothing fire situation, so that when the real thing occurs he will move rapidly and with certainty.

Make sure that no glowing sparks remain in clothing. Smoldering clothing should be removed gently. Unburned clothing should not be removed unnecessarily. Charred cloth in direct contact with burned areas of the skin should not be removed where it adheres to the flesh. Charred cloth which does not adhere should be carefully cut away.

Cold water in copious quantities, applied very gently, is the best treatment for burns. Then the patient should be covered to prevent chilling. Prompt medical attention is essential for serious burns.

Runaway Reactions

The second fire situation is that involving explosion or the threat of explosion. The chemist handling an exothermic or potentially runaway reaction must know when to stop trying to control things and do something positive for both his own protection and that of others. All reactions of this type, as well as any which may involve flammable or toxic vapors or gases, should be run in a ventilated hood with the doors closed. If the doors are not already closed, they should be closed at the first indication of trouble.

The next step is evacuation of the room. Persons must be trained to use the nearest exit which does not lead them through the immediate danger area. This training is especially important, since the usual reaction is to use the normal entrance door. Occasional evacuation drills using the emergency doors or chutes and avoiding the normal entrance are therefore desirable.

Rescue

In some situations there may be need to enter a burning room or one in which persons have been overcome by toxic gases. Such a situation requires the utmost in good judgment on the part of the emergency man. There is no merit in losing a rescuer or incurring great risk to remove a person already dead.

Unless there is good evidence that the room atmosphere is not seriously contaminated, rescue should be attempted only with self-contained breathing apparatus (all laboratory personnel should be trained in the use of this equipment). It is possible in some instances for a man to effect a rescue by not breathing during the attempt. Obviously this will work only when the distance to the injured person is only a few feet and the injured person is not trapped.

Small rescue types of self-contained apparatus may be used to good advantage. Whenever possible, the rescuer should work with a "buddy." If a rope is available, someone outside the room may use it to drag the rescuer out if he is overcome.

In entering a burning room, keep low to avoid superheated air. A wool jacket over the head will provide some insulation from heat. The rescuer must keep an escape path between himself and the door.

If help is available, a hose line with spray nozzle may be used to protect the rescuer. Water spray not only will protect from heat; it will also help to disperse toxic vapors.

Fire Notification

Next in importance to rescue in any fire situation is notification. Usually there will be enough personnel available so that someone may turn in the alarm while rescue is still going on.

The manual fire alarm box, if provided, should be actuated. One person should remain at the box to direct firemen to the fire location. A fire call may also be made by telephone. The telephone number of the fire department or the institutional fire center should be plainly indicated on every telephone.

Fire fighting operations are outlined in Chapter 22. First aid procedures are covered in Chapter 29.

Building Evacuation

Regular building evacuation drills are recommended. In any large laboratory or institution, the fire drill procedure should provide for searchers, also known as wardens or drill captains, to thoroughly check the area and to report to a central location.

Immediate roll-call checks are undependable. It is not easily possible to account for absences due to vacations and illness or to provide for visitors and service personnel. Search should include toilets, stockrooms, and such service areas as pipe chases or rooftops where men may be working. Management must always be aware of special work above suspended ceilings or in ducts, so that the men involved can be evacuated in any emergency.

The emergency man must determine the need for evacuation. Any leak of an appreciable amount of irritating gas usually requires evacuation. On the other hand, the burnout of an electric motor, even when the smoke and odor are quite noticeable, seldom requires evacuation.

Shutdown

Certain laboratory operations may create an additional building hazard if left running when the building is evacuated; others may be hazardous if power or heat is turned off. A card should be prepared for hazardous operations which will indicate the action to be taken in an emergency. One typical situation is a reaction which requires agitation.

In general, power and other services to a laboratory building should not be shut off in emergencies. Ventilating fans should be kept running unless fire fighting forces determine that ventilation contributes to the fire. Electric power should not be shut off except in the rare instance of a serious flammable liquid spill or gas leak; in this case, power should be shut off from a point *outside* the building. This situation requires preplanning.

Pilot plants require special attention. Plans should be made in advance to determine whether it is best to shut down a pilot operation; the shutdown procedure should be carefully detailed.

Public Relations

Arrangements should be made in advance to meet representatives of the press in a safe location. Correct information should be made available as quickly as possible and from a single qualified source.

In general, the press will respond well to the establishment of a "press" room and the information furnished by a qualified spokesman. When it is safe to do so, reporters and photographers may be admitted to the area.

In talking with the press, no attempt should be made to avoid the word "explosion" and use such ambiguous terms as "rapid evolution of gases." If classified material is involved, the press will respect the necessity to keep secret the names of materials or the project involved. If there is a satisfactory open meeting with the press, it will do much to prevent their getting and printing information from unauthorized sources.

Notification of Kin

Every effort must be made to notify the next of kin of any death or any injury resulting in hospitalization before any information reaches the press. If this cannot be done before the press arrives, simply inform reporters that you are withholding names until the next of kin are notified. Do not hold out too long, however, if someone is missing; it may be better to notify the next of kin and then inform the press, even though that person may turn up later.

LARGE LABORATORIES AND INSTITUTIONS

In larger organizations, more elaborate emergency plans are required. To set up such plans, an emergency committee may be convened. This group should include those in charge of the medical, safety, fire protection, security, maintenance, and transportation functions. Representation by the public relations function may also be desirable. This group should provide definite plans for the following items.

Communications

Telephone "fan out" lists should be prepared in detail, so that persons helpful in an emergency may be reached at home. A list for one function will usually include ten or more names. The person calling will start at the top of the list. The first man reached may call the rest of the list from his home, or may often delegate further calls to his wife while he goes to the scene. A regularly updated call list is needed in each person's home.

Because the institution switchboard may be jammed with incoming calls, it is desirable to have an unlisted line which key personnel may use to reach the communication center.

No matter how simple the plan, arrangements must be provided so that these calls may be made automatically and not require any action on the part of the emergency man himself other than a first call to the communications center.

The call lists should be arranged in order of decreasing importance, and should include the community emergency medical organization, the community fire department, and local rescue services. Large organizations can furnish their own team of rescue services, including maintenance equipment, cranes and earth movers, a pipe fitting crew, electricians, and so on. When these services are provided by outside sources, emergency numbers should be available. A man should always be designated to meet outside rescue personnel at the gate or door and direct them properly.

COMMAND CENTER

In any serious emergency, it is desirable to designate a "command post" or headquarters, which may be any convenient office or emergency car or truck, always in a safe location.

The command post should have maps and plans that show the area involved and such essential services as water mains and underground gas and electric lines. There should be good communication in the form of telephones and radio. In small laboratories, citizen's band "walkie-talkies" can be used. Larger organizations will usually have an existing plant radio system. It is especially desirable to keep in touch with exploratory teams which enter areas that may have hazardous atmospheres.

The emergency man in charge should be available at the command post unless there is a real need for him at the actual scene. In any event, he should be in touch with headquarters by radio.

COORDINATION WITH COMMUNITY SERVICES

Fire Department

Officers of the fire department should be invited to inspect the laboratories periodically. Safety personnel should become personally acquainted with the fire officers, so that there will be a smooth working relationship. Special hazards and protective devices should be made known and emergency plans made in advance. It is particularly important to point out to fire officials possible sources of toxic gases and materials which may react violently with water or when heated.

Rescue Services

Some communities have volunteer rescue teams which can render valuable service. These teams should always be guided by a laboratory or institution person familiar with the area and its special problems.

Medical Services

Many Communities have arrangements for making hospital beds available. Arrangements should be made in advance for any emergencies which may involve a number of persons.

Police

Advance emergency planning should be made with the local police department, who can usually assist best by keeping a route clear for ambulances and by keeping unwanted spectators out of the danger area. Meetings between the police and emergency personnel should be arranged to assure mutual acquaintance. It is also important to provide identification cards for emergency personnel so that the police can assist them as necessary in getting into the area.

28

ACCIDENT INVESTIGATION

Any incident occurring in a laboratory which results in injury, fire, or loss, such as damage to equipment, or which might have caused such results, should be carefully investigated. The principal purpose of such accident investigation is to learn how to prevent future accidents. Thus, the near-miss situation, in which, by good fortune, no one is hurt or there is no loss, may be far more significant and far more important to investigate carefully than one which actually results in injury but has also an easily understood cause.

While every incident should be investigated to some degree, it is obvious that one which results in severe injury, death, great loss or a near miss of great significance should be investigated with particular care, from both the legal and the prevention standpoints.

TYPE OF INVESTIGATION

The type of investigation therefore varies appreciably with the severity or potential severity of the incident. In ordinary circumstances, where the injury may be lacerations, bruises, or relatively minor chemical burns, the investigation is usually best made by a single person and the report form used should be simple (Figure 28-1). The form can be kept in pads on the supervisor's or instructor's desk, so that it is a simple matter to fill in

NA 3832 REV. 8-70 ☎ 8-70

ACCIDENT INVESTIGATION REPORT

CYANAMID

AMERICAN CYANAMID COMPANY

Investigate disabling injuries promptly. Issue report within 48 hours after accident. Refer to Cyanamid Safety and Loss Prevention Bulletins No. 5 & 11.

PLANT	DEPARTMENT		DEPT. NO.	SUPERVISOR	

NAME OF INJURED	AGE	CHECK NO	OCCUPATION OF INJURED	DATE ASGD. THIS OCCUPATION	DOING REGULAR WORK ☐ YES ☐ NO

DATE OF ACCIDENT	HOUR	☐ A.M. ☐ P.M.	DATE STARTED LOSING TIME	EXPECTED LENGTH OF DISAB	PERMANENT PARTIAL DISAB. ☐ YES ☐ NO	DAYS TO BE CHGD. (ANSI Z16.1)

NATURE OF INJURY	CONSECUTIVE DAYS WORKED SINCE LAST DAY OFF	HOURS WORKED FOR 30 DAYS (PREC. ACC.)

DESCRIPTION OF ACCIDENT (DESCRIBE FULLY WHAT INJURED WAS DOING AT TIME OF ACCIDENT.)

CAUSE OF ACCIDENT

STEPS TAKEN TO PREVENT A RECURRENCE: (IF PREVENTIVE PLANS NOT COMPLETE, FOLLOW WITH SUPPLEMENT)

NAMES OF INVESTIGATING COMMITTEE. (SIGNATURES NOT NECESSARY)

SIGNATURE (NAME OF PERSON MAKING REPORT)	DATE	APPROVED BY (SIGNATURE & TITLE	DATE

Fig. 28-1. Accident investigation report.

the form whenever a person is treated for a minor injury. Copies should always be sent to the staff unit responsible for safety, so that a proper follow-up may be made if necessary.

Even in small laboratories in relatively nonhazardous work, provision should be made for investigation of all injuries. Although an individual injury may not appear to be significant, the summation of a number of investigations may indicate trends or may justify more elaborate equipment. Serious injury, especially eye injury, or fatal injury is possible even in a high school laboratory. Setting up procedures for investigating a serious accident may uncover hazard situations that need correction. In the event of a serious incident, the existence of proper procedures for investigation will be of great assistance.

MINOR INJURY INVESTIGATION

Treatment of the injured person comes first, even if little more is required than a finger bandage. The investigation should include a statement by the injured person in his own words (Fig. 28-2). Every effort should be made to put the injured person at ease and to indicate that the whole purpose of the investigation is the prevention of future injuries. To avoid excessive formality, the nurse or other person treating the injury may simply ask how it happened and fill out the form later. Subsequent investigation should be made by the injured person's supervisor (Fig. 28-3). It is essential that minor injuries be treated and reported so that any questioning or filling out of forms in the presence of the injured person is avoided.

Statements of eyewitnesses are valuable, but the whole atmosphere should be one of informality. No attempt should be made to fix responsibility. Later, responsibility may be assigned if this is felt desirable.

Every investigation should, however, list the measures to be taken to prevent repetition. If it is agreed that nothing could be done, this fact should be stated.

SERIOUS INJURIES

Any injury which causes a permanent disability or a loss of time of a day or more requires a careful and more formal investigation. Usually this is best done by the small group of persons most concerned. Such a group is usually appointed for each incident, rather than a committee set up in advance (Figs. 28-4, 28-4a, 28-5, and 28-5a). This group should include the direct supervisor, and others directly concerned with the laboratory. It is important to investigate quickly, before anything is disturbed except for

M I N O R   I N J U R Y   R E P O R T

EMPLOYING UNIT	MINOR INJURY NUMBER
DEPARTMENT	DATE OF INJURY
INJURED EMPLOYEE	TIME OF INJURY A.M. P.M.
NATURE OF INJURY	REPORTED FOR MEDICAL ATTENTION DATE A.M. P.M.
TREATMENT	
SIGNATURE OF FIRST AID ATTENDANT	

LOCATION (NEAR OR IN WHAT BUILDING)	OCCUPATION
NATURE OF WORK	LENGTH OF COMPANY SERVICE
	LENGTH OF JOB EXPERIENCE

DESCRIPTION OF INCIDENT

WHAT INSTRUCTIONS WERE GIVEN BEFORE WORK WAS BEGUN?

WHAT PROTECTIVE EQUIPMENT WAS USED?

UNDERLYING CAUSES OF INJURY

BASIC CAUSE

WHAT IS BEING DONE TO PREVENT SIMILAR INJURY SITUATIONS?

INVESTIGATED BY	DATE
	AUTHORIZED SIGNATURE

Fig. 28-2. Minor injury report.

GOF 2827 REV. 1-60 PRINTED IN U.S.A. 1-60

SUPERVISORS ACCIDENT REPORT

Refer to Safety & Loss Prevention Bulletin No. 22 for the preparation of this report

DEPARTMENT	NO. OF BLDG.	DATE OF ACCIDENT	TIME OF ACCIDENT
			☐ A.M. ☐ P.M.

WHO WAS INJURED OR WHAT WAS DAMAGED

NATURE OF INJURY OR ACCIDENT

DESCRIBE THE ACCIDENT

CAUSE – DESCRIBE UNSAFE ACT AND/OR UNSAFE CONDITION

CORRECTIVE ACTION TAKEN

SUPERVISOR	DATE
REVIEWED BY (DEPT. HEAD)	DATE

COMMENTS

REVIEWED BY	DATE

COMMENTS

Fig. 28-3. Supervisor's accident report.

M TAB SETS→
G-111 REV. 1-68

DU PONT

v   v   v

SUB-MAJOR INJURY REPORT

NO._____

EMPLOYING UNIT (NAME PLANT, LAB., WHSE., CONST., ETC.)	DATE OF INJURY	DAY OF WEEK	HOUR ____A.M. ____P.M.
NAME AND No. OF BUILDING IN OR NEAR WHICH ACCIDENT OCCURRED			CHECK ONE ☐ INDOORS ☐ OUTDOORS
NAME OF INJURED AND SOCIAL SECURITY NUMBER		SECTION OF EMPLOYING UNIT	
OCCUPATION	LENGTH OF SERVICE—THIS UNIT	LENGTH OF SERVICE—THIS COMPANY	DATE OF BIRTH

NATURE OF INJURY

WHAT FIRST AID	DATE	HOUR ____ A.M. ____P.M.
WHERE GIVEN	BY WHOM	EST. DURATION OF TREATMENT

| WHERE SENT CHECK ONE | ☐ HOSPITAL ☐ SURGEON | ☐ HOME ☐ OTHER (SPECIFY) | RETURNED _____ (DATE) TO WORK _____ (HOUR) |

DESCRIPTION OF ACCIDENT (GIVE FULL DETAILS BEING PARTICULAR TO DESCRIBE CONDITIONS PRECEDING THE ACCIDENT, WORK IN PROGRESS, STAGE OF PROCESS, ACTUAL ACTS OF INJURED AND FELLOW WORKMEN. ETC., SO THAT A CLEAR PICTURE OF THE ACCIDENT IS GIVEN — USE ANOTHER SHEET OF PAPER IF NECESSARY).

ASSIGNMENT OF NEW WORK (BE SPECIFIC)

ESTIMATED DATE ABLE TO RETURN TO PREVIOUS ASSIGNMENT: _____

DO STATEMENTS OF INJURED AND WITNESSES AGREE WITH ABOVE? YES ☐ NO ☐ IF NOT, STATE WHEREIN THEY DIFFER.

IN HOSPITALIZATION-FOR-OBSERVATION CASES ATTACH PHYSICIAN'S STATEMENT. (SEE SAFETY BULLETIN No. 422, SECTION 2.1.3.2.3)

Fig. 28-4. Sub-major injury report.

M TAB SETS→   V   V

CONDITIONS AT TIME OF ACCIDENT (CROSS OUT THOSE NOT APPLYING)—NORMAL OPERATION—MINOR REPAIRS—MAJOR REPAIRS—ADJUSTMENTS—OILING—EXPERIMENTAL—BREAKDOWN—SERIOUS EMERGENCY—SPACE INSUFFICIENT—OBSTRUCTIONS—ARTIFICIAL ILLUMINATION—POOR LIGHT—EXCESSIVE HEAT—COLD—FUMES—STEAM—SLIPPERY—POOR HOUSEKEEPING—OTHER

ADDITIONAL INFORMATION

WHAT GUARDS, SAFETY DEVICES, ETC., WERE APPLICABLE?

WERE THESE PROVIDED?	IN USE?	IN GOOD CONDITION?
DU PONT STANDARD?	EFFECTIVE?	

IF THEY DID NOT FUNCTION PROPERLY, EXPLAIN WHY.

HOW LONG WAS INJURED FAMILIAR WITH THE PROCESS AND LOCAL CONDITIONS WHERE ACCIDENT OCCURRED?

HOW LONG WAS INJURED EXPERIENCED IN WORK HE WAS DOING?	WHERE WAS PERSON IN DIRECT CHARGE OF THIS WORK?

WHAT SPECIFIC INSTRUCTIONS HAD HE GIVEN?

IF FELLOW WORKMAN CONTRIBUTED TO ACCIDENT, GIVE HIS EXPERIENCE IN THIS CLASS OF WORK.

IF INJURED OR FELLOW EMPLOYEE THROUGH FAILURE TO OBEY RULES OR INSTRUCTIONS CONTRIBUTED TO ACCIDENT, STATE CLEARLY HOW.

IF LACK OF EXPERIENCE OR SKILL OF INJURED OR FELLOW EMPLOYEE CONTRIBUTED TO ACCIDENT, STATE CLEARLY HOW.

IF DEFECTIVE CONSTRUCTION OR EQUIPMENT OR USE OF IMPROPER APPLIANCES CONTRIBUTED TO ACCIDENT, STATE CLEARLY HOW.

COULD CLOSER SUPERVISION OR MORE DETAILED INSTRUCTIONS HAVE PREVENTED THIS ACCIDENT? YES ☐ NO ☐ IF SO, HOW?

FROM YOUR INVESTIGATION OF CONDITIONS ASSIGN THE CAUSE CODE (SEE SAFETY BULLETIN, No. 422) AND GIVE REASONS FOR THE ASSIGNMENT.

WHAT CHANGES HAVE BEEN RECOMMENDED TO PREVENT RECURRENCE?

WHAT ACTION ON THESE?

DATE OF REPORT	THIS ACCIDENT WAS INVESTIGATED BY

_____ MGR.

Fig. 28-4. (continued)

M TAB SETS——→

G-105   REV. 1-68

DUPONT
REG. U.S. PAT. OFF.

MAJOR INJURY REPORT

P. I. NO. _____

EMPLOYING UNIT (NAME PLANT, LAB., WHSE., CONST.)	DATE OF INJURY	DAY OF WEEK	HOUR _____ A.M. _____ P.M.
NAME AND No. OF BUILDING IN OR NEAR WHICH ACCIDENT OCCURRED			CHECK ☐ INDOORS ONE ☐ OUTDOORS
NAME OF INJURED AND SOCIAL SECURITY No.		SECTION OF EMPLOYING UNIT	
OCCUPATION	MARITAL STATUS	RATE PER HOUR	WORKING HOURS PER WEEK
LENGTH OF SERVICE—THIS UNIT	LENGTH OF SERVICE—THIS COMPANY		DATE OF BIRTH

NATURE OF INJURY

WHAT FIRST AID	DATE	HOUR _____ A.M. _____ P.M.
WHERE GIVEN	BY WHOM	EST. LOSS OF TIME
WHERE SENT ☐ HOSPITAL ☐ HOME CHECK ONE ☐ SURGEON ☐ OTHER (SPECIFY)	MEANS OF CONVEYANCE	

DESCRIPTION OF ACCIDENT (GIVE FULL DETAILS BEING PARTICULAR TO DESCRIBE CONDITIONS PRECEDING THE ACCIDENT, WORK IN PROGRESS, STAGE OF PROCESS, ACTUAL ACTS OF INJURED AND FELLOW WORKMEN, ETC., SO THAT A CLEAR PICTURE OF THE ACCIDENT IS GIVEN—USE ANOTHER SHEET OF PAPER IF NECESSARY.)

DO STATEMENTS OF INJURED AND WITNESSES AGREE WITH ABOVE? YES ☐ NO ☐ IF NOT, STATE WHEREIN THEY DIFFER.

IN FATAL CASES ATTACH A STATEMENT GIVING NAMES AND RELATIONSHIP OF DEPENDENTS, AND IN ALL SERIOUS CASES OBTAIN AND ATTACH SIGNED STATEMENTS OF WITNESSES.

Fig. 28-5. Major injury report.

M TAB SETS——▶   V   V

CONDITIONS AT TIME OF ACCIDENT (CROSS OUT THOSE NOT APPLYING)—NORMAL OPERATION—MINOR REPAIRS—MAJOR REPAIRS—ADJUSTMENTS—OILING—EXPERIMENTAL—BREAKDOWN—SERIOUS EMERGENCY—SPACE INSUFFICIENT—OBSTRUCTIONS—ARTIFICAL ILLUMINATION—POOR LIGHT—EXCESSIVE HEAT—COLD—FUMES—STEAM—SLIPPERY—POOR HOUSEKEEPING—OTHER

ADDITIONAL INFORMATION		
WHAT GUARDS, SAFETY DEVICES, ETC., WERE APPLICABLE?		
WERE THESE PROVIDED?	IN USE?	IN GOOD CONDITION?
DU PONT STANDARDS?	EFFECTIVE?	
IF THEY DID NOT FUNCTION PROPERLY, EXPLAIN WHY.		
HOW LONG WAS INJURED FAMILIAR WITH THE PROCESS AND LOCAL CONDITIONS WHERE ACCIDENT OCCURRED?		
HOW LONG WAS INJURED EXPERIENCED IN WORK HE WAS DOING?		WHERE WAS PERSON IN DIRECT CHARGE OF THIS WORK?
WHAT SPECIFIC INSTRUCTIONS HAD HE GIVEN?		
IF FELLOW WORKMAN CONTRIBUTED TO ACCIDENT, GIVE HIS EXPERIENCE IN THIS CLASS OF WORK.		
IF INJURED OR FELLOW EMPLOYEE THROUGH FAILURE TO OBEY RULES OR INSTRUCTIONS CONTRIBUTED TO ACCIDENT, STATE CLEARLY HOW.		
IF LACK OF EXPERIENCE OR SKILL OF INJURED OR FELLOW EMPLOYEE CONTRIBUTED TO ACCIDENT, STATE CLEARLY HOW.		
IF DEFECTIVE CONSTRUCTION OR EQUIPMENT OR USE OF IMPROPER APPLIANCES CONTRIBUTED TO ACCIDENT, STATE CLEARLY HOW.		
COULD CLOSER SUPERVISION OR MORE DETAILED INSTRUCTIONS HAVE PREVENTED THIS ACCIDENT? YES ☐ NO ☐ IF SO, HOW?		
FROM YOUR INVESTIGATION OF CONDITIONS ASSIGN THE CAUSE CODE (SEE SAFETY BULLETIN, No. 422) AND GIVE REASONS FOR THE ASSIGNMENT.		
WHAT CHANGES HAVE BEEN RECOMMENDED TO PREVENT RECURRENCE?		
WHAT ACTION ON THESE?		
*GIVE DATE COMPENSATION BEGINS	AT WHAT WEEKLY RATE	
DATE OF REPORT	THIS ACCIDENT WAS INVESTIGATED BY	

*IN FATALITY CASES ATTACH LIST OF
DEPENDENTS, GIVING NAME OF DEPENDENT,
RELATIONSHIP, DATE OF BIRTH, AND SEX.

_____MGR.

Fig. 28-5. (continued)

such obvious necessities as restoring damaged services. There should be no delay such as waiting for an insurance representative or for police or fire authorities; the institution should make its own investigation immediately.

Any delay permits persons involved to think about the incident and to begin to formulate ideas about its occurrence, perhaps even unconsciously, that may not be in accord with the facts.

INVESTIGATION INTERVIEW FOR SERIOUS INCIDENTS

The investigating group should meet in a convenient room and call in witnesses individually. It must be pointed out to each witness that the purpose of the investigation is not to place responsibility but to bring out facts that will determine the cause and help in preventing a repetition. Let each man tell his own story, but it is advisable not to have a secretary present or to record it mechanically. The chairman may make rough notes. Later, if necessary, a statement can be written out for the witness to sign if he is in full agreement (Figs. 28-6 and 28-6a).

Usually, stories by witnesses will vary slightly. No attempt should be made to get these stories to agree in the presence of witnesses. Later, all stories may be carefully reviewed to try to determine what actually happened, but a statement by a witness should not be altered.

In this brief space, it is not possible to go into detail on procedures or investigators' methods. In a serious or complex situation, it may be desirable to bring in an experienced industrial safety man, who can assist in the investigation and, if needed, find authorities on fire, explosion, chemical reactions, and so on. A few basic suggestions that may be helpful for several of the more common investigations follow.

Fire

The point of origin of a Class A (ordinary combustibles) fire can often be pinpointed, since such fires usually burn upward from the starting point. A charred or scorched wall, side of a bench, or the like which shows a low point with the charred area spreading outward above this point usually shows the point of origin clearly. It should be remembered that charred wood curls toward the side exposed to the most severe fire.

Most flammable vapors, being heavier than air, will fall to floor level and may travel some distance to reach a point of ignition. Hydrocarbon liquids which contain a great deal of carbon, such as benzene, leave a greasy soot which forms a trail to the point of ignition. Some vapors, such as carbon bisulfide, may be ignited by metal surfaces at about 212°F. Flammable vapors may be ignited by static sparks. The witness may recall seeing, hearing, or feeling a static spark or discharge as he reached toward a piece of equipment. However, most laboratory fires are ignited by other means, such as sparking motors, open electric switches, and open flames. All ignition sources should be carefully checked.

Pressure Effects

Ignition of unconfined flammable vapors usually results in a momentary increase in air pressure as a flash or puff. When a vapor-air mixture is near the middle of the explosive range or the expansion is restricted by equipment or building walls, the effect may properly be called an explosion and there will be some building damage, such as the blowing out of doors and windows. Checking of these effects is helpful in pinpointing the area of origin.

Reactions or explosions in laboratory glassware may result in glass projectiles being thrown out. By noting where these have struck, especially those pieces that remain stuck in doors and walls, it is possible to locate the source accurately. It is possible to determine which of several flasks in a close grouping may have exploded. Projectiles of course travel in straight lines; by plotting these lines on a scale drawing of the laboratory, the point of origin can be found.

This technique may be used in investigating pilot-scale laboratory explosions. It would ordinarily be advisable in such an incident to call in persons familiar with explosion investigation. The point of origin may sometimes be determined precisely even when the building and equipment have been destroyed.

Injuries

The injured persons should receive medical care immediately and, in general, should not be questioned. Seriously injured persons are in a state of shock and cannot remember the circumstances accurately.

It is important, however, to determine where the person was standing and the position of his hands, head, and other parts of the body. Since the injured person will have been removed before the investigation starts, it will be helpful to use such tangible evidences as blood spots, clothing items, and the like to help fix the location. Good photographs are essential.

FORM GEN-799 REV. 2-62   INVESTIGATION OF LOST TIME INJURY

INSTRUCTIONS

1. SKETCH OR DIAGRAM, IF REQUIRED, SHOULD BE ATTACHED.
2. SEND COMPLETED REPORT TO: DIRECTOR, SAFETY AND PLANT PROTECTION, CHARLOTTE OFFICE.

PLANT			REPORT NO.		DATE	
NAME OF INJURED			OCCUPATION		☐ MALE ☐ FEMALE	
AGE	SERVICE WITH CORP.	TIME ON PRESENT JOB	DATE & TIME OF ACCIDENT	DATE LOST TIME BEGAN	LOST TIME	EST. LOST TIME

A.M./P.M.

DEPARTMENT OR LOCATION OF ACCIDENT

NATURE OF INJURY AND PART OF BODY INJURED

ATTENDING PHYSICIAN

EQUIPMENT AND MATERIAL

	YES	NO
DID THIS ACCIDENT INVOLVE ANY DOWNTIME OF EQUIPMENT	☐	☐
DID THIS ACCIDENT INVOLVE ANY LOSS OF PRODUCT	☐	☐

EVENTS LEADING UP TO ACCIDENT

DETAILS OF OCCURRENCE OF ACCIDENT

Fig. 28-6. Report of investigation of lost time due to injury.

EVENTS FOLLOWING ACCIDENT

CONCLUSION

RECOMMENDATIONS

COMPLETION DATE OF RECOMMENDATIONS

INVESTIGATED BY: DATE

SUPERINTENDENT _____

FOREMAN _____

SUPERVISOR OF SAFETY AND PLANT PROTECTION _____

ENGINEERING MANAGER OR PRODUCTION MANAGER _____

REVIEWED BY: PLANT MANAGER _____

NOTE: USE OF THIS FORM IN NO WAY AFFECTS METHOD OF REPORTING INJURIES ON REGULAR WORKMEN'S COMPENSATION INSURANCE FORM, OR FORWARDING OF SUCH FORMS TO INSURANCE CARRIER, OR TO OTHER GOVERNMENT AGENCIES.

Fig. 28-6. (continued)

Deaths

In case of death, the official medical investigator or coroner must be notified. Although it is seldom necessary or advisable to leave a body on the accident scene, it is required by some state laws. The body should be covered and permission requested of local authorities to remove it. Care should be taken to search for wedding rings and other personal items, so that these can be returned to relatives.

The investigation report should be made available to the police and medical examiner as quickly as possible.

REPORTING

In any incident involving death, serious injury, or severe equipment damage, the press and local governmental authorities will be concerned. It is desirable to set up in advance a suitable press-relation plan.

Because the preliminary investigation should precede any reporting by the press, it is essential this investigation be conducted promptly. In reporting, known facts should be stated. Theorizing about the cause should be avoided; but if the cause cannot be determined, the two or three most likely causes may be mentioned as possibilities. Some bad public relations have developed from fear of using certain words, such as "explosion." To the average person, anything which causes building failure or damage from pressure effects is an explosion.

Final Report

After careful review of the initial report and statements of witnesses, a final report should be prepared for the institution. This report should be made available to insurance investigators, governmental agencies, and other interested parties. It is also suggested that an account of any serious incident of a chemical nature be sent to the Manufacturing Chemists Association for use in the *Accident Case Histories*. This report to MCA should not contain names.

Classified Material

An accident may involve government classified material or other work regarded as secret. The first released written report in this case should not mention such materials by name or the name of the project. Governmental investigators will understand the need for such secrecy. The press will also usually respect such needs, especially on government defense work.

GOF. 3483 REV. 1-60 PRINTED U.S.A. 1-60
FIRE AND DAMAGE REPORT

CYANAMID

AMERICAN CYANAMID COMPANY

Refer to Safety and Loss Prevention Bulletin No. 24 for the preparation of this report. Use back for additional information.

PLANT	DEPARTMENT	LOCATION (BLDG. FLOOR, ROOM)	SUPERVISOR

DATE AND TIME OF OCCURRENCE	DISCOVERED BY	REPORTED BY	TIME ALARM SOUNDED	TIME FIRE OUT

ALARM BOX NO.	OTHER ALARM USED	INSURANCE CO. REPORTED TO	OFFICE	DATE	FIRE EQUIPMENT RESTORED TO USE

☐ YES   ☐ NO

WHAT FIRE EXTINGUISHMENT EQUIPMENT WAS USED

OUTSIDE ASSISTANCE, IF ANY

DESCRIBE ANY FAILURE OF FIRE PROTECTIVE EQUIPMENT

DESCRIPTION OF OCCURRENCE (TELL WHAT HAPPENED)

CAUSE OF OCCURRENCE (EXPLAIN HOW AND WHY)

PREVENTIVE ACTION TAKEN

INJURIES TO PERSONNEL

REPORTED BY-SIGNATURE	DATE	REVIEWED BY-SIGNATURE	DATE

PRELIMINARY LOSS REPORT- *Do not hold this report for detailed loss data. Forward within ▮▮▮▮ 7 days.*

DESCRIPTION OF DAMAGE TO BUILDING

ESTIMATED LOSS
$

DESCRIPTION OF DAMAGE TO EQUIPMENT

ESTIMATED LOSS

DESCRIPTION OF DAMAGE TO PRODUCT

ESTIMATED DAMAGE

ESTIMATE OF USE & OCCUPANY LOSS IN DOLLARS	ESTIMATE OF TOTAL PROPERTY LOSS IN DOLLARS

ESTIMATE OF QUANTITY OF PRODUCTION LOST

ESTIMATE OF DOWNTIME BEFORE RETURN TO 100% OPERATION	ESTIMATE OF TOTAL PROPERTY & U & O LOSS IN DOLLARS

ESTIMATES BY-SIGNATURE	DATE	APPROVED BY-SIGNATURE	DATE

Fig. 28-7. Fire and damage report.

A full report, with all necessary information, should be prepared within government security requirements and be accessible only to authorized persons.

The Report

The report of an accident is best written in straight narrative form. Complicated blanks, with questions to be answered, should be avoided (Fig. 28-7).

If the cause appears clear, state it in straightforward language without equivocation. If, however, the cause is not absolutely clear, state the most logical explanation with a qualification, such as "it appears that" or "the most reasonable explanation is." If there is more than one possibility, state all of them in the order of probability. If it is obvious that the institution is responsible through failure to provide safe equipment, a work-safe environment, or suitable protective equipment, say so; any good lawyer can find the weak points in any attempt to cover up or minimize responsibility.

In indicating responsibility, make clear that the intent is to prevent a future accident, not to place blame or to furnish reason for disciplinary action. If there was deliberate intent or gross negligence, disciplinary action is of course needed; but such cases are extremely rare.

29

LABORATORY FIRST AID AND SUGGESTIONS FOR MEDICAL TREATMENT

Hazards common to chemical laboratories call for somewhat specialized phases of first aid, and it is with these that this discussion deals. General directions for first aid are not included, and can best be found in such a source as the *Textbook on First Aid* of the American Red Cross.

Burns, eye injuries, and poisoning are the injuries with which laboratory personnel must be most concerned. First emphasis in the laboratory, as in any place where dangerous materials are handled, should be on preventing accidents. This means observing all recognized safe practices, using necessary personal protective equipment, and exercising proper control over poisonous substances at the source of exposure.

In first aid and subsequent treatment, there are short cuts called "antidotes." Chemical antidotes are usually to be avoided, since body tissues are poor test tubes for reactions which may result from their use. One's first thought should be to remove the victim from contact with the material—not "neutralization." A few well-recognized specific treatments, used only under the direction of a physician, are listed at the end of this chapter.

It should be definitely understood that this section contains considerable material not intended as a guide for first aid attendants, but, as indicated in the text, given as suggestions for medical treatment. Only a physician should put these suggestions for medical treatment into practice.

The attending physician is of course the one to determine the treatment to be used and the best judge as to what is indicated. However, because many physicians have little occasion to treat conditions due to some chemical exposures, where specific types of treatment have been developed, these are given in case the physician may wish to use them.

So that a physician can be summoned promptly, every laboratory should have posted the names, telephone numbers, and addresses of doctors to be called in an emergency requiring medical care. Telephone numbers for hospitals and ambulance services should also be readily available.

To help the attending physician, a patient suffering from a chemical injury who must be sent to a hospital should be "tagged" with a label giving the following information: name and address of patient and of employer; name or type of hazardous material to which he is believed to have been exposed; and specific drugs or treatment that have been administered before transfer.

BURNS

Thermal Burns

When possible, immerse burned area in very cold or ice water as quickly as possible. Continue immersion until pain is relieved and does not return when burn is removed from cold water. If burn cannot be immersed, ice cold compresses may be applied with good effect. Prompt application of cold eases the pain and severe tissue reaction associated with burns, and tends to reduce the ultimate severity of the burn. A physician should direct further treatment after first aid immersion.

Where there are extensive burns, look out for shock. Notify physician at once. Keep patient as quiet as possible. Remove clothing to uncover burn, being careful not to contaminate burned area any more than necessary. Do not attempt to cleanse burned area or apply ointments. Cover burned area with sterile gauze or a clean sheet. Whenever possible, get prompt medical attention, but do not delay transfer to hospital as quickly as possible. Send patient in care of an attendant, and notify hospital in advance.

Chemical Burns

A chemical burn is a severe injury involving destruction of tissue following contact with strong acids, alkalies, or oxidizing materials. First aid calls for removal from contact as promptly and completely as possible. When clothing has been contaminated, this means prompt removal of all such clothing, under a shower if possible, not overlooking shoes, socks, garters, and so on.

Affected areas of skin should be promptly and freely flushed with water, by shower, hose, or whatever means may be quickly available. This should be done thoroughly. This copious flushing is necessary regardless of the solubility in water of the material involved. The object is to remove mechanically or by solution, as quickly as possible, all injurious material. No neutralizing or buffering agents should be used. During the flushing, do not remove goggles from patient until his head and face areas have been thoroughly flushed.

Do not consider chemical antidotes. Reactions producing further injury may be set up in this way. If later an antidote is to be applied, it should be only as directed by the physician. After flushing to remove all possible irritants, treat as indicated for thermal burns.

EYE INJURIES

Foreign Bodies in the Eye

Loose, unattached foreign bodies—under or on the lid—may often be safely removed with a wet piece of clean cotton on an applicator. If the particle is on the cornea or attached to the surface of the eye or embedded in it, the case should be referred to a physician, preferably an ophthalmologist. Serious injury may follow "digging" or "picking" at a foreign body as a first aid measure. An associate or supervisor must not attempt to remove a foreign body unless specifically authorized by the physician.

Chemical Burns of the Eye

Splashes of irritant chemicals in the eye, or even exposure to vapor or mist of some chemicals, may lead to serious eye injury. Those who may be exposed to such chemicals should always use proper protective goggles or face shields.

Seconds count. First aid should be immediate, and consists of a thorough flushing of the eye with tap water, using eye bath fountain if available, a gentle stream of water from a hose, or any other means by which the eye may be freely flushed. Lids should be forcibly held apart so that the entire surface of the eye may be flushed. Under most circumstances this flushing should be continued for at least fifteen minutes. Contact lenses should not be worn in chemical laboratories because of the added difficulty they cause in eye irrigation.

The patient should then be referred to a physician, preferably an ophthalmologist with experience in handling chemical burns of the eye. Neutralizing solutions should never be used for first aid, since experience has demonstrated that they often aggravate the injury. Ointments are not recommended for first aid use.

When referring a case of chemical injury to a physician, if possible, always try to tell him the chemical name of the material causing the injury. Some compounds cause little immediate damage but have a delayed action very evident after a few hours or even days. The extent of damage caused by an acid, for instance, is fairly evident at once, but an alkali may cause markedly progressive damage. Alkali burns are not as painful as acid burns, so that the amount of pain is no gauge of the severity of the burn.

In most laboratories the hazard to the eyes is well recognized in the case of such acids as sulfuric, nitric, hydrochloric, and hydrofluoric, but may be overlooked in handling some of the anhydrides and chlorides. Similarly, most persons are aware of the inherent danger in such alkalies as sodium or potassium hydroxide, lime, or ammonia, but may not realize that similar severe effects may be caused by many of the amines. The possibility of delayed damage may be overlooked in instances of exposure to vapors or mists of such materials as hydrogen sulfide, methyl silicate, and hydroquinone.

Suggestions as to Medical Treatment

All cases of eye burns should be referred to a physician, preferably an ophthalmologist. A most efficient method of treatment of chemical eye burns, especially from alkaline compounds, is the "denuding" technique. This has been in use for some years and has saved many eyes, but should of course be carried out only by an opthalmologist experienced in this class of work, and only in the type of eye burn for which it has been found most suitable.

POISONING

Poisoning is a very general, and loosely used, term commonly connected with swallowing of toxic substances. In this chapter some local irritants have been included, although "poisoning," properly interpreted, should apply only to systemic effects. As encountered in the laboratory, poisoning by swallowing is a rare occurrence. Most important in the laboratory are the hazards from inhalation and skin absorption.

No attempt has been made to give a long list of poisonous substances with their "antidotes." There are a few compounds for which reasonably specific antidotes are available, the more important of which *are* mentioned. In general, certain principles of first aid apply. Most important of all is to prevent any hazardous exposure through any route—inhalation, skin contact, or swallowing.

Poisoning by Inhalation

This may occur through inhalation of gases, vapors, fumes, mists, or dusts. Some of these are so acutely irritating or have such marked odors that these warning properties make dangerous exposure unlikely, unless the exposed person is physically unable to leave the area of exposure. Such substances include chlorine, bromine, hydrochloric acid, sulfur dioxide, formaldehyde, acrolein, and ammonia. The action of materials in this group is essentially that of an extreme local irritant, the effects of which cannot properly be classed as "poisoning."

Other compounds without such prompt irritant action, even though usually having a characteristic odor, may be present in dangerous concentrations before this is realized. Among these are the halogenated hydrocarbons, particularly tetrachloroethane, carbon tetrachloride, methyl bromide, and ethylene chlorhydrin, as well as such compounds as oxides of nitrogen, carbon disulfide, and benzene.

Others may have very slight or no odor, even in dangerous concentrations, as in the case of carbon monoxide, methyl chloride, aniline, arsine, and mercury.

Some in higher concentrations, for example, carbon monoxide, hydrogen sulfide, hydrogen cyanide, cause unconsciousness almost immediately. Others usually have a delayed action, symptoms due to a dangerous exposure not coming on for perhaps some hours. Among these are oxides of nitrogen, phosgene, cadmium fumes, aniline, chlorine, and hydrochloric acid.

Many compounds are dangerous because of the chronic long-term exposures to concentrations too low to cause acute symptoms. These include, among others, benzene, carbon tetrachloride and other chlorinated hydrocarbons, mercury, lead, and some dusts.

First Aid. Remove patient from exposure as quickly as possible. When he is in a closely confined space, rescuers may need personal protective equipment such as air-supplied or self-contained breathing apparatus. When exposure has been severe or patient is unconscious, call a physician at once, giving location of patient and, whenever possible, the identity of the toxic material. Keep patient warm and lying down. If breathing has stopped, do not wait for mechanical equipment but start artificial respiration at once. The preferred method is mouth-to-mouth resuscitation as described below.

If there is foreign matter visible in the mouth, wipe it out quickly with your fingers or a cloth wrapped around your fingers.

Tilt the head back so the chin is pointing upward (Fig. 29-1). Pull or push the jaw into a jutting-out position (Figs. 29-2 and 29-3).

These maneuvers should relieve obstruction of the airway by moving the base of the tongue away from the back of the throat.

Open your mouth wide and place it tightly over the victim's mouth. At the same time pinch the victim's nostrils shut (Fig. 29-4) or close the nostrils with your cheek (Fig. 29-5). Or, close the victim's mouth and place your mouth over the nose (Fig. 29–6). Blow into the victim's

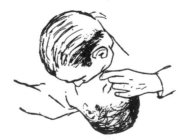

Fig. 29-1

Fig. 29-2   Fig. 29-3

Fig. 29-4   Fig. 29-5

Fig. 29-6   Fig. 29-7

mouth or nose. (Air may be blown through the victim's teeth, even though they may be clenched.)

The first blowing efforts should determine whether or not obstruction exists.

Remove your mouth, turn your head to the side, and listen for the return rush of air that indicates air exchange. Repeat the blowing effort.

For an adult, blow vigorously at the rate of about twelve breaths per minute. For a child, take relatively shallow breaths appropriate for the child's size, at the rate of about twenty per minute.

If you are not getting air exchange, recheck the head and jaw position (Figs. 29-1, or 29-2 and 29-3). If you still do not get air exchange, quickly turn the victim on his side and administer several sharp blows between the shoulder blades in the hope of dislodging foreign matter (Fig. 29-7).

Again sweep your fingers through the victim's mouth to remove foreign matter.

Those who do not wish to come in contact with the person may hold a cloth over the victim's mouth or nose and breathe through it. The cloth does not greatly affect the exchange of air.

Where mouth-to-mouth resuscitation cannot be used the alternative method is the arm-lift back-pressure method, described as follows:

The victim is placed in the face-down or prone position, with the arms folded and the hands placed on top of each other (Fig. 29-8). The face is then placed on the hands. The operator kneels on either knee, or on both knees, at the victim's head and places his hands under the arms just above the elbows. He lifts them upward at the same time that he rocks backward, drawing the arms toward himself until he meets firm resistance and tension (Fig. 29-9). This is the

Fig. 29-8

Fig. 29-9   **Fig. 29-10**

inspiratory phase. The arms are then replaced on the floor, and the operator moves his hands to the midback just below the shoulder blades and rocks forward, keeping his arms stiff, and exerting pressure almost vertically downward to cause active expiration (Fig. 29-10). This expiratory phase may be used to start the procedure if desired. The complete maneuver is repeated twelve times per minute. Of course the usual precautions should be taken to see that the tongue, dentures, or other materials are not blocking the victim's airway.

Whenever artificial respiration is applied, there is usually a definite advantage in the administration of oxygen at the same time. This is especially true in the case of carbon monoxide poisoning. Pure oxygen is usually administered alone, and is preferred, although some use oxygen with 7% carbon dioxide.

Oxygen may be given, preferably through the use of an inhalator or resuscitator. It should be administered only by someone familar with the procedure and with the equipment available.

After severe exposure to carbon monoxide or irritant gases, such as ammonia or chlorine, and especially those known to have a delayed action, such as oxides of nitrogen or phosgene, pure oxygen should be administered as soon as possible.

In the case of exposure to nitrogen oxides or phosgene, it is very important that the victim be observed during the next twelve-hour period. Pulmonary edema frequently occurs from six to eight hours after exposure—thus after the victim has gone home.

Note to physicians: In most exposures, administration of pure oxygen at atmospheric pressures has been found to be adequate. This is best accomplished by use of a face mask having a reservoir bag of the non-rebreathing type.

Some authorities believe that superior results in relief of symptoms are obtained when exposures to lung irritants are treated with oxygen under an exhalation pressure not exceeding 4 cm water. These same authorities believe that breathing oxygen under pressure is also useful as an aid in the prevention of the pulmonary edema which may occur after breathing an irritant chemical.

Masks providing for such exhalation pressures are available.

In the event of an exposure causing symptoms, the patient may be treated with oxygen for several hours. Oxygen inhalation must be continued as long as necessary to maintain the normal color of the skin and mucous membranes. Treatment should be continued until symptoms subside; if there are no signs of lung congestion and if breathing is easy, oxygen inhalation may be discontinued.

Recently, pulmonary edema has been treated successfully by administration of oxygen through intermittent positive pressure breathing equipment (IPPB). In addition to providing oxygen under positive pressure, this equipment has the value of allowing simultaneous administration of aerosolized bronchodilators and/or foam suppressants as the clinical conditions of the patient may indicate.

The administration of a chemically neutralizing material by inhalation is not recommended as a first aid measure. Stimulants should be given only as recommended by the attending physician. Adrenalin should not be used in cases involving exposure to chlorinated hydrocarbons, such as carbon tetrachloride.

REMEMBER: Never give anything by mouth to an unconscious patient.

Poisoning by Skin Contact and Absorption through the Skin

Skin contact with various chemicals may result in changes ranging from a defatting action from solvents not otherwise irritating, to rapid and often deep destruction of tissue in the case of the stronger acids and alkalies. Defatting may only make the skin more sensitive to irritation or secondary infection.

Between this and the direct destruction of tissues, there is a large group of effects described generally as contact dermatitis. These conditions are so varied, even in different individuals exposed to the same materials, and include the complex question of sensitization, that their consideration belongs in the field of dermatology rather than in that of laboratory safety.

Skin absorption refers to the systemic effects resulting from absorption through the skin. In the case of some chemicals, the effects due to absorption through the skin are approximately as severe, dose for dose, as from inhalation or swallowing. Such chemicals include allyl alcohol, aniline, ethylene chlorhydrin, and ethylenimine.

For a considerable group of chemicals, toxicity by skin absorption is very definitely a hazard. A few of these are:

acrylonitrile	nitrobenzene
allyl chloride	phenol
antimony trichloride	tetrachlorethane
arsenic trichloride	tetraethyl lead
cresols	toluidine
cyanides	xylidine
nitroaniline	

There are economic poisons, such as the following.

benzene hexachloride
dimethyl bromide
nicotine sulfate
parathion
tetraethyl pyrophosphate

Many chemicals which are definitely hazardous because of toxicity through skin absorption present relatively little or no evidence of local skin effects. Examples are acrylonitrile, allyl alcohol, aniline, and ethylene chlorhydrin. Others, such as cresol and phenol, have severe local as well as systemic effects.

First Aid.   The primary consideration is the prompt removal of the chemical from contact with the skin. This is true whether or not the material has local action. All contaminated clothing should be removed at once, preferably under a shower, and the contacted areas freely flushed with water, preferably with plenty of soap, and under a shower or running water. If exposure has been severe, call a physician, telling him the location of the patient and chemical involved.

The copious use of water to remove as far as possible all traces of the chemical is the most available and effective first aid measure. This applies whether or not the material is water-soluble. Chemical antidotes, such as alkalies for acid contacts and vice versa, or solvents such as alcohol for phenol, should *not* be used as first aid measures.

After thorough removal of the chemical, the patient should be kept warm and preferably lying down. Further treatment should be as directed by the physician.

Poisoning by Swallowing

Nearly all chemicals handled in a laboratory are "poisonous" if swallowed; it is a question of degree. A broad definition of a poison is "a substance which will injure the body if too much is taken for too long." If every harmful material were labeled "poison," the term would soon lose all value as a warning. All chemicals should be treated with due respect according to their properties, but the term "poison" should be reserved only for those which are highly toxic.

First Aid.   A physician should be called at once and be told of the location of the patient and If possible the chemical swallowed. If a chemical of known or suspected toxicity has been swallowed, the first step is to remove it from the stomach before it is absorbed. If patient is vomiting, give him warm water to drink freely, to aid in vomiting and to dilute any chemical retained in the stomach.

If vomiting is not spontaneous, give lukewarm water freely. It will help to induce vomiting if one or two tablespoonfuls of common salt are dissolved in each glass of water or if a soapy solution is given. One to three teaspoonfuls of powdered mustard may be stirred into each glass of water for the same purpose. Vomiting may also be induced by using a strip of paper or a finger to tickle the throat. Vomiting should be encouraged until vomited liquid is clear. Vomiting should not be induced if a petroleum distillate or certain halogenated hydrocarbons have been swallowed.

Use of a stomach tube should be limited to a physician only. In the case of corrosive materials, its use may be dangerous.

After the stomach has been evacuated by free vomiting, the patient should be kept warm, preferably lying down, and watched for shock. In cases of severe poisoning, the patient must also be watched to be sure breathing has not stopped. If breathing has stopped, artificial respiration should be started immediately. Before starting, make sure no dentures, tobacco, gum, or other substances are in the mouth. Administration of oxygen is often of help and should be given in severe cases and when breathing has been restored. Further treatment, including use of antidotes and stimulants should be only as directed by a physician.

REMEMBER: Never give anything by mouth to an unconscious patient.

SPECIFIC ANTIDOTES AND TREATMENT SUGGESTED FOR USE BY PHYSICIANS

Cyanides

In any case of cyanide poisoning, call a physician immediately. In all areas in which cyanide compounds are manufactured or handled, a first aid kit containing the following items should be readily available.

Two boxes (two dozen) of amyl nitrite pearls (ampules).
Two sterile ampules of sodium nitrite solution (10 cc of a 3% solution in each).
Two sterile ampules of sodium thiosulfate solution (50 cc of a 25% solution in each).
One 10 cc and one 50 cc sterile glass syringe with sterile intravenous needles.
One tourniquet.
One stomach tube.
One dozen gauze pads and one small bottle of 70% alcohol.
Two 1-pint bottles of 1% sodium thiosulfate solution.

The kit should be conveniently located and checked at regular intervals by a responsible person.

First Aid.   After inhalation and skin contact, the patient should be allowed to breathe the vapors of the contents of amyl nitrite pearls, one every five minutes for twenty minutes. If breathing has ceased, an assistant should administer the amyl nitrite while the patient is receiving artificial respiration. The pearls are to be wrapped lightly in a handkerchief or gauze pad, then broken in the handkerchief or pad, and the latter held about one inch from the patient's mouth and nostrils until the strength of the pearl is spent.

WARNING: Those giving first aid should be careful to keep the broken pearls away from their own mouths and noses; otherwise, they may inhale the amyl nitrite, become dizzy, and be rendered incompetent to give proper assistance to the poisoned person.

The vapor of amyl nitrite is flammable and may present a fire or explosion hazard if a source of ignition is present.

If cyanide compound has been swallowed, and the patient is conscious, induce vomiting, using as an emetic warm salt water (one tablespoonful of common salt to each cup of water) or one pint of 1% solution of sodium thiosulfate in water. Then continue treatment as above.

REMEMBER: Never give anything by mouth to an unconscious patient.

Suggestions for Medical Treatment.   In cases of cyanide poisoning, gastric lavage may be performed by a physician. In addition, the physician may administer sodium nitrite and sodium thiosulfate intravenously. The details of this treatment are described in an article by Chen, Rose, and Clowes, "The Modern Treatment of Cyanide Poisoning," *Journal of the Indiana State Medical Association, 37* (7), 344–350 (July 1944). A summary of the treatment is as follows:

The physician shows an assistant how to break, one at a time, pearls of amyl nitrite in a handkerchief and hold the latter over the victim's nose for fifteen to thirty seconds per minute. At the same time the physician quickly loads his syringes: the 10 cc syringe with a 3% solution of sodium nitrite, and the 50 cc syringe with a 25% solution of sodium thiosulfate.

The physician discontinues the administration of amyl nitrite and injects intravenously 0.3 g (10 cc of a 3% solution) of sodium nitrite at the rate of 2.5 to 5.0 cc/min.

The physician injects by the same needle and vein, or by a larger needle and a new vein, 12.5 g (50 cc of a 25% solution) of sodium thiosulfate.

The patient should be watched for at least twenty-four hours or even forty-eight hours. If signs of poisoning reappear, the injection of both sodium nitrite and sodium thiosulfate should be repeated, but each in one-half of the above dose. Even if the patient appears perfectly well, the medication may be given for prophylactic purposes two hours after the first injections.

WARNING: A patient who has been treated with amyl nitrite, with or without injection of sodium nitrite and sodium thiosulfate, should not be treated with methylene blue.

Hydrofluoric Acid

First Aid.   In cases of contact with HF, remove contaminated clothing immediately, preferably under a shower. Wash thoroughly all contacted areas with plenty of water. Following this, an iced aqueous or alcoholic solution, 0.13% (1:750) of benzalkonium chloride ("Zephiran" Chloride); an iced 70% alcohol solution; or an ice-cold saturated solution of magnesium sulfate (Epsom Salt) should be applied for at least thirty minutes or longer if pain persists.

If the burn is in such an area that it is impracticable to immerse the part, then the iced alcohol or the iced magnesium sulfate should be applied with saturated compresses, which should be changed at least every two minutes; this treatment should be continued for thirty minutes or until the skin looks normal.

The physician should be on hand to administer treatment before the completion of the iced solution treatment. If, however, he has not arrived by that time, it is then permissible to apply a generous quantity of paste made from powdered magnesium oxide and glycerine (U.S.P., and, preferably, freshly prepared). (Oils or greases should not be applied except under instructions from a physician.)

For eyes, flush freely with water for at least fifteen minutes. Further treatment should be administered by physician.

Suggestions for Medical Treatment.   If the physician sees the patient soon after contact with a diluted (1–20%) solution of hydrofluoric acid, there may be little or no evidence of injury. Evidence of the injury may not appear for several hours.

If after adequate iced-solution soaking there is evidence of penetration of acid beneath the skin, calcium gluconate solution (standard ampule of 10% intravenous solution) with a local anesthetic may be injected by infiltrating the skin and subcutaneous tissues in the same manner as a local anesthetic is injected subcutaneously. All the skin which has been exposed to the acid should be infiltrated including at least ¼-½ in. around the area. This treatment will usually prevent the development of severe burns.

After the affected area has been treated with either iced alcohol or iced magnesium sulfate or after the injection of calcium gluconate solution, magnesium oxide-glycerine paste, A & D ointment, or topical steroid-antibiotic ointment, should be applied. Then the areas should be well padded with gauze and a pressure dressing applied to the area in the same manner as for thermal burns.

In severe burns, blisters filled with a sero-purulent fluid develop and the skin assumes a blanched appearance. Ten percent calcium gluconate should be immediately injected into and around the affected areas. Blisters should be cut away completely and magnesium oxide and glycerine paste applied to the denuded area for a period of twenty-four to forty-eight hours. At the end of this time, if there is no further evidence of extension of the burned area, magnesium oxide ointment should be applied.

Treatment of Burns around Fingernails. Burns involving finger and toe nails may show no visual effects as the material apparently passes through the nail without altering its appearance. The nail should be immediately removed under regional anesthesia and treatment pursued as described previously.

Contact with Eyes. Eye exposure should be followed immediately by prolonged, gentle irrigation with copious amounts of cool tap water. A topical ophthalmic anesthetic should be used for pain. Persistent pain usually indicates a need for additional gentle irrigation. Fluorescein should be instilled in the eye and, if staining occurs, intermittent irrigation should be continued for twenty-four hours. Antibiotic-steroid ointments may be useful. Ophthalmologic consultation should be promptly sought but irrigation must not be delayed.

Gastric Lavage. Lavage with lime water should be instituted promptly by a physician only. If lime water is not available, milk may be used as a substitute. Soluble calcium inactivates the fluoride ion. In addition to lavage, 10 cc of a 10% solution of calcium gluconate should be injected intravenously. Respiratory depression should be combatted with oxygen and stimulants if necessary; and artificial respiration should be used if needed.

Arsenic Compounds

Induce vomiting by sticking a finger down the throat or giving strong salt or soapy water. Give a strong laxative dose of magnesium sulfate. Force fluids.

Dimercaprol (BAL) is indicated except in the presence of known liver damage. The following information relative to this preparation, is from *New and Nonofficial Remedies 1952*, p. 427:

The toxicity of dimercaprol is less in patients suffering from arsenic, gold, or mercury poisoning, but doses of 300 mg (5 mg/kg of body weight) may produce nausea, vomiting and headache, a burning sensation of the lips, mouth, throat and eyes, generalized muscular aches with burning and tingling of the extremities, and a sense of constriction in the chest. The symptoms usually subside in 30 to 90 min.

Dosage: In the treatment of arsenic or gold poisoning, 3 mg of dimercaprol per kg (as a 10% solution in oil) should be administered by intramuscular injection every four hours for the first two days; four injections should be given on the third day; and two injections daily thereafter for ten days, or until complete recovery. In milder cases, the dose may be reduced to 2.5 mg per kg.

The following dosage form is identical to Dimercaprol Injection (U.S.P.): Hynson,Wescott & Dunning, Inc.—Solution BAL in Oil: 3 cc ampules, a solution in peanut oil containing 10% dimercaprol and 20% benzyl benzoate.

Mercury Compounds

Induce vomiting followed by gastric lavage with 5% sodium formaldehyde sulfoxylate solution, allowing a small amount to remain in the stomach. Dimercaprol (BAL) is the treatment of choice in mercury poisoning. It should be given as soon as possible to prevent serious kidney injury. See "Dimercaprol" under "Arsenic" for treatment schedule.

Parathion and Other Organic Phosphate Pesticides

First Aid.   Call a physician at once in all cases of suspected parathion poisoning. If symptoms or signs of poisoning include blurred vision, abdominal cramps, and tightness in the chest, don't wait for a doctor but give two atropine tablets (each 1/100 grains) at once. Remove contaminated clothing and wash the skin clean with plenty of soap and water to remove all traces of parathion. If swallowed, induce vomiting by giving warm salty or soapy water.

REMEMBER: Never give anything by mouth to an unconscious person.

Physician's note: Warning symptoms include weakness, headache, tightness in the chest, blurred vision, nonreactive pinpoint pupils, salivation, sweating, nausea, vomiting, diarrhea, and abdominal cramps.

Treatment.   Give 2 mg atropine intravenously every eight to ten minutes until patient is atropinized, as shown by dilated pupils, flushed face, dry skin and rapid pulse. Maintain atropinization by appropriate dosage intramuscularly or intravenously for as long as needed to maintain atropinization as long as needed. In severe poisoning, as much as 30 to 40 mg of atropine may be needed within the first forty-eight hours. Never give morphine. Clear chest by postural drainage. Artificial respiration or oxygen administration may be necessary. Observe patient continuously for forty-eight hours. Repeated exposure to cholinesterase inhibitors may, without warning, cause prolonged susceptibility to very small doses of any cholinesterase inhibitor. Allow no further exposure until time for cholinesterase regeneration has been allowed as determined by blood tests.

If available, pralidoxime (Protopam) chloride may be used in *addition* to atropine for the purpose of stimulating reversal of cholinesterase inhibition. Give 2.5 g in 100 ml of sterile water or 5% glucose solution intravenously in not less than fifteen minutes. Protopam is an adjunct to atropine therapy; it should not be used in place of atropine.

Aromatic Nitro and Amino Compounds

The aromatic nitro and amino compounds are generally active in causing methemoglobinemia, which is recognized by varying degrees of cyanosis. Concentrations of methemoglobin up to 20% usually cause no symptoms, and the cyanosis is hard to identify. Above 20% and up to 50 or 60%, methemoglobin concentration, symptoms of headache, dizziness, and breathlessness become evident and more severe as the methemoglobin concentration increases. Concentrations above 50% are classed as medical emergencies requiring energetic treatment.

Aromatic nitro and amino compounds may enter the body by swallowing, inhalation, or skin absorption. They penetrate the skin readily, and this means of entry to the body is a common problem. If swallowed induce vomiting by strong salt or soapy water. Repeat until vomit is clear. If inhaled, remove patient to fresh air. Skin contact must be flushed at once with water followed by washing with soap and water. Transport patient to medical care. Do not permit him to walk. At the medical department, the bathing must be repeated using warm water and plenty of soap, making sure that hairy areas and fingernail areas are thoroughly cleaned. Methemoglobin determinations should be made at once in all cases of exposure and repeated at hourly intervals until there is conclusive evidence that the concentration is falling.

Treatment.   Treatment is generally determined by the methemoglobin concentration. In mild poisoning (methemoglobin concentration of 20% or less) cleanse individual thoroughly and keep under observation for several hours to assure that methemoglobin concentration is not rising. No specific therapy is needed. Make sure that the patient does not wear contaminated clothing, including hat, gloves, and shoes.

In moderate poisoning (methemoglobin concentration of 20% to 50%) give oxygen to relieve symptoms. Keep under observation at bed rest. Some authorities believe that intravenous dextrose solution may hasten the rate of conversion of methemoglobin to hemoglobin. Its use may be desirable when concentrations of methemoglobin approach 40% or more. Determine methemoglobin concentrations at hourly intervals to follow the course of the disease.

Severe poisoning (methemoglobin concentrations of 50% or more) constitutes a medical emergency. Start patient on oxygen and prepare to administer methylene blue intravenously when the methemoglobin concentration reaches 60%. Methylene blue is given as a 1% aqueous solution. The usual dose is 1 mg per kg of body weight. The solution is injected slowly over a period of eight to ten minutes.

In all cases, the methemoglobin level must be followed at hourly intervals until there is evidence that the concentration is falling. Once the concentration starts declining, it will continue to decrease, with normal hemoglobin values usually being established within forty-eight hours. No further exposure to aromatic nitro or amino compounds should be permitted until normal hemoglobin levels have been established.

APPENDIX 1

BIBLIOGRAPHY—GENERAL

The following is a partial list of technical and trade associations, insurance companies and associations, government agencies, authors and publishers together with their full names and addresses and from which are available films, and posters, chemical and safety publications, etc. Most organizations offer a complete list of their publications upon request.

1. Data Sheets and Manuals

Source	*Literature Available*
Manufacturing Chemists Association (MCA) 1825 Connecticut Avenue, N.W. Washington, D.C. 20009	More than ninety chemical safety data sheets covering specific chemicals; also, *Guide to Precautionary Labeling of Hazardous Chemicals*
American Conference of Governmental Industrial Hygienists (ACGIH) 1014 Broadway Cincinnati, Ohio 45202	Pamphlet giving threshold limit values for toxic dusts, fumes, gases, vapors and mists
Air Conditioning and Refrigeration Institute 1815 N. Fort Myer Drive Arlington, Va. 22209	Standards for air conditioning and refrigeration equipment including safety provisions

American Insurance Association
85 John Street
New York, New York 10038

Handbook of Industrial
Safety Standards;
National Building Code
Data sheets on specific chemicals

American Industrial Hygiene
Association (AIHA)
210 Haddon Avenue
Westmont, New Jersey 08108

Hygienic Guides for about 150
chemicals

American National Standards
Institute (ANSI)
1430 Broadway
New York, N.Y. 10016

Standards and guides
relating to safety

Factory Mutual Engineering
Corporation
1151 Boston-Providence Turnpike
Norwood, Mass. 02062

*Properties of Flammable Liquids,
Gases, and Solids*

American Mutual Insurance
Alliance
20 N. Wacker Drive
Chicago, Ill. 60606

*Handbook of Organic
Industrial Solvents,* 3rd ed.

National Council on
Radiation Protection and
Measurements (NCRP&M)
4000 Brandywine Street, N. W.
Washington, D.C. 20016

Reports: "'Control and Removal of
Radioactive Contamination in Labora-
tories," "Safe Handling of Radioactive
Materials"; these and other reports are
available from the National Council on
Radiation Protection and Measurements,
Publication Sales Section, Box 4867,
Washington, D.C. 20008. In addition,
many of the NCRP reports are available
as National Bureau of Standards hand-
books from the Superintendent of
Documents, Government Printing Office,
Washington, D.C. 20025

National Electrical
Manufacturers Association
155 E. 44th Street
New York, N.Y. 10017

Standards for electrical equipment

National Fire Protection
Association (NFPA)
60 Batterymarch Street
Boston, Mass. 02110

Fire Protection Guide
on
Hazardous Materials

National Safety Council (NSC)
425 North Michigan Avenue
Chicago, Ill. 60611

Data sheets for more than 60 chemicals.
Accident Prevention Manual for
Industrial Operations

Underwriters' Laboratories,
Inc. (UL)

Standards on refrigerators, x-ray
equipment, motors, generators, lighting

207 E. Ohio Street
Chicago, Ill. 60611

fixtures, and other electrical equipment for use in hazardous locations; extinguishers, automatic sprinklers, fire detection equipment; electrical equipment and appliances for use in ordinary locations; lists of devices, materials, and systems inspected under these standards

U.S. Bureau of Mines
4800 Forbes Avenue
Pittsburgh, Pa. 15213

Bibliography of Bureau of Mines Health and Safety Publications

Superintendent of Documents
U.S. Government Printing Office
Washington, D.C. 20403

Occupational Safety and Health Administration
Occupational Safety and Health Regulations

Other Publications

Fawcett, H. H., and W. S. Wood, *Safety and Accident Prevention in Chemical Operations,* Interscience Publishers, a division of John Wiley & Sons, Inc., New York, 1965.

Lewis, H. F., *Laboratory Planning for Chemistry and Chemical Engineering,* Van Nostrand Reinhold Co., New York, 1962.

Quam, G. N.,*Safety Practice for Chemical Laboratories,* Villanova Press, Villanova, Pa.

Safety in the Laboratory, Union Carbide Corporation, Chemicals Division, 270 Park Ave., New York.

Manual of Laboratory Safety, Fisher Scientific Company, 717 Forbes Ave., Pittsburgh, Pa.

Toxic Eye Hazards, National Society for the Prevention of Blindness, 1790 Broadway, New York.

Patty, F. A., *Industrial Hygiene and Toxicology,* 2nd ed., Interscience Publishers, a division of John Wiley & Sons, New York, 1962.

Fairhall, L. T., *Industrial Toxicology,* Williams and Wilkins Co., Baltimore, Md.

Sax, N. I., *Dangerous Properties of Industrial Materials* 3rd ed., Van Nostrand Reinhold Co., New York, 1968.

Steere, N. V., Handbook of Laboratory Safety, Chemical Rubber Co., Cleveland, Ohio.

2. Case Histories of Accidents

Manufacturing Chemists Association, Washington, D.C.
Case Histories of Accidents in the Chemical Industry, Vol. 1, 1962, Vol. 2, 1966; Vol. 3, 1970. Contains classified index and includes many case histories of laboratory accidents.

Alpha Chi Sigma Fraternity
Committee on Safety
5503 East Washington Street
Indianapolis, Indiana

3. Safety Films and Slides

Some of the safety films and slide presentations available are listed below. Others are listed in *Film Guide,* available from the Manufacturing Chemists Association; in *Safety Film News,* available from Engineering and Safety Department, American Insurance Association; and in the *National Directory of Safety Films,* available from the National Safety Council.

Safety in the Chemical Laboratory
16 mm. sound, color film, 20 min; Manufacturing Chemists Association, Washington, D.C. Order from Edward Feil Productions, 1514 Prospect Avenue, Cleveland, Ohio 44115.

Chemical Boobytraps
10 1/4 min color film; free rental; General Electric Company, Research Laboratory, P.O. Box 1088, Schenectady, New York.

Eye and Face Protection in Chemical Laboratories; National Society for the Prevention of Blindness, 79 Madison Ave., New York, N.Y. 10016.

4. Safety Posters

Chemical Laboratory Safety Posters—set of 12, 12 in. x 18 in.; Manufacturing Chemists Association, Washington, D.C.; available to U.S. purchasers only.

Laboratory Emergency Chart, Fisher Scientific Company, Pittsburgh, Pa.

First Aid Chart and Spillages of Hazardous Chemicals Chart published by British Drug Houses Ltd., BDH Laboratory Chemicals Division, Poole, Dorset, England are available from Gallard-Schlesinger Chemical Mfg. Corp., 584 Mineola Ave., Carle Place, L.I., New York. Minimum order—ten charts in any combination. In Canada order from British Drug Houses (Canada) Ltd., Barclay Ave., Toronto, Ont., Canada.

APPENDIX 2

BIBLIOGRAPHY—ALPHABETICAL

The following list is adapted from References to Representative Sources of Information on Chemicals by H. H. Fawcett, National Academy of Sciences, Washington, D.C.

Copies of publications or further information can be secured from these organizations. The numerical references included In the listing are interpreted as follows:

(1) MCA Chemical Safety Data Sheet, Manufacturing Chemists Association, 1825 Connecticut Ave., N.W., Washington, D.C. 20009.

(2) MCA Chem-Card.

(3) MCA Cargo Information Card.

(4) NSC Safety Data Sheet, National Safety Council, 425 North Michigan Ave., Chicago, Ill. 60611.

(5) AIHA Hygienic Guide, American Industrial Hygiene Association, 210 Haddon Avenue, Westmont, New Jersey 08108.
The letters "HG" appearing under the heading "AIHA" indicate that material on the particular subject can be found in the American Industrial Hygiene Associations' *Hygiene Guide*. Under the other headings, the publisher's number for the specific publication is given.

(6) Compressed Gas Association, 500 Fifth Avenue, New York, N.Y. 10036.

(7) The threshold limit values, TLV, are set by the American Conference of Governmental Industrial Hygienists. The full list of threshold limit values is revised annually by the ACGIH. Single copies of the list are available from the secretary of the ACGIH, 1014 Broadway, Cincinnati, Ohio, 45202.

(8) Penna. Short Term Limits (STL) and hygienic Information Guides (HIG) are from the Pennsylvania Department of Health, P.O. Box 90, Harrisburg, Pa. Cited references are to Chapter 4, Article 432, Regulations Establishing Threshold Limits in Places of Employment (revised Aug. 17, 1965). For justifi-

cation see, "Short Term Limits for Exposure to Air-Borne Contaminants, a documentation."

(9) NAS–NRC Short Term Limits (STL) were established by NAR–NRC Committee on Toxicology, and were published in H. F. Smyth, Jr., "Military and Space Short-Term Inhalation Standards," *Arch. Environ. Health* **12**, 488–490 (April 1966).

(10) NAS–NRC #1465: A Tentative Guide to Evaluation of the Hazard of Industrial Chemicals for Bulk Water Transportation, Committee on Hazardous Materials, National Academy of Sciences, 2101 Constitution Ave., N.W. Washington, D.C. 20418.

(11) CED are Commodity and Equipment Data Sheets, published by the National Tank Truck Carriers, Inc., 1616 P St., N.W., Washington, D.C. 20036.

(12) CG–388: Chemical Data Guide for Bulk Shipment by water, 1969 U.S. Government Printing Office, Washington, D.C. 20402.

(13) NFPA No. 49, Hazardous Chemicals Data, Published by National Fire Protection Association, 60 Batterymarch St., Boston, Mass. 02110.

(14) Sax, N.I., *Dangerous Properties of Industrial Materials,* 3rd ed., Van Nostrand Reinhold Co., New York.

(15) Laboratory Waste Disposal Manual, May 1970, Manufacturing Chemists Association. (See Appendix 4).

In the table, (R) denotes a registered trade mark.

Acetaldehyde   (1) SD–43; (2) CC–49; (3) CIC-1 (7); (10); (12); (13); (14); (15)2

Acetic acid   (1) SD–41; (3) CIC–2 (4) 410; (5); (7); (8) STL; (10); (11) 1; (12); (13); (14); (15)3

Acetic anhydride   (1) SD–15 (3) CIC–3; (7); (10); (12); (13); (14); (15)4

Acetone   (1) SD–87; (2) CC–23; (4) 398; (5); (7); (8) STL, HIG; (10); (11) 36; (12); (13); (14); (15)5

Acetone cyanhydrin   (2) CC–34; (3) CIC–4; (10); (12); (13); (14); (15)6

Acetonitrile   (2) CC–64; (3) CIC–5; (5); (7); (8)STL; (10); (12) (13); (14); (15)7

Acetylene   (1) SD–7); (4) 494; (6) G–1; (13); (14); (15) 10

Acetylene tetrabromide   (7); (13); (14); (15) 1004

Acids, mixed   (1) SD–65; (2) CC–65; (14)

Acrolein   (1) SD–85; (2) CC–76; (4) 436; (7); (8) STL HIG; (10); (13); (14); (15) 13

Acrylic esters (monomers)   (11) 48; (14)

Acrylonitrile   (1) SD–31; (2) CC–15; (3) CIC–6; (8) STL; (10); (11) 64; (13); (14); (15) 16

Adipic acid   (4) 438; (14); (15) 17

Adiponitrile   (3) CIC–7; (8); (10); (14); (15) 18

Aldrin   (7); (8) HIG; (14); (15) 19

Allyl alcohol   (2) CC–35; (3) CIC–8; (7); (8) STL; (10); (12); (13); (14); (15) 23

Allyl chloride   (2) CC–63; (3) CIC–9; (7); (10); (12); (14); (15) 26

Allyl glycidyl ether   (7); (8) STL; (14); (15) 28

Alum   (11) 49; (14); (15) 31

sec-**Butyl alcohol** (2) CC–80; (10); (12); (14); (15) 163

i-**Butyl alcohol** (10); (12); (14); (15) 162

tert-**Butyl alcohol** (2) CC–80; (7); (8) STL; (10); (12); (14); (15) 164

Butylamine (5); (7); (8) STL; (14); (15) 165

tert-**Butyl chromate** (7); (8) STL; (14); (15) 172

n-**Butyl Glycidyl ether** (7); (14); (15) 175

Butyllithium (2) CC–28; (14); (15) 177

n-**Butyllithium** (1) SD–91; (14); (15) 177

p-tert-**Butyl toluene** (7); (14); (15) 186

Butyl mercaptan (7); (14); (15) 178

n-**Butyraldehyde** (1) SD–78; (2) CC–50; (3) CIC–17; (10); (12); (13); (14); (15) 189

i-**Butyraldehyde** (3) CIC–56; (12); (13); (14); (15) 190

Cadmium (4) 312; (5); (8) HIG; (14); (15) 197

Cadmium oxide fume (7); (8) STL; (14); (15) 199

Calcium arsenate (7); (14); (15) 201

Calcium carbide (1) SD–23; (13); (14); (15) 202

Calcium chloride (11) 59; (14); (15) 205

Calcium oxide (7); (8) STL; (13); (14); (15) 210

Camphor oil (light) (3) CIC–18; (7); (10); (14);

Carbaryl (sevin) (R) (7); (14);

Carbolic acid (phenol) (1) SD–4; (2) CC–48; (3) CIC–19; (4) 405; (5); (7); (8) HIG; (10); (11) 29; (12); (13); (14); (15) 803

Carbon dioxide (4) 397; (6) G–6; (7); (8) HIG; (14);

Carbon disulfide (1) SD–12; (3) CIC–20; (4) 341; (5); (7); (8) STL, HIG; (9); (10); (12); (13); (14); (15) 216

Carbon monoxide (4) 415; (5); (7); (8) STL HIG; (9); (14); (15) 217

Carbon tetrachloride (1) SD–3; (3) CIC–21; (5); (7); (8) STL HIG; (10); (11) 50; (12); (14); (15) 219

Casinghead, natural gasoline (10); (11) 12; (15) 527

Caustic potash, liquid and solid (1) SD–10; (2) CC–32; (3) CIC–22; (10); (12); (13); (14); (15) 223

Caustic soda, sodium hydroxide' (1) SD–9; (2) CC–33; (3) CIC–23; (4) 373; (7); (8) HIG; (10); (11) 13; (12); (13); (14); (15) 224

Cellosolve (12); (14); (15) 225

Chlorates (1) SD–42; (4) 371; (14); (15) 846, 937

Chlordane (7) skin; (8) HIG; (14); (15) 231

Chloroacetaldehyde (7); (8) STL; (14); (15) 236

Chlorinated camphene (7); (14);

Chlorinated diphenyl oxide (7); (14);

Chlorine (1) SD–80; (2) CC–53; (3) CIC–24; (4) 207; (7); (8) STL HIG; (10); (12); (13); (14); (15) 233

Chlorine dioxide (4) 525; (5); (7); (14); (15) 234

Chlorine trifluoride (2) CC–2; (7); (8) STL; (14); (15) 235

Chlorobenzene (3) CIC–25; (5); (7); (10); (12); (13); (14); (15) 245

Chlorobromo methane (7); (14); (15) 247

Chlorodiphenyl (42% and 54% chlorine) (7); (14); (15) 252

Chloroform (1) SD–89; (3) 26; (7); (8) STL; (10); (11) 67; (12); (14); (15) 253

Chlorohydrin crude (3) CIC–27; (10); (12); (14)

Dicyclopentadiene (10); (14); (15) 367

Dieldrin (7) skin; (14); (15) 368

Diethanolamine (3) CIC–34; (10); (11) 70; (12); (14); (15) 369

Diethylamine (1) SD–97; (2) CC–27; (5); (7); (8) STL, HIG; (10); (14); (15) 373

Diethylbenzene (10); (12); (14);

Diethylene glycol (10); (12); (14); (15) 377

Diethylene triamine (1) SD–46; (2) CC–72; (3) CIC–35; (10); (11) 71; (12); (13); (14); (15) 379

Difluorodibromomethane (7); (14);

Diglycidyl ether (7); (8) STL; (14); (15) 387

Diisobutyl ketone (5); (7); (8) STL; (10); (14); (15) 389

Diisobutylene (10); (14);

Dimethylamine (3) CIC–37; (7); (10); (12); (13); (14); (15) 399

Dimethylaniline (7); (14); (15) 402

Dimethyl-1,2-dibromo-2,2-dichloroethyl phosphate (Dibrom) (R) (7); (14); (15) 409

Dimethyl ether (2) CC–61; (14); (15) 665

Dimethyl formamide (5); (7); (14); (15) 410

1,1-Dimethylhydrazine (5); (7) skin; (14); (15) 1090

Dimethyl sulfate (1) SD–19; (2) CC–55; (7); (8) HIG; (11) 72; (14); (15) 415

o,m,p-Dinitrobenzene (5); (7); (14); (5); (15) 419–421

Dinitro-o-Cresol (7); (14); (15) 422

2,4-Dinitrophenol (5); (14); (15) 424

Dinitrotoluene (1) SD–93; (7) skin; (8) HIG; (14); (15) 426

Dioxane (5); (7) skin; (14); (15) 429

Dipentene (10); (14); (15) 430

Dipropylene glycol (10); (14);

Dipropylene glycol methyl ether (7) skin (14); (15) 435

Di-sec-Octyl phthalate (di-2-ethylhexyl phthalate) (7); (14); (15) 428

Dry ice (4) 397; (6) P–2; (14);

Endrin (R) (7) skin; (14); (15) 439

Epichlorohydrin (2) CC–29; (3) CIC–38; (5); (7); (10); (12); (14); (15) 440

EPN (R) (7) skin; (8) HIG; (14); (15) 441

Epoxy resin systems (4) 533; (5); (8) HIG; (14); (15) 444

Ethanolamine (5); (7); (14); (15) 449

2-Ethanoxyethyl acetate (cellosolve acetate) (7) skin; (14); (15) 450

2-Ethoxyethanol (cellosolve) (7) skin; (10); (12); (14); (15) 225

Ethyl acetate (1) SD–51; (2) CC–18; (5); (7); (8) STL HIG; (10); (11) 89; (12); (13); (14); (15) 953

Ethyl acrylate (1) SD–79; (2) CC–85; (3) CIC–39; (5); (7) skin; (8) STL; (10); (11) 73; (12); (13); (14); (15) 455

Ethyl alcohol (ethanol) (2) CC–70; (4) 391; (5); (7); (10); (11) 7, 84; (12); (13); (14); (15) 448

Ethylamine (7); (11) 83; (14); (15) 456

Ethyl benzene (5); (7); (8) STL HIG; (10); (12); (13); (14); (15) 459

Ethyl bromide (7); (8) HIG; (14); (15) 145

Ethyl chloride (1) SD–50; (2) CC–24; (3) CIC–40; (5); (7); (10); (12); (13); (14); (15) 464

Ethyl ether (diethyl ether or diethyl oxide) (1) SD–29; (2) CC–16; (3) CIC–46; (4) 396; (5); (7); (8) HIG; (10); (12); (13); (14); (15) 478

Ethyl formate (7); (14); (15) 481

2-Ethyl hexanol (10); (12); (14); (15) 482

Hexone (MIBK) (7); (8) STL; (13); (14); (15) 180

sec-Hexylacetate (7); (14); (15) 557

Hexylene glycol (10); (12); (14)

Hydrazine (anhydrous) (2) CC–7 & CC–9; (8); STL; (13); (14); (15) 561

Hydrochloric acid (Hydrogen chloride; aqueous and anhydrous) (1) SD–39; (2) CC–82; (3) CIC–51 and 53; (5); (7); (8) STL HIG; (9); (10); (11) 2; (12); (13); (14); (15) 566

Hydrocyanic acid (1) SD–67; (5); (7); (8) STL HIG; (14); (15) 567

Hydrofluoric acid (hydrogen fluoride, anhydrous and aqueous) (1) SD–25; (2) CC–42; (3) CIC 52 and 54; (4) 459; (5); (7); (8) HIG; (9); (10); (12); (13); (14); (15) 568

Hydrogen gas (6) G–5; (12); (14); (15) 569

Hydrogen liquid (2) CC–12; (12); (14);

Hydrogen bromide (7); (14); (15) 565

Hydrogen peroxide (1) SD–53; (2) CC–14; (5); (7); (13); (14); (15) 570

Hydrogen selenide (5); (7); (14); (15) 572

Hydrogen sulfide (1) SD–36; (4) 284; (5); (7); (8) STL HIG; (9); (13); (14); (15) 573

Hydroquinone (7); (14); (15) 574

Iodine (4) 457; (7); (8) STL HIG; (14); (15) 583

Iron oxide fume (7); (8) HIG; (14);

Isoamyl alcohol (7); (14); (15) 58

Isobutyl acetate (10); (14); (15) 158

Isobutyl alcohol (10); (14); (15) 162

Isobutyraldehyde (3) CIC–56; (10); (14); (15) 190

Isocyanates (TDI and MDI) (1) SD–73; (4) 489; (7); (14); (15) 1045

Isodecaldehyde (10); (12); (14);

Isooctanol (10); (14); (15) 764

Isooctylaldehyde (10); (14); (15) 320

Isoprene (3) CIC–57; (10); (12); (14); (15) 588

Isopropanol (2) CC–71; (7); (10); (12); (13); (14); (15) 872

Isopropyl acetate (10); (12); (14); (15) 869

Isopropylamine (1) SD–72; (2) CC–56; (7); (11) 85; (14); (15) 874

Isopropylbenzene (cumene) (5); (12); (14); (15) 296

Isopropyl ether (2) CC–58; (7); (10); (14); (15) 883

Isopropyl glycidyl ether (7); (14); (15) 886

Jet Fuel, JP–3 JP–4 & JP–5 (10); (11) 17; (14); (15) 589

Kerosene (10); (11) 18; (14); (15) 590

Ketene (7); (14); (15) 591

Latex (11) 19; (14); (15) 596

Lead (4) 433; (5); (7); (8) HIG; (14); (15) 599

Lead arsenate (5); (7); (8) HIG; (14); (15) 601

Lead oxides (1) SD–64; (5); (14); (15) 604

Lime (4) 24; (13); (14); (15) 210

Lindane (R) (7) skin; (14); (15) 607

Lithium hydride (7); (13); (14); (15) 613

LPG (liquefied petroleum gas) (7); (11) 20; (14); (15) 608

Magnesium (4) 426; (5); (7); (13); (14); (15) 616

Malathion (R) (7); (14); (15) 622

Maleic anhydride (1) SD–88; (13); (14); (15) 624

Manganese (4) 306; (7); (8) STL HIG; (14); (15) 625

Mercury (4) 203; (5); (7); (8) HIG; (14); (15) 630

Nickel carbonyl   (5); (7); (8) HIG; (14); (15) 717

Nickel, metal and soluble   (7); (14);

Nicotine   (7) skin; (8) HIG; (14); (15) 720

Nitrate-nitrite salt baths   (4) 270; (14); (15) 722

Nitric acid   (1) SD–5; (2) CC–47; (5); (7); (8) STL HIG; (12); (13); (14); (15) 723

Nitric acid (red fuming)   (2) CC–3; (14); (15) 734

p-Nitroaniline   (1) SD–94; (7); (13); (14); (15) 728

Nitrobenzene   (1) SD–21; (2) CC–79; (5); (7); (13); (14); (15) 729

p-Nitrochlorobenzene   (7); (14); (15) 259

Nitroethane   (5); (7); (14); (15) 731

Nitrogen dioxide   (7); (8) HIG; (12); (14); (15) 734

Nitrogen fertilizer solutions   (11) 23; (15) 735

Nitrogen liquid   (2) CC–6; (14);

Nitrogen tetroxide   (2) CC–1; (12); (13); (14); (15) 734

Nitrogen trifluoride   (7); (14); (15) 736

Nitroglycerin   (5); (7); (14);

Nitromethane*   (5); (7); (13); (14); (15) 738

Nitropropane   (5); (7); (10); (13); (14); (15) 743

2-Nitropropane   (7); (10); (13); (14); (15) 744

n-Nitrosodimethylamine   (dimethylnitrosoamine)   (7); (14); (15) 747

Nitrotoluene   (7); (14); (15) 753–5

Nitrous oxide   (4) 206; (7); (10); (12); (14);

Nonyl phenol   (10); (12); (14); (15) 756

Octane   (7); (14); (15) 761

Oil, lubricating   (11) 27; (15) 767

Oil mist mineral   (7); (8) HIG; (15) 768

Oil vegetable   (11) 44; (15) 772

Oleum   (2) CC–68; (3) ClC–67; (10); (12); (14);

o-Dichlorobenzene   (1) SD–54; (7); (14); (15) 349

Osmium tetroxide   (7); (14); (15) 775

Oxalic acid   (4) 406; (7); (14); (15) 776

Oxygen   (4) 472; (6) G–4; (8) HIG; (12); (14); (15) 778

Oxygen difluoride   (7); (9); (12); (14); (15) 779

Oxygen liquid   (2) CC–13; (4) 283; (14);

Ozone   (5); (7); (8) STL HIG; (14); (15) 780

Paint   (11) 28; (15) 781

Paraformaldehyde   (1) SD–6; (14); (15) 783

Parathion   (5); (7); (8) HIG; (13); (14); (15) 785

Pentaborane   (1) SD–84; (2) CC–5; (4) 508; (5); (7); (14); (15) 786

Pentachloronaphthalene   (7); (14);

Pentachlorophenol   (5); (7); (14);

n-Pentane   (5); (7); (10); (14); (15) 790

Isopentane   (10); (12); (14); (15) 791

Perchloric acid   (1) SD–11; (4) 311; (13); (14); (15) 796

Perchloroethylene (tetrachloroethylene)   (1) SD–24; (5); (7); (8) STL HIG; (10); (11) 78; (12); (14); (15) 1009

Perchloromethyl mercaptan   (7); (14); (15) 797

---

*Nitromethane has detonated in bulk when subjected to shock and/or heat (see NFPA Hazardous Chemicals Data 49–121, 1966, and also Bureau of Mines Publications).

Perchloryl fluoride (2) CC–4; (7); (14); (15) 798

Petroleum ether (7); (10); (14); (15) 799

Phenol (1) SD–4; (2) CC–48; (3) CIC–68; (4) 405; (5); (7) skin; (8); (10); (11) 29; (12); (13); (14); (15) 803

p-Phenylene diamine (7); (14); (15) 806

Phenyl glycidyl ether (PGE) (7); (14); (15) 442

Phenyl hydrazine (7) skin; (14); (15) 810

Phosdrin (Mevinphos) (R) (7) skin; (14); (15) 816

Phosgene (1) SD–95; (5); (7); (8) STL HIG; (14); (15) 818

Phosphine (5); (7); (14);

Phosphoric acid (1) SD–70; (3) CIC–69; (5); (7); (10); (11)4; (12); (14); (15) 820

Phosphoric anhydride (1) SD–28; (5); (14); (15) 821

Phosphorus, red (1) SD–16; (4) 282; (10); (14); (15) 823

Phosphorus, white or yellow in water (2) CC–37; (3) CIC–70; (7); (8) HIG; (12); (13); (14); (15) 822

Phosphorus oxychloride (1) SD–26; (2) CC–39; (14); (15) 824

Phosphorous pentachloride (7); (14); (15) 825

Phosphorus pentasulfide (1) SD–71; (2) CC–41; (7); (13); (14); (15) 826

Phosphorus trichloride (1) SD–27; (2) CC–40; (13); (14); (15) 829

Phthalic anhydride (1) SD–61; (5); (7); (13); (14); (15) 830

Picric acid (4) 351A; (7); (13); (14); (15) 833

Platinum (soluble salts) (7); (14); (15) 839

Polonium-210 (5); (14);

Polyvinyl acetate emulsion (11) 79; (14); (15) 841

Propane, commercial (7); (10); (12); (14); (15) 859

β-Propiolactone (7); (10); (12); (14); (15) 560

Propionaldehyde (10); (12); (13); (14); (15) 865

Propionic acid (3) CC–71; (10); (12); (13); (14); (15) 866

i-Propyl acetate (12); (13); (14); (15) 869

n-Propyl acetate (7); (10); (12); (13); (14); (15) 870

i-Propyl alcohol (5); (8) STL; (12); (13); (14); (15) 872

n-Propyl nitrate (7); (13); (14); (15) 887

Propylene (1) SD–59; (10); (12); (13); (14); (15) 879

Propylene dichloride (5); (7); (11) 80; (12); (14); (15) 363

Propylene glycol (10); (11) 45; (12); (13); (14); (15) 861

Propylenimine (7) skin; (14); (15) 882

Propylene oxide (3) CIC–72; (5); (7); (10); (12); (13); (14); (15) 443

Propargyl bromide (propyne, 3-bromo) Has detonated in mixtures with other chemicals.*

Pyrethrum (7); (14); (15) 890

Pyridine (4) 310A; (5); (7); (10); (13); (14); (15) 891

Quinone (5); (7); (14); (15) 898

Radon and daughters (5); (14);

*For reference to a similar compound, see "Attenuation of the Explosive Decomposition of Dicetylene by Inert Diluents," F. B. Moshkovich, R. Ya. Mushii and V. P. Kostyuk, KHIM PROM., **41** (2), 137–139 (1965). All double and triple bond compounds must be suspect until data is available to the contrary.

APPENDIX 3

HAZARDOUS REACTIONS

The following is a list of 265 explosive situations with literature references for further study. Very careful reading of this material is recommended for anyone responsible for the operation of a chemical laboratory, pilot plant, or manufacturing facility. It can help prevent unexpected explosions either because the exact counterpart of a proposed set-up has been implicated in an explosion or because a set-up you are about to go ahead with is analogous to one which has been so implicated. Either of these situations should give pause for thought, re-evaluation, or at the very least, the application of safety shields and other devices to mitigate the effects of a possible explosion.

The list below is divided into the following categories: *explosions involving*

1. acids
2. acid anhydrides
3. acid halides
4. alcohols
5. aldehydes
6. amides
7. amines
8. azo and diazo compounds
9. esters
10. ethers
11. halogen compounds
12. hydrocarbons
13. ketones
14. nitrates
15. nitriles
16. nitro compounds
17. nitroso compounds
18. peracids and perchlorates
19. peroxides
20. phenols
21. sulfonyl chlorides
22. unsaturated compounds
23. miscellaneous causes

Cross references are indicated numerically as follows: 4–2, this would indicate that the cross reference refers to category #4, item 2.

1. Explosions Involving Acids

1. *Cyanoacetic acid,* see 4–2.
2. *Caro's acid* (H_2SO_5).[1] Peroxymonosulfuric acid is prepared by reacting chloro-

[1]C.A. **49**, 14325d: J. O. Edwards (Brown Univ. Providence, R.I.) *Chem. Eng. News* **33**, 3336 (1955).

sulfonic acid and 90% H_2O_2 at low temperatures. One sample was stored overnight at 0°C, then removed to a test tube rack. In ten minutes it exploded.

3. *Chloric acid* $(HClO_3)$.[2] The recorded explosions of $HClO_3$ have been due to the formation of explosive compounds with Bi, Sb, NH_3, and organic substances.

4. *3,5-Dinitro-4-hydroxybenzenearsonic Acid,* see 16–7.

5. *Caro's acid* (H_2SO_5).[3] An explosion which occurred during the wet washing of a polymer sample was caused by the inadvertent addition of acetone to a solution thought to have contained Caro's acid. Explosive reactions of Caro's acid with alcohols and spontaneous explosions without the addition of organic material have been reported.

6. *Formic Acid and 2-furfuryl alcohol polymerization,* see 4–6.

7. *Chromic acid* (CrO_3) *in a container.*[4] A container with 50 kg CrO_3 exploded when laid on the ground. The container may have been contaminated with an oxidizable substance.

8. *Picric acid dust,* see 16–17.

9. *Picric acid crystal* see 16–18.

10. *Hydrocyanic acid (HCN).*[5] 100 g of 95–96% HCN in a glass-stoppered bottle was stored in a small case not exposed to sunlight. After two months' storage, it exploded with extreme violence.

11. *Storage of cleaning solution.*[6] A screw-cap acid bottle filled with the standard H_2O_4-dichromate cleaning solution exploded. Best guess is that the cleaning solution had been used and that dissolved organic matter was present, building up pressure.

12. *Etching reagent.*[7] A well- known etching reagent for nickel has been known to explode from 1 1/2 to 6 hrs after it is mixed. The composition of this reagent is equal parts of acetone, concentrated nitric acid, and 75% acetic acid.

13. *5-Ethyl-2-methyl pyridine and nitric acid.*[8] These materials were mixed in a small autoclave and heated and stirred for 40 min. The emergency vent was opened due to a sudden pressure rise and a violent explosion occurred 90 sec. later.

2. Explosions Involving Acid Anhydrides

1. *Purification of acetic anhydride.*[9] 10 g chromium trioxide was added to 500 ml of reagent grade acetic anhydride in a fractionating column to remove oxidizable aldehydes. After the initial heat of reaction subsided, heat was applied and at 30°C a detonation occurred.

2. *Maleic anhydride.*[10] An exothermic decomposition of maleic anhydride initiated by caustic soda resulted in an explosion that killed two men. The anhydride was in an 800-gal vessel and the caustic leaked in from a connecting pipe to another process unit.

3. *Maleic anhydride and tertiary amines form dangerous mix.*[11] Basic tertiary

[2]C.A. **46**, 2805c: V. Majer, *Chemie (Prague)* **3**, 90–1 (1948).
[3]Private communication.
[4]*C.A.* **31**, 4010⁴: Goertz, *Arbeitsshutz* **1935**, 323; *Chim. Ind.* **36**, 511.
[5]*C.A.* **6**, 2748⁸ G. Salomone. Pinerolo, *Gazz. Chim. Ital.* **42**, 1, 617–22.
[6]*Chem. Eng. News*, May 2, 1955, p. 1844.
[7]*Chem. Eng. News*, Oct. 24, 1960, p. 46.
[8]*Chem. Eng. News* **30**, 3348 (1952).
[9]Merck Safety Report, July 1960.
[10]Merck Safety Report, June 1960.
[11]*Chem. Eng. News*, Feb. 20, 1964, p. 41.

amines catalyze the decomposition of molten maleic anhydride. For example, 0.1% of pyridine in maleic anhydride is enough to cause an exothermic decomposition at 185°C with rapid evolution of gas. In one experiment, flask contents hit the ceiling.

4. *Acetic anhydride and CrO₃.*[12] When ½ mole quantities of these reagents were mixed in a flask, the mixture became warm during 45 min. and on being moved, exploded with considerable violence.

3. Explosions Involving Acid Halides

1. *o-Nitrobenzoyl chloride,* see 16–11.
2. *o-Nitrophenylacetyl chloride (O₂NC₆H₄CH₂COCl) in air.*[13] Two explosions with the compound took place. In both instances, *o*-nitrophenylacetic acid was suspended in chloroform, and thionyl chloride added. The mixture was refluxed for two to three hours and solvent removed under vacuum. In one case the solvent-free residue decomposed violently. In the other case, the residue exploded as soon as the solvent was evaporated.
3. *Anisyl chloride (CH₃OC₆H₄COCl).*[14] A 5-lb bottle containing anisyl chloride on a laboratory shelf exploded during the night. After several weeks of storage in a desiccator at room temperature, a 200-g bottle of commercial anisyl chloride exploded.

4. Explosions Involving Alcohols

1. *Cumene hydroperoxide,* see 19–9.
2. *Esterification of cyanoacetic acid with furfuryl alcohol (C₄H₃O·CH₂OH +NCCH₂ COOH→ NCCH₂COOCH₂·OC₄H₃).*[15] Cyanoacetic acid was reacted with furfuryl alcohol in an attempt to form furfuryl cyanoacetate. The reagents were mixed and transferred to a 1-liter flask. Three to four minutes after the agitator was turned on and heat applied, the explosion occurred.
3. *β-Phenylhydroxylamine hydrochloride,* see 7–3.
4. *Phenol-formaldehyde polymerization,* see 5–3.
5. *Ethylene oxide, alcohol, and mercaptan mixture,* see 10–9.
6. *Polymerization of furfuryl alcohol.*[16] When mixed with approximately equimolar amounts of formic acid (98–100%) at about 30°C, 2-furfuryl alcohol may polymerize with explosive violence.
7. *Reduction of hexahydroxyplatinate.*[17] Two attempts at preparing diacetoplatinum by the reduction of a solution of hexahydroxyplatinate in glacial acetic acid resulted in explosions that completely shattered the glass flasks.
8. *Furfuryl alcohol and formic acid.*[18] During an attempt to prepare furfuryl formate from furfuryl alcohol and concentrated formic acid, an explosion occurred.

5. Explosions Involving Aldehydes

1. *Formaldehyde and nitrogen dioxide.*[19] The slow reaction between NO₂ and HCHO becomes explosive in the region of 180°C.

[12]*Chem. Ind. (London)* **1964** (23), 973.
[13]S. Hayao (Miles Laboratories, Elkhart, Ind.). *Chem. Eng. News,* March 30, 1964, p. 39.
[14]*Chem. Eng. News* **38** (34), 40 (1960); **38** (43), 5 (1960).
[15]M.C.A. Case History No. 858.
[16]*C.A.* **34,** 4571⁸.
[17]*Chem. Ind.* **7,** 306 (1966).
[18]*Chem. Eng. News* **18,** 72 (1940).
[19]*C.A.* **44,** 25g; F. H. Pollard and P. Woodward. *Trans. Faraday Soc.* **45,** 767–70 (1949),

2. *Butadiene-crotonaldehyde,* see 22–7.
3. *Phenol-formaldehyde polymerization.*[20] Three employees were killed when an explosion occurred in a kettle in which a phenol-formaldehyde polymerization was taking place. The addition of the catalyst at the wrong time is a possible explanation.

6. Explosions Involving Amides

1. *Synthesis of dinitrosodimethyloxamide.*[21] An extremely violent explosion occurred while distilling CCl_4 from $Me(ON)NC(:O)_2$ at atmospheric pressure on a hot water bath.
2. $C_2H_5CN + COCH_2CH_2CONBr$, see 15–4.
3. *Dinitrosodimethyloxamide,* see 17–4.
4. *Perfluorosuccinamide and lithium aluminum hydride,*[22] see 11–21.
5. *Cyclohexyl-2-bromoethylcyanamide.*[23] A crude sample of this material was being distilled at 0.4 mm pressure and 160°C pot temperature when the column began to flood. The liquid darkened rapidly and soon exploded.

7. Explosions Involving Amines

1. *Methyldichloroamine (CH_3NCl_2).*[24] CH_3NCl_2 (2 g) exploded violently on addition of 1.6 g Na_2S. CH_3NCl_2 exploded when distilled over $CaOCl_2$.
2. *Aminoguanidine nitrate ($CH_6N_4 \cdot HNO_3$).*[25] Aminoguanidine nitrate (25g) in water solution exploded violently while being evaporated to dryness in vacuo on the steam bath.
3. *An amine salt ($C_6H_5NH \cdot OH \cdot HCl$).*[26] 700 g of β-phenylhydroxylamine hydrochloride had been sitting in a brown bottle on a shelf for two weeks when it exploded with considerable force.
4. *Benzoyl peroxide and dimethyl aniline,* see 19–12.
5. *Cyanoethylation of pyrrole.*[27] 1.5 moles of practical grade pyrrole was dissolved in 400 ml acrylonitrile. Three drops of Triton B (benzotrimethylammonium hydroxide) were added to the reaction mixture. After a slight rise in temperature and while an ice bath was being placed under the flask, a violent explosion occurred.
6. *β-Aminopropionitrile,* see 15–5.
7. *Preparation of diethylaminoethyl dinitrate,* see 14–7.
8. *Dimethylamine ampoules.*[28] Dimethylamine which is obtained in sealed, soda-glass ampoules containing 100 g boils at 7°C at atmospheric pressure. When a research worker was immersing an ampoule in an acetone-"Dricold" mixture in preparation for opening it, it exploded. One eye was seriously damaged. Several months later a similar explosion occurred.

[20]Quarterly Report of the National Fire Protection Assoc., Pittsfield, Mass., General Electric Co.
[21]*C.A.* **59**, 4967d; R. Preussmann, *Angew. Chem.* **75**, (13), 642 (1963).
[22]*Chem. Eng. News* **29:** 3042 (1951).
[23]Quarterly Safety Summary of the British Chemical Inc., Safety Council, April–June, 1964.
[24]*C.A.* **54**, 17887b; Biul. Wojshowej, *Aked. Tech. Im. J. Dabrowskiego (Warsaw)* **8** (48), 75–9 (1959).
[25]*C.A.* **51**, 10061b; H. Koopman (Central Research Lab. N.V. Philips-Roxane, Weesp, Neth.) *Chem. Weekblad.* **53**, 97–8 (1957).
[26]Merck Safety Report, July, 1961.
[27]Private communication.
[28]Quarterly Safety Summary, Assoc. of Britsh C.M. **31**, 124 (1960). *J. Roy. Inst. Chem.* **84**, 451 (1960).

9. *Guanidine nitrate.*[29] Guanidine nitrate as prepared from ammonium thiocyanate $NH_4SCN + Pb(NO_3)_2 + 2NH_3 \rightarrow PbS + CN_3H_5 \cdot HNO_3 + NH_4NO_3$ (H. Gockel, *C.A.* **29,** *65754)* demolished an autoclave built for 50 atm.

10. *Autoclave.*[30] While students were engaged in the manufacture of $PhNH_2$ and $PhNH_3Cl$ by autoclaving at 240–260°C for 24 hr, an explosion, killed one and seriously injured another graduate student.

11. *Maleic anhydride and tertiary amines form dangerous mixtures,* see 2–3.

12. *N-Chloroethylenimine,* see 11–43.

13. *p-Bromodimethylaniline,* see 11–44.

14. *Preparation of an oxime with hydroxylamine hydrochloride.*[31] In the preparation of an oxime,, pyridine, sodium acetate, and hydroxylamine·HCl were heated in a stainless steel autoclave. At 90°C the reaction temperature suddenly increased very rapidly causing the 5000-psi rupture disc to fail. The reaction had previously been run successfully under identical conditions at one-tenth the scale in a glass-lined autoclave.

15. *A decomposition occurred in the preparation of an oxime from an aldehyde.*[32] The reaction mixture of aldehyde, hydroxylamine·HCl, pyridine, and ethanol was diluted with water, extracted with ether, and the ether solution washed and dried. It was then concentrated by vacuum distillation at a bath temperature of 70–80°C. Near the completion of the distillation a violent decomposition occurred.

16. *Hydroxylamine·HCl mother liquors.*[33] In the manufacture of hydroxylamine·HCl, mother liquors have decomposed violently on several occasions when concentrated by distillation.

17. *Hydrolyzed oxime ($C_3H_8CNOH + HOH \xrightarrow{H+} C_3H_8CO + H_2NOH$).*[34] After methyl ethyl ketoxime had been hydrolyzed with sulfuric acid, the methyl ethyl ketone was distilled off at 12 mm and a bath temperature of 110–115°C. Three minutes after the bath was removed and vacuum released an explosive decomposition took place.

18. *Tri-n-bromomelamine and allyl alcohol.*[35] This mixture exploded fifteen minutes after mixing at room temperature.

19. *Phenylhydroxylamine·HCl.*[36] In an attempt to stabilize some phenylhydroxylamine for safekeeping, 700 g were made into the hydrochloride but after it had been in the laboratory about two weeks, it exploded with considerable force.

8. Explosions Involving Azo and Diazo Compounds

1. *Reaction of diazocylopentadiene with triphenyl phosphine.*[37] Diazocyclopentadiene, $C_5H_9N_2$, should be handled cautiously since during one preparation a violent explosion took place after distillation.

[29]*C.A.* **30,** 2004[8]; C. Schopf and H. Klapproth, *Angew. Chem.* **49,** 23 (1936).
[30]*C.A.* **17,** 471[2]; Anon. *Chem. Met. Eng.* **27,** 1044 (1922).
[31]Private communication.
[32]Private communication.
[33]Private communication.
[34]Private communication.
[35]*Chem. Eng. News* **30,** 1916 (1952).
[36]Wallace, 1966.
[37]*C.A.* **53,** 10088f; F. Ranairez and S. Levy (Columbia Univ.), *J. Org. Chem.* **23,** 2036–7 (1958).

2. *Preparation of diazoacetonitrile,* see 15–2.
3. *Zinc benzenediazonium chloride.*[38] Precipitated Zn benzenediazonium chloride dried by washing with dry acetone unexpectedly exploded after 15 hr of storage in a vacuum desiccator.
4. *Diazomethane ($H_2NCO \cdot NNO \cdot CH_3$ + KOH → CH_2N_2).*[39] Diazomethane was being prepared by portionwise additions of nitrosomethyl urea to a flask containing 40% KOH and methylene chloride. Immediately before or after the fourth addition a loud detonation occurred.
5. *Diazomethane precursor when heated above the melting point.*[40] 1-Methyl-3-nitro-1-nitrosoguanidine should not be heated in closed systems.
6. *α,α'-Azodiisobutyronitrile.*[41] An explosion occurred when a solution of $(NCCMe_2N:)_2$ in acetone was concentrated in a glass-lined steam-jacketed vessel.
7. *Diazotized sulfanilic acid.*[42] An employee took some diazotized sulfanilic acid from a powder bottle in the refrigerator with a metal spatula. An explosion occurred and the powder bottle was shattered.
8. *Diazomethane.*[43] The hazards associated with diazomethane are discussed.
9. *Ethyl azodicarboxylate ($C_2H_5O_2CN=NCO_2C_2H_5$).*[44] Following the directions in the reference, an explosion occurred while dismantling the equipment after distillation of the product had been completed.
10. *Aminotriazole.*[45] After the diazotisation of 3-amino-5-phenyl-1,2,4-triazole a solid was noticed in the reaction solution. This solid was filtered off and on transferring the solid to a filter paper, a violent explosion occurred.
11. *3-Pyridyldiazonium fluoroborate.*[46] After preparing this material the salt was spread out on aluminum foil to dry. A short time later the material detonated.

9. Explosions Involving Esters

1. *Esterification,* see 4–2.
2. *Lithium aluminum hydride decomposition.*[47] 10 g of a mixture of chlorinated organic compounds (major component was probably a chlorine-substituted tetrahydrofuran) was subjected to reductive dechlorination with 3 g LiAlH₄. After the reaction, ethyl acetate was added in small increments to decompose excess LiAlH₄. After a few drops of ethyl acetate had been added, a violent explosion occurred.
3. *Explosive polymerization reactions,* see 22–10.
4. *Ethyl azodicarboxylate,* see 8–9.

[38]*C.A.* **50**, 9021b; G. D. Muir (B.D.H. Lab. Chemicals Group, Poole, Dorset, England), *Chem. Ind.* **1956**, 58–59.
[39]Merck Safety Report, Oct. 1960.
[40]*C.A.* **47**, 11737b; J.N. Eisendrath, *Chem. Eng. News* **31**, 3016 (1953).
[41]*C.A.* **43**, 8682a; P. J. Carlisle, *Chem. Eng. News* **27**, 150 (1949).
[42]Quarterly Safety Summary Assoc. of British C.M. **30**, 120 (1959). *Chemierbeit,* 1959, Jahrgang XI, Nr. 9, 59.
[43]C.D. Gutsche, *Org. Reactions* **8**, 391–394 (1954).
[44]*Org. Synthesis, Coll. Vol.* **3**, 375, (1955).
[45]*Chem. Ind.* **1965**, (19), 812.
[46]*Chem Eng. News,* Oct. 16, 1967, p. 44.
[47]*Chem. Ind.* **14**, 432 (1957).

10. Explosions Involving Ethers

1. *Distillation of tetrahydrofuran.*[48] Tetrahydrofuran should be considered as a form of peroxides comparable to ether.

2. *Tetrahydrofuran can cause fire when used as solvent for LiAlH₄.*[49] Peroxides of tetrahydrofuran or their reaction products probably caused a vigorous reaction with $LiAlH_4$ and subsequent fire.

3. *Tetrahydrofuran.*[50] Mother liquors dissolved in tetrahydrofuran were being concentrated in a 100-ml flask over a steam cone. When about 50 ml remained in the flask, an explosion occurred.

4. *Isopropyl ether peroxide.*[51] A chemist suffered fatal injuries when a pint bottle of isopropyl ether at least three years old exploded as he was opening it.

5. *Dipropargyl ether (CH≡CCH₂)₂O).*[52] An explosion in a 50-gal stainless steel still occurred during the distillation of dipropargyl ether.

6. *Ether peroxide,* see 19–7.

7. *Lithium aluminum hydride not safe for drying methyl ethers.*[53] Explosions are attributed to CO_2 solubility in Me ethers. High concentration of peroxides were found present.

8. *2-Chloromethylfuran (CH:CH·CCL:CCH₃·O).*[54] A small sample (20 ml) of 2-chloromethylfuran had been made and distilled and allowed to stand over the weekend before use. During the week-end the material exploded.

9. *Ethylene oxide hazard (CH₂CH₂O).*[55] Violent explosions have occurred by the unexpected reaction of ethylene oxide with certain mercaptans and with an alcohol.

10. *Lithium aluminum hydride decomposition,* see 9–2.

11. *Ether.*[56] An operator lost his sight when 20 ml of ether exploded violently during distillation. It was estimated that the ether must have contained about 1.2 g of peroxide, calculated as H_2O_2.

12. *Ethers.*[57] All ethers studied form explosive peroxides on storage, but the rate of formation is variable. Danger occurs whenever an ether is dry distilled. Peroxides are measured by adding acidified KI·EtOH and titrating the I liberated; they may be destroyed by adding FeSO₄, NaHSO₃, acidified KI, or Na₂SO₃. The ether may be safely distilled with steam. Oxidation inhibitors such as NaHPh₂, α-and β-C₁₀H₇OH, and quinol when added to the ether retard development of peroxides.

13. *Diisopropyl ether.*[58] Two explosions are reported. In the first instance the explosion occurred as a flask of diisopropyl ether was being heated in a water bath and shaken gently. In the second instance the explosion occurred after practically all the ether had been distilled.

[48]*C.A.* **50**, 7097i; J. Schurz and H. Stubchen (Univ. Graz, Austria), *Angew. Chem.* 68, 182 (1956).
[49]C.A. **49**, 2073h; R. E. Moffett and B. D. Aspergren *Chem. Eng.* **32**, 4328 (1954).
[50]M.C.A. Case History 77.
[51]M.C.A. Case History 603.
[52]Merck Safety Report, Oct. 1958.
[53]*C.A.* **47**, 8371i; R. M. Adams (Callery Chem. Co., Callery, Pa.). *Chem. Eng. News* **31**, 2334 (1953).
[54]Quarterly Safety Summary, Assoc. of British C. M. **32**, 128 (1961).
[55]*C.A.* **37**, 2616; D.P. Meigs, *Chem. Eng. News* **20**, 1318 (1942).
[56]*C.A.* **32**, 9503⁴; J. Tandberg, *Tek. Trd. Repring,* **1938** (24), 4 pp.
[57]*C.A.* **31**, 4123³; E. C. Williams, *Chem. Ind.* **1936**, 580–581.
[58]*C.A.* **30**, 5416⁴; G. T. Morgan and R. H. Pickard, *Chem. Ind.* **1936**, 421–422.

14. *Diethyl ether and liquid air.*[59] A mixture of liquid air and diethyl ether exploded spontaneously.
15. *Diglyme and lithium aluminum hydride.*[60] During the distillation of diglyme (diethylene glycol dimethyl ether) from $LiAlH_4$ (used as drying agent) the flask exploded.

11. Explosions Involving Halogen Compounds

1. *Uncatalyzed addition of trichlorobromomethane to ethylene.*[61] Freshly distilled $BrCCl_3$ was shaken with C_2H_4 at 120°C and 50 atm initial pressure. After 1.5 hr a violent explosion occurred.
2. *Methyldichloroamine,* see 7–1.
3. *Chloroaziridine.*[62] After storage of 20 ml of 1-chloroaziridine in a flask for several months at 0°C, the stopper had frozen and the flask was shattered by dropping it into a disposal pit. A nitroglycerin-type explosion resulted.
4. *Peroxides of polyhalo compounds.*[63] Addition of Br to a mixture of chlorotrifluoroethylene and O_2 causes an explosion. Liquid $CF_2{:}CFCl$ and O_2 give CF_2ClCOF and chlorotrifluoroethylene peroxide, which explodes when heated.
5. *Perchloryl fluoride (FClO₃).*[64] While making $AcCF_2CO_2ET$, a mixture of $FClO_3$ and MeOH vapor exploded when the final portion of $NaOCH_3$ was added and came in contact with the $FClO_3$.
6. *Benzenesulfenyl chloride,* see 21–1.
7. *Nitrosyl chloride and acetone in the presence of platinum,* see 21–2.
8. *Zinc benzenediazonium chloride,* see 8–3.
9. *Condensation of 2–iodo-3,5-dinitrobiphenyl with sodium acetoacetic ester,* see 16–1.
10. *Perfluoroalkyl derivatives of sulfur.*[65] Electrolysis of $MeSO_3H$ in anhydrous HF produces OF_2 which explodes during the reaction.
11. *Explosion of a steel pressure vessel during a double decomposition reaction,* see 22–2.
12. *Chlorination.*[66] During the chlorination step of methyl parathion production, the chlorination exploded without warning, killing eight employees. Chlorine feed to the batch was controlled automatically by means of a temperature recorder connected to a thermocouple located in a thermowell in the chlorinator. Failure of the instrument allowed chlorine to be added at a rate too fast to control the reaction.
13. *Trichlorethylene in presence of strong alkali.*[67] Trichlorethylene, $CHCl{:}CCl_2$, reacts with strong alkalies such as caustic soda to form flammable and explosive mixtures.
14. *Synthesis of trichlorophenol salt (C₆H₂Cl₄ + NaOH → C₆H₂Cl₃ONa).*[68] The sodium

[59]M.C.A. Case History 616 (1960).
[60]*Chem. Ind.,* **1964** (1b), 665.
[61]*C.A.* **57**, 9638f.
[62]*C.A.* **53**, 3695b; A. F. Graefe, *Chem. Eng. News* **36** (43), 52 (1958).
[63]*C.A.* **53**, 12858h; R. N. Haszeldine and F. Nyman (Univ. Cambridge, England), *J. Chem. Soc.* **1959**, 1004–1090.
[64]*C.A.* **53**, 22954i; V. Papesch, *Chem. Eng. News* **37** (28), 60 (1959).
[65]*C.A.* **50**, 12808a; T. Gramstad and R. N. Haszeldine (Univ. Chem. Lab., Cambridge, Eng.), *J. Chem. Soc.* **1956,** 173–180.
[66]M.C.A. Case History No. 371.
[67]M.C.A. Case History No. 495.
[68]Merck Safety Report, May 1960.

salt of trichlorophenol is manufactured in an autoclave. The autoclave was charged with NaOH, methanol, and tetrachlorobenzene and the mixture heated. In the process of heating up, the temperature and pressure suddenly increased rapidly and an explosion occurred.

15. *Perchloric acid,* see 18–6.

16. *Reducing fluorinated compounds with LiAlH$_4$.*[69] Violent reactions are described involving two different fluoro compounds.

17. *Friedel-Crafts Mixture,* see 16–2.

18. *Halogenated oxime (ClCH$_2$CHO + H$_2$NOH·HCl → ClCH$_2$CH:NOH).*[70] Stirring 200 ml 50% ClCH$_2$CHO with 120 g H$_2$NOH·HCl for ½ hr, dissolving the oily layer in ether, and distilling the residue of the dried ether solution in vacuo gives ClCH$_2$CH:NOH. The distillation must not be carried out too far lest a violent explosion occur.

19. *Trichlorethylene.*[71] A violent explosion occurred when trichlorethylene CHCl:CCl$_2$ was being distilled from a vessel at atmospheric pressure.

20. *The polycondensation of benzyl chloride.*[72] During a distillation of technical-grade PhCH$_2$Cl, HCl was evolved. Therefore, air was bubbled through the solution to remove it. Suddenly an explosion occurred.

21. *Perfluorosuccinamide and LiAlH$_4$.*[73] The perfluorosuccinamide (38 g., 0.2 mol) was added to the ether (1 liter) solution of LiAlH$_4$ (42.5 g, 1.12 mol) under a nitrogen atmosphere. After the reaction occurred, hydrolysis of the mixture was attempted by the dropwise addition of water. When the second drop was added, a violent explosion resulted.

22. *Ammonium periodate (NH$_4$IO$_4$).*[74] A serious explosion occurred on scooping NH$_4$IO$_4$ from one container into another.

23. *Solid brominating agents.*[75] Tri-N-bromomelamine added to allyl alcohol exploded after 15 min at room temperature.

24. *Aromatic fluorine compound (3,4,6,2-F$_3$(O$_2$N)C$_6$HNH$_2$ + HCl + NaNO$_2$ → C$_6$HF$_2$N$_3$O$_3$).*[76] Diazotization of 5 g 3,4,6,2-F$_3$(O$_2$N)C$_6$HNH$_2$ in 10 ml concn. HCl with 3 g NaHO$_2$ in 5 ml H$_2$O gave C$_6$HF$_2$N$_3$O$_3$. The recrystallized compound exploded violently on impact.

25. *Aluminum chloride nitromethane catalyst.*[77] An explosion is reported of an autoclave containing AlCl$_3$·MeNO$_2$ complex and a gaseous olefin.

26. *Aluminum chloride.*[78] Owing to spontaneous decomposition upon long storage in closed containers, AlCl$_3$ often releases pressure when the container is opened.

27. *Decomposition of ethylene,* see 22–8.

[69]*C.A.* **49,** 9278h; W. Karo (Monomer-Polymer, Leominster, Mass.), *Chem. Eng. News* **33,** 1368 (1955).

[70]*C.A.* **47,** 1589g.

[71]*C.A.* **47,** 11737c; C. Brade (Zqickau, Ger.), *Chem. Tech.* (Berlin) **4,** 506–507 (1952).

[72]*C.A.* **47,** 12277b; F. Oehme (Kothen, Ger.), *Chem. Tech. (Berlin)* **4,** 404 (1952).

[73]*Chem Eng. News* **29,** 3042 (1951).

[74]*C.A.* **46,** 3279c; G. Frederick Smith (Univ. of Illinois, Urbana), *Chem. Eng. News* **29,** 1770 (1951).

[75]*C.A.* **46,** 7329i; J. A. Vona and P. C. Merker, *Chem. Eng. News* **30,** 1916 (1952).

[76]*C.A.* **45,** 7034a: G. C. Finger, F. H. Reed, D. M. Burness, D. M. Fort & R.R. Blough (Illinois State Geol. Survey, Urbana), *J. Am. Chem. Soc.* **73,** 145–9 (1951).

[77]*C.A.* **42,** 7045b: F. M. Cowen and O. Rorso (American Cynamid Co. Stamford, Conn.), *Chem. Eng. News* **26,** 2257 (1948).

[78]*C.A.* **41,** 6723d; P.V. Popov, *Zavodshaya Lab.* **13,** 127 (1946).

28. *Autoclave (CBrCl₃ + H₂C=CH₂ → BrCH₂CH₂CCl₃).*[79] A violent explosion (shattering the autoclave) occurred during the uncatalyzed addition of $CBrCl_3$ to ethylene.
29. *2-Chloromethylfuran,* see 10–8.
30. *Nitrogen trifluoride, NF₃, and tetrafluorohydrazine, N₂F₄.*[80] Several hundred grams of crude reaction mixture involving NF_3 and N_2F_4 had been collected in a small stainless steel cylinder. During the opening of valves to measure the cylinder's pressure, it exploded, killing one man and injuring another.
31. *Reaction of chlorinated rubber with zinc oxide.*[81] An undescribed exothermic reaction of chlorinated rubber with zinc oxide has been pegged as the cause of the explosion that leveled a manufacturing area.
32. *Aluminum and chlorinated solvents.*[82] Experiments have shown that aluminum powder in contact with methylchloride, carbon tetrachloride, or with a carbon tetrachloride-chloroform mixture is capable of exploding. Aluminum in contact with methyl chloride may form aluminum methyl which is spontaneously combustible.
33. *Tert-butyl hypochlorite ((CH₃)₃COCl).*[83] A sealed glass ampoule containing approximately 10 g *tert*-butyl hypochlorite exploded violently when it was exposed either to fluorescent or ordinary daylight at room temperature.
34. *Chlorination of alkyl isothioureas to prepare alkyl sulfonyl chlorides.*[84] Reports are given on explosions which occurred in the study of the chlorination of formamidmethiolacetic acid HCl (HO₂CCH₂SC(:NH)NH₂·HCl) and of *S*-ethylisothiourea sulfate.
35. *Lithium aluminum hydride decomposition,* see 9–2.
36. *3-Chlorocyclopentene explosion.*[85] 3-Chlorocyclopentene was prepared by a published procedure. A portion was used immediately and the remainder (35 g) was set aside to be disposed of later. The following day it exploded.
37. *Dinitrofluoroethane distillation,* see 16–13.
38. *Unstable nitroso chloride derivative,* see 17–3.
39. *Trichloroethylene and potassium nitrate.*[86] A batch of 3257 g boron, 9362 g KNO_3, 989 g laminac, and 500 g Cl₂C=CClH had been mixing for 5 min when an explosion occurred.
40. *Sodium methylate and chloroform.*[87] A chemist mixed 270 g $NaOCH_3$ with 1 liter of methanol and 200 g chloroform in a 2-liter flask. Very shortly thereafter the reaction began to boil and then a violent explosion occurred.
41. *Chloracetone.*[88] An explosion of chloracetone is described.

[79]Quarterly Safety Summary, Assoc. of British C.M. **33,** 131 (1962); Elsner and Saure, *Angew. Chem. Intern Ed.* **1962** (1), 218.
[80]M.C.A. Case History No. 683.
[81]*Chem. Eng. News,* Sept. 10, 1962, p. 79.
[82]Quarterly Safety Summary, Assoc. of British C.M., **34,** 134, (1963). *F.P.A. Journal* **59,** 110 (1963).
[83]Quarterly Safety Summary, Assoc. of British C.M. **34,** 134, (1963). *Chem. Eng. News* **40** (43), 62 (1962).
[84]*C.A.* **36,** 1179[8]; K. Folkers, A. Russell, and R. N. Bost, *J. Am. Chem. Soc.* **63,** 3530–2 (1941).
[85]Merck Safety Report, April 1962.
[86]M.C.A. Case History No. 745.
[87]M.C.A. Case History No. 693, March 1961.
[88]*C.A.* **25,** 4404[9]; C. F. H. Allen and W. A. L. Trig, *Ind. Eng. Chem., News Ed.* **9,** 184 (1931). G. E. Ewe, *Ibid.,* 229.

42. *Triethyl aluminum and carbn tetrachloride mixture.*[89] Mixtures of $Al_2Cl_2Et_3$ and CCl_4 were prepared under cooling. On warming to room temperature, a mild explosion occurred. A mixture of $AlEt_3$ and CCl_4 in a molecular proportion of 1:3 was prepared with ice cooling. After removing the ice cooling, a violent explosion occurred, even before room temperature was reached.

43. *N-Chloroethylenimine $(CH_2CH_2NH+NaOCl \rightarrow CH_2CH_2NCl)$.*[90] A sample of N-cholorethylenimine, prepared from ethylenimine and sodium hypochlorite in dry, oxygen-free ether [*JACS, 80,* 3939 (1958)] spontaneously exploded. The compound was distilled immediately after it was prepared, then allowed to stand for 10 days before redistillation. After redistillation, a 50-g portion was stored in an amber bottle at room temperature for three months when it suddenly exploded. A black residue indicated polymerization.

44. *p-Bromodimethylaniline $(BrC_6^{\cdot}H_4N(CH_3)_2)$.*[91] During a vacuum distillation of *p*-bromodimethylaniline the contents of the flask began to heat up and could not be controlled. The resulting explosion injured two chemists.

45. *Liquid chlorine and carbon bisulfide and iron.*[92] An employee was attempting to mix liquid chlorine and carbon bisulfide for use with infrared equipment to determine chlorine impurities. An explosion occurred when liquid chlorine was added to a 1700 ml metal cylinder containing 70 ml of carbon bisulfide.

46. *Sodium chlorate.*[93] Approximately 200 kg of sodium chlorate, stored in a basement vault for more than 20 yr was being removed for disposal when an explosion occurred that killed four employees.

47. *2,6-Dibromo-p-benzoquinone-4-p-benzoquinone-4-chlorimine.*[94] While drying thin-layer chromatograms with a hot-air dryer, a capped, 25-g bottle of 2,6-dibromo-*p*-benzoquinone-4-*p*-benzoquinone-4-chlorimine, 1–2 ft away on the bench exploded.

48. *p-Chlorophenyl Isocyanate.*[95] A violent explosion occurred in a laboratory during vacuum distillation of *p*-chlorophenyl isocyanate prepared by the Curtius reaction of *p*-chlorobenzoylazide.

49. *Experimental fluorination.*[96] During the preparation of perfluoropropyl hypofluorite, the addition of 160 g fluorine proceeded uneventfully over a 10-hr period. At the end of this period with no warning there was a sudden explosion.

50. *Chlorination of S-ethyl isothiourea sulfate.*[97] An explosion occurred during the chlorination of S-ethyl isothiourea sulfate and formamidine thiolacetic acid HCl.

51. *Sulfamic acid and chlorine.*[98] An explosion occurred when chlorine was being passed at room temperature into a reaction mixture which included sulfamic acid and water.

52. *Nitrogen trichloride.*[99] Nitrogen trichloride was prepared by bubbling chlorine into an aqueous solution of ammonium sulfate and 100 ml of di-*n*-butyl ether.

[89]Reinheckel, H. (German Academy of Science), *Agnew. Chem.* **75** (24), 1205 (1963).
[90]*Chem. Eng. News,* Feb. 24, 1964, p. 41.
[91]*Chem. Eng. News,* March 27, 1961, p. 37.
[92]Merck Safety Report, June 1964.
[93]*J. Chem. Educ.* **44,** 320 (1967).
[94]*Chem. Ind.* **37,** 1551 (1967).
[95]*Chem. Ind.* **38,** 1625 (1965).
[96]M.C.A. Case History No. 1045.
[97]*J. Am. Chem. Soc.* **63,** 3530–3532 (1941).
[98]Private communication.
[99]*Chem. Eng. News* **44,** (31), 46 (1966).

Then 50 ml of the ether-NCl_3 mixture (over ammonium sulfate solution) was stored in a refrigerator and 5 min later the mixture exploded violently.

53. *Methanesulfonyl chloride, MeSOCl.*[100] A sample of methanesulfonyl chloride in a sealed ampoule exploded after standing several months on the shelf.

54. *Sodium hypochlorite and oxalic acid.*[101] The two chemicals were placed in a beaker in preparation of a bleach solution. Just as water was first added, the mixture exploded.

55. *Calcium hypochlorite and sulfur.*[102] A mixture of equal parts of calcium hypochlorite and finely divided sulfur exploded when heated in a closed vessel.

12. Explosions Involving Hydrocarbons

1. *Hydrocarbon flash.*[103] A small laboratory centrifuge was being used to separate a slurry from a flammable hydrocarbon when a flash fire occurred. It is assumed that a spark from the centrifuge motor ignited the material.

2. *Tetranitromethane-hydrocarbon mixtures,* see 16–12.

13. Explosions Involving Ketones

1. *p-Urazine, (CO:(NH·NH)₂:CO) by-product.*[104] A chemist was preparing *p*-urazine. During a manipulation the material in the glass container, probably a nitrogen-containing by-product, exploded, shattering the vessel he was holding.

2. *Distillation of 2-acetyl-3-methylthiophenone.*[105] A vacuum distillation of 2-acetyl-3-methylthiophenone was being performed on a laboratory bench when suddenly it exploded. One of the injured graduate students died 20 days later.

3. *Brominated ketone ($C_6H_5COC(CH_3)_2$ + Br_2 →).*[106] To a solution of isobutyrophenone in CCl_4 was added bromine dropwise at 20–31°C. After bromine was all added and reaction was complete, the flask was packed in ice. After 15 min, the flask exploded with a sharp report.

4. *4-Methylcyclohexanone.*[107] Assuming that 4-methylcyclohexanone would be easier to oxidize than its corresponding alcohol and using a technique which had previously been successfully applied with 4-methylcyclohexanol led to an explosion. An inherent difference in the reactivity of nitric acid with cyclic ketones and cyclic alcohols was not emphasized in the literature. The ketone was added to nitric acid at a temperature of 69 to 77°C over 1 hr. It detonated at 76°C.

5. *Chloroacetone,* see 11–41.

6. *Etching reagent,* see 1–12.

14. Explosions Involving Nitrates

1. *A reaction of cellulose nitrate with butylamine, and the reaction product.*[108] Small amounts of butylamine used on cellulose nitrate reacted explosively.

2. *Aminoguanidine nitrate,* see 7–2.

[100]*J. Org. Chem.* **29** (4), 951–2 (1964).
[101]M.C.A. Case History 839 (1962).
[102]*Mellor* **2,** 254–62 (1946–1947).
[103]M.C.A. Case History 417.
[104]M.C.A. Case History 144.
[105]Merck Safety Report, Feb. 1960.
[106]Private communication, Parke, Davis & Co.
[107]*Chem. Eng. News* **37** (35), 48 (1959).
[108]*C.A.* **52,** 19121f; Ermo Kaila, *Paperi ja pun* **40,** 339–340 (1958) (in English).

3. *Nitrate salt.*[109] An explosion occurred following the rearrangement of an organic nitrate salt with 98% sulfuric acid in an agitated 500-gal glass-lined reactor equipped with a 2-in. free vent. About 2¾ hr following a 30-min aging period at 90°C the batch began to react vigorously and within a few moments a violent explosion occurred.

4. *Ammonium nitrate.*[110] Ammonium nitrate exploded while it was being evaporated in an open kettle.

5. *Complex salt.*[111] A small amount of thoroughly washed, dry trihydrazine nickel nitrate $Ni(NO_3) \cdot 3N_2H_4$, unexpectedly exploded about 10 min after exposure to the atmosphere.

6. *Decomposition of triethylammonium nitrate-dinitrogen tetroxide mixtures.*[112] In the preparation of triethylammonium nitrate (I) from triethylammonium chloride and dinitrogen tetroxide (II) a pale yellow solid is obtained by ether extraction at −50°C having the composition 1 mol (II) to 2 moles (I.) This compound exploded violently when ether was removed at room temperature.

7. *Preparation of diethylaminoethyl dinitrate.*[113] In the preparation of O_2NOCH_2-$CH_2NEt_2 \cdot HNO_3$ according to the method of Barbiere (*C.A.* **40**, 2110[6]), 23.4 g $HOCH_2CH_2NEt_2$ is stirred into 75.6 g fuming HNO_3 in an ice bath and the excess HNO_3 evaporated in vacuo at a bath temperature of less than 40°C. When most of the HNO_3 is evaporated, the mixture invariably explodes.

8. *Acetyl nitrate ($CH_3CO_2NO_3$).*[114] After standing for 2–3 days a capped bottle which contained about 80–100g redistilled acetyl nitrate was to be opened under the hood. While applying every precaution, it detonated in the hands of the student with enormous explosive power. The student lost both hands.

9. *Sodium nitrite and sodium thiosulfate.*[115] When a sodium nitrite and thiosulfate mixture was heated to dryness, a violent explosion occurred.

10. *Mercuric nitrate.*[116] An explosion occurred in the use of mercuric nitrate for determining sulfur in Ball's reaction.

11. *Potassium nitrate and titanium disulfide.*[117] A mixture of titanium disulfide and potassium nitrate detonated when heated.

15. Explosions Involving Nitriles

1. *Oxidation by dinitrogen tetroxide in the presence of indium.*[118] Indium (5 g) was immersed in 20 ml of an acetonitrile-N_2O_4 mixture. A slow steady reaction occurred. After 2 days at room temperature, much of the original tetroxide was consumed, and a further 10 ml of acetonitrile-N_2O_4 mixture was added. On shaking to mix, a violent detonation occurred.

[109]M.C.A. Case History No. 526.
[110]Merck Safety Report, Feb. 1957.
[111]*C.A.* **49**, 6607g; H. Ellern and D. E. Olander (Universal Match Corp,, St. Louis, Mo.), *J. Chem. Educ.* **32**, 24 (1955).
[112]*C.A.* **48**, 3691f; C. C. Addison and N. Hodge (Univ. of Nottingham, England), *Chem. Ind.* **1953**, 1315.
[113]*C.A.* **46**, 7522h; J. Fakstorp and J. Christiansen (Pharmacia, Copenhagen), *Acta Chem. Scand.* **5**, 968–969 (1951).
[114]*Angew. Chem.* **67**, (5), 157 (1955).
[115]Mellor **10**, 501 (1946–1947).
[116]*Chem. Eng. News* **26**, 3300 (1948).
[117]Mellor **7**, 91 (1946–1947).
[118]*C.A.* **53**, 1715d; C. C. Addison, J. C. Sheldon, and B. C. Smith (Univ. of Nottingham, England), *Chem. Ind.* (London) **1958**, 1004–1005.

2. *Preparation of diazoacetonitrile.*[119] It is important that the nitrile be used only in dilute solution; it is highly explosive when concentrated.
3. *Cyanoethylation of pyrrole,* see 7–5.
4. *Propionitrile and N-bromosuccinimide ($C_2H_5CN + COCH_2CH_2CONBr \rightarrow$).*[120] 300 g propionitrile and 534 g *N*-bromosuccinimide were refluxed at 105°C all day and overnight. Two boiling chips were in the flask and heat was supplied by an infrared lamp. The next morning the flask exploded.
5. *β-Aminopropionitrile.*[121] A bottle of distilled *β*-aminopropionitrile, which contained no polymerization inhibitors (diphenylamine or hydroquinone), suddenly exploded after standing on a shelf for several months.
6. *α,α′-Azodiisobutyronitrile,* see 8–6.
7. *Acetyl nitrate ($CH_3CO_2NO_2$).*[122] In an experiment on nitration of pyrocatecholsulfonic acid ethylene ether with $AcNO_3$ in 20% oleum, there resulted a violent explosion, probably of $AcNO_3$.
8. *Mercury oxycyanide ($Hg_2CN)_2O$).*[123] Several instances are cited where explosions have occurred in handling or manipulating this substance. Rubbing the material is a frequent cause of the explosions.
9. *Iminodipropionitrile.*[124] Acrylonitrile was reacted with aqueous ammonia solution to give a mixture of *β*-aminopropionitrile, iminodipropionitrile, and the tertiary trinitrile. The mixed products were distilled to give a fraction corresponding to substantially pure iminodipropionitrile which was put into two bottles fitted with bakelite screw caps. After 18 mo storage, a violent explosion occurred which completely shattered both bottles.
10. *Phosphorus tricyanide ($3AgCN + PCl_3 \rightarrow P(CN)_3 + 3AgCl$).*[125] Phosphorus tricyanide may be prepared by the reaction of silver cyanide with phosphorus trichloride, the final product to be purified by vacuum sublimation at about 100°C. On three occasions explosions have occurred during the final sublimation.
11. *Acetone cyanhydrin (($CH_3)_2COHCN$).*[126] During addition of sulfuric acid to a vat of acetone cyanhydrin, pressure produced by rapid reaction ruptured the vessel explosively.
12. *Glycolonitrile polymerization.*[127] Two weeks after glycolonitrile was distilled, it started to polymerize, generating enough heat and pressure to explode the bottle.
13. *Acrylonitrile and bromine.*[128] Bromine was added dropwise to acrylonitrile while the reaction to mixture was alternately allowed to warm to room temperature and cool to 0–5°C. After about one-half of the bromine had been added, the temperature rose rapidly to 70° C and before the ice bath could be put under the flask it exploded.

[119]*C.A.* **51,** 718g; D. D. Phillips and W. C. Champion (Cornell Univ., Ithaca, N.Y,), *J, Am, Chem. Soc.* **78,** 5452 (1956).
[120]Private communication, Upjohn Co.
[121]*Merck Safety Report,* Oct. 1953.
[122]*C.A.* **38,** 2627³; E.A.M.F. Dahmen and P.M. Heertjes, *Chem, Weekblad* **39,** 447–448 (1942).
[123]*C.A.* **16,** 2010³ and *C.A.* **11,** 300⁷; E. Merck, *Pharm. Ztg.* **67,** 284 (1922).
[124]Quarterly Safety Summary, Assoc. of British C.M. **38,** No. 152 (1967).
[125]*Chem. Ind.* **38,** 1593 (1967).
[126]Occupancy Fire Record FR 57–5:5 (1957).
[127]*Chem. Eng. News,* Nov. 28, 1966, p. 50.
[128]Merck Safety Report, Aug. 1966.

16. Explosions Involving Nitro Compounds

1. *Condensation of 2-iodo-3,5-dinitrobiphenyl with sodium acetoacetic ester.*[129] The condensation of 2-halo-3,5-dinitrobiphenyl (I) with sodium acetoacetic ester should be carried out with only 5–6 g of (I) per batch, for larger amounts lead to explosions.

2. *Friedel-Crafts mixture.*[130] $AlCl_3$ added to nitrobenzene containing 5% phenol rose in temperature and exploded violently.

3. *Diazomethane precursor,* see 8–5.

4. *2,4-Dinitrobenzenesulfenyl chloride, see* 21–3.

5. *Distilling 2,4-dinitrochlorobenzene.*[131] An explosion occurred while distilling 2,4-dinitrochlorobenzene residue at 1mm pressure.

6. *Trinitro compounds.*[132] Explosions were encountered during distillations of both $(O_2N)_3CH$ and $(O_2N)_3CCH_2OH$.

7. *3,5-Dinitro-4-hydroxybenzenearsonic acid.*[133] $4,3,5,1\text{-}OHC_6H_2(NO_2)_2AsO_3H_2$ and its metallic salts may be as explosive as its close relative, picric acid. When a wet cake of the acid was heated an explosion occurred which was accompanied by liberation of As or AsH_3.

8. *Nitrotoluene $(O_2NC_6H_4CH_3)$.*[134] A charge of p-nitrotoluene was being dissolved in 93% H_2SO_4 in a steam-jacketed kettle. Failure in the temperature controller allowed a rise in temperature to 135°C and a violent explosion followed.

9. *Tetranitromethane $((CH_3CO)_2) + HNO_3 \rightarrow (NO_2)_4C)$.*[135] A violent explosion occurred in a laboratory in which tetranitromethane was being made by adding acetic anhydride to anhydrous nitric acid in a cooled, steel mixing tank.

10. *Nitrobenzene and nitric acid.*[136] An explosion claimed fifteen lives and injured 200. It has been attributed to detonation of a mixture of nitrobenzene, nitric acid, and water.

11. *o-Nitrobenzoyl chloride.*[137] An explosion of o-nitrobenzoyl chloride $(O_2NC_6H_4\text{-}COCl)$ is reported. It is considered dangerous to heat any quantity of the material above 100°C.

12. *Tetranitromethane-hydrocarbon mixtures.*[138] Ten students were killed and twenty severely wounded by an explosion of 10 g of a tetranitromethane-toluene mixture. Tetranitromethane forms with hydrocarbons, such as benzene and toluene, the most brisant, destructive explosive mixture possible to prepare today. It is always formed as a troublesome by-product in the nitration of aromatic hydrocarbons.

13. *Dinitrofluoroethane distillation.*[139] A fractional distillation at reduced pressure of dinitrofluoroethane was being carried out when an explosion occurred killing one man and injuring two others.

[129]*C.A.* **50**, 10688c; S. H. Zahur and I. K. Kacker (Central Labs. Sci, Ind, Research, Hyderabad-Deccan), *J. Indian Chem. Soc.* **32**, 491 (1955).
[130]*C.A.* **48**, 1684c; Anon., *Chem. Eng. News* **31**, 4915 (1953).
[131]*C.A.* **46**, 3279e; B. D. Halpern, *Chem. Eng. News* **29**, 2666 (1951).
[132]*C.A.* **45**, 4642f; N. S. Marans and R. P. Zelinski (DePaul Univ., Chicago), *J. Am. Chem. Soc.* **72**, 5329–5330 (1950).
[133]*C.A.* **43**, 2437a; M. A. Phillips, Chem. Ind. **1947**, 61.
[134]*C.A.* **43**, 8681a; J. K. Hunt, *Chem. Eng. News* **27**, 2504 (1949).
[135]Quarterly of the National Fire Protection Assoc.
[136]*Chem. Eng. News,* Oct. 17, 1960, p. 39; *Chem. Processing,* Jan. 1961, p. 122.
[137]*C.A.* **40**, 21295; J. Am. Soc. **68**, 344–5 (1946)
[138]*C.A.* **36**, 43394; A. Stettbacher, *Tech. Ind. Schweiz Chem. Ztg.* **24**, 265–71 (1941).
[139]M.C.A. Case History No. 784, Feb. 1962.

14. *Trichloroethylene and potassium nitrate,* see 11–39.
15. *Catalytic reduction of nitroanisole.*[140] The explosion of an autoclave in the catalytic reduction of 400 g nitroanisole according to Brown, Etzel, and Henke (*C.A.* **22,** 1891) is reported.
16. *Sodium dinitrophenate ($Na^+(NO_2)_2C_6H_3O$).*[141] An explosion occurred during the manufacture of sodium dinitrophenate, killing two people.
17. *Picric Acid (($NO_2)_3C_6H_2OH$).*[142] An explosion at a picric acid works was caused by picric acid in the form of dust. No detonating body was present.
18. *Picric acid (($NO_2)_3C_6H_2OH$).*[143] An explosion occurred during the grinding of dry picric acid crystals and was as severe as a detonation. It is believed to have been an explosion of picric acid dust mixed with air.
19. *Action of sodium ethylate on tetranitromethane.*[144] NaOEt dissolved in alcohol was added portionwise to $(NO_2)_4C$ in a cold-water bath. After the reaction, the Na salt separated and the mixture was allowed to stand for some time in the cold-water bath. On addition of another small portion of NaOEt, there was an instant, violent explosion.
20. *Beaker containing lead styphnate (lead 2,4,6-trinitroresorcinate).*[145] An employee was removing a beaker of lead styphnate from a laboratory oven. He grasped the beaker with his left hand and, as he turned he apparently bumped the beaker on the side or bottom of the oven opening and a detonation occurred.
21. *o-Nitrophenylacetyl chloride* see 3–2.
22. *o-Nitrophenylacetyl chloride.*[146] In two instances wherein o-nitrophenylacetic acid was suspended in chloroform and thionyl chloride added and the mixture refluxed for 2–3 hr and solvent removed under vacuum, explosions occurred. In one case, the solvent-free residue decomposed violently. In the other case, the residue exploded as soon as the solvent was evaporated.
23. *o-Nitrobenzoyl chloride.*[147] During an attempted distillation of o-nitrobenzoyl chloride it detonated violently.
24. *Sodium o-nitrothiophenoxide.*[148] A violent and unexpected explosion occurred during the preparation of sodium o-nitrothiophenoxide from o-nitrothiophenol and sodium methoxide in methanol.
25. *Hydrazine nitroformate.*[149] Two violent decompositions into nitrogen followed by explosions occurred during the manufacture of hydrazine nitroformate.
26. *Chloronitrotoluenes.*[150] An explosion resulted when the feed stream into a tank of mixed chloronitrotoluenes became inadvertently contaminated with caustic.
27. *Dipotassium nitroacetate and water.*[151] Dipotassium nitroacetate exploded when dry salt was moistened with a little water.

[140]*C.A.* **25,** 3838[7]; T. S. Carswell, *J. Am. Chem. Soc.* **53,** 2417–2418 (1931).
[141]*C.A.* **21,** 323[5]; Anon., 241 *Chem. Age* (London) **15,** 522 (1926).
[142]*C.A.* **9,** 716[7]; Anon., *Chem. Trade J.* **55,** 542 (1914).
[143]*C.A.* **9,** 965[8]; A. Copper Key, *Home Offic. Rept.* **211,** 9 (1914).
[144]*C.A.* **7,** 3973[7]; A. K. Macbeth, *Ber.* **46,** 2537–2538.
[145]M.C.A. Case History No. 957.
[146]*Chem. Eng. News,* March 30, 1964, p. 39.
[147]*J. Am. Chem. Soc.* **68,** 344 (1946).
[148]*Chem. Ind.* **6,** 257 (1966).
[149]M.C.A. Case History No. 1010.
[150]M.C.A. Case History No. 907, 1963.
[151]*Chem. Eng. News* **27,** 1473.

28. *Ethylene and nitromethane-aluminum chloride catalyst.*[152] A mixture of ethylene with nitromethane-aluminum chloride catalyst in an autoclave exploded at a temperature below 40°C.

29. *2,6,-di-t-Butyl-4-nitrophenol.*[153] Two grams of this nitro compound exploded violently after warming on a steam bath for 2–3 min.

30. *p-Nitrotoluene ($NO_2C_6H_4CH_3$).*[154] *p-Nitrotoluene and sulfuric acid mixture exploded at 80°C.*

31. *Sodium nitromethane ($NaCH_2NO_2$).*[155] Sodium or potassium salts of nitromethane exploded when the dry salt was moistened with a little water.

17. Explosions Involving Nitroso Compounds

1. *Synthesis of dinitrosodimethyloxamide,* see 6–1.
2. *Diazomethane precursor,* see 8–5.
3. *Unstable nitroso chloride derivative.*[156] The nitroso chloride of α-methylstyrene was prepared by treating a mixture of the olefin and amyl nitrate with concn. HCl in the usual manner. After isolating and drying the product, it was put in a wide-mouth screw-capped bottle. The following morning the material decomposed with a loud report, filling the room with white fumes.
4. *Dinitroso dimethyloxamide.*[157] In the preparation of *N,N'-dinitroso-N,N'-dimethyloxamide*, CCl_4, the solvent used to introduce the nitroso group, was not evaporated in vacuo, but under atmospheric pressure on a hot water bath. Toward the end of the distillation, a violent explosion occurred.

18. Explosions Involving Peracids and Perchlorates

1. *Perchlorate.*[158] Nine grams of *cis*-dichlorobis(1,2-diaminopropane) chromium (III) perchlorate in 190 ml of 71% $HClO_4$ exploded while being stirred at 22°C.
2. *Tropylium perchlorate.*[159] Agitation of 80 g of tropylium perchlorate with a stirring rod caused a violent detonation.
3. *Silver perchlorate ($AgClO_4$).*[160] Approximately 17 g of AgClO, filter cake exploded as it was being pulverized in a mortar.
4. *Spontaneous combustion of nitrosyl perchlorate ($NOClO_4$).*[161] Decomposition of nitrosyl perchlorate, $NOClO_4$, begins just below 100°C. Above 100°C (115–120°C) low-order explosions occur.
5. *Perchlorate.*[162] A 10-g mixture of an inorganic perchlorate, a glycol, and an organic polymer was heated to 265–270°C in a test tube inserted in a nichrome wire-heated silicone oil bath. The reaction was completed and about 5 g of product had been withdrawn when the explosion occurred.
6. *Perchloric acid.*[163] A mixture of boron trichloride absorbed in dioxane and

[152]*Chem. Eng. News* **26**, 2257 (1948).
[153]*ASESB Expl. Rept.* **24** (1961).
[154]*Chem. Eng. News* **27**, 2504.
[155]*Chem. Eng. News* **28**, 1473 (1950).
[156]M.C.A. Case History No. 747.
[157]R. Preussman (U. of Frieburg I. Br.), *Angew. Chem.* **75** (13), 642 (1963).
[158]*C.A.* **59**, 11177b; Anon., *Chem. Eng. News* **41** (27), 47 (1963).
[159]*C.A.* **57**, 10091a; P. G. Ferrini and A. Marder (Ciba Aktiengesellschaft, Basel, Switz.), *Angew. Chem.* **74**, 488–489 (1962).
[160]*C.A.* **51**, 10061g; F. Hein, *Chem. Tech. (Berlin)* **9,** 97 (1957).
[161]*C.A.* **50**, 10409f; H. Gerding and W. F. Haak, *Chem. Weekblad* **52**, 282–283 (1956).
[162]M.C.A. Case History No. 464.
[163]Merck Safety Report, Feb. 1955.

water was being evaporated in a beaker on a hot plate. Nitric acid was added three times followed by a treatment with 70% perchloric acid, and the boiling continued for 30 min. While the chemist was at lunch, a violent explosion occurred.

7. *Performic acid.*[164] A graduate student was preparing 5 ml of 90% performic acid when the receiving flask exploded tearing off his right hand. The danger period had been thought to be over by the time the detonation occurred.

8. *Hydrazine nickel (II) perchlorate: an extremely dangerous substance* $(Ni(ClO_4)_2 + N_2H_4 \rightarrow Ni(N_2H_4)_2 \cdot (ClO_4)_2)$.[165] In an attempt to prepare $(Ni(N_2H_4)_2 \cdot (ClO_4)_2$, 0.2 mole $Ni(ClO_4)_2$, 0.17 mole N_2H_4, and 150 ml H_2O were mixed. After 5 days a blue precipitate had formed and 150 ml water was added. When a glass stirring rod was introduced into the suspension, a violent explosion resulted.

9. *Sodium perchlorate.*[166] While concentrating a solution containing Na perchlorate, a few decigrams exploded during transfer of the crucible from a small gas flame to a water bath.

10. *Pyridinium perchlorate* $(C_6H_5N \cdot HClO_4)$.[167] Explosions occurred during purification of pyridine by the method of Arndt and Nachtwey (*C.A.* **20**, 2163).

11. *Purification of pyridine.*[168] Pyridine was purified with perchloric acid following the method of Arndt and Natchway (*C.A.* **20**, 2163). The liberated pyridine is distilled under diminished pressure on the water bath. An experienced chemist was half-way through the final distillation when an extremely violent explosion took place. Two men were very seriously injured, one losing an eye.

12. *Perchloric acid (HClO_4).*[169] An explosion of a tank containing 150 gal perchloric acid and 50–60 gal acetic anhydride in an electroplating plant in Los Angeles killed seventeen persons and wrecked 116 buildings. The explosion was attributed to contamination of the solution with some easily oxidizable material.

13. *Magnesium perchlorate.*[170] Anhydrous $Mg(ClO_4)_2$, used in drying unsaturated hydrocarbons, exploded from unknown cause on heating to 220°C.

14. *Instability of silver perchlorate.*[171] A case of the explosion of $AgClO_4$ crystallized from benzene is reported.

15. *Phosphonium perchlorate* $(2PH_3 \cdot 3HClO_4)$.[172] Serious accidents resulted from the violent explosions of phosphonium perchlorate in spite of every precaution.

16. *Cobalt (III) pentamine hypophosphite perchlorate detonation.*[173] A detonation occurred in a sintered filter funnel containing $[Co(NH_3)_5(H_2PO_2)](ClO_4)_2$. This complex had been precipitated from solution, then washed with ethyl alcohol and diethyl ether, and air dried on the sintered filter funnel.

[164]Merck Safety Report, Sept. 1952.
[165]*C.A.* **46**, 3280c; B. Maissen and G. Schwarzenbach (Univ. Zurich, Switz.), *Heh. Chem. Acta* **34**, 2084–2085 (1951).
[166]*C.A.* **45**, 5929h; H. Moureu and H. Munch (Lab. munic., Paris), *Arch. maladies profess. med. travail. et securite sociale* **12**, 157–61 (1951).
[167]*C.A.* **44**, 6129e; R. Kuhn and W. Otting (Kaiser Wilhelm-Inst., Heidelberg, Germany), *Chem. Ztg.* **74**, 139–140 (1950).
[168]*C.A.* **42**, 6537h; M. K. Zacherl, *Mikrochemie ver. Mikrochim. Acta* **33**, 387–388 (1948).
[169]*C.A.* **41**, 5308i; J. H. Kuney, *Chem. Eng. News* **25**, 1658–1659 (1947).
[170]*C.A.* **36**, 5349⁹; P. M. Keertjes and J. P. W. Houtman, *Chemweekblad,* **38**, 85 (1941).
[171]*C.A.* **35**, 897³; S. R. Brinkley, Jr., *J. Am. Chem, Soc,* **62**, 3524 (1940).
[172]*C.A.* **28**, 4325⁹; F. Fichter and Hermann, *Arni. Helv. Chim. Acta* **17**, 222–224 (1934).
[173]Division of Operational Safety, U.S. Atomic Energy Comm., Washington, D.C. 20545.

17. *Magnesium perchlorate.*[174] During vacuum distillation of dimethyl sulfoxide from magnesium perchlorate, an explosion occurred.
18. *Magnesium perchlorate.*[175] Drying gaseous ethylene oxide with magnesium perchlorate resulted in an explosion.
19. *Magnesium perchlorate.*[176] When magnesium perchlorate was used as a drying agent for unsaturated hydrocarbons, an explosion occurred.
20. *Nitrosyl perchlorate (NO·ClO$_4$).*[177] Nitrosyl perchlorate ignites and explodes with acetone, primary amines, and diethyl ether.
21. *Benzene II-cyclopentadienyl-iron perchlorate.*[178] This material recrystallized from alcohol detonated violently when touched with a spatula.

19. Explosions Involving Peroxides

1. *Organic peroxides in the laboratory.*[179] Proper and safe methods for handling, storage, and disposal of laboratory quantities of organic peroxides are discussed. The safety characteristics of some peroxides are tabulated from data derived from the specialized tests described. Methods are given for the detection and removal of residue or trace quantities of peroxides as well as precautionary measurements to be followed in their purification.
2. *Recrystallizing benzoyl peroxide.*[180] Explosions during the recrystallization of benzoyl peroxide have been reported. For purification, benzoyl peroxide should be dissolved in cold CHCl$_3$ and precipitated with cold MeOH.
3. *Cumene hydroperoxide.*[181] Cumene hydroperoxide, C$_6$H$_5$C(CH$_3$)$_2$OOH, at concentrations of 91 and 95% could not be detonated at room temperature but burned easily. It decomposed violently at a temperature of about 150° C.
4. *Nitrogen peroxide-cyclohexane mixture (NO$_2$ + C$_6$H$_{12}$ → Explosion).*[182] Through an error, liquid instead of gaseous nitrogen peroxide was fed into a nitration column containing hot cyclohexane. An explosion resulted.
5. *Melting point determination.*[183] The mineral oil used in a Thiele-Denis tube had been heated by a gas flame to 260°C. when an explosion occurred. The violence of the explosion indicated that it may have been caused by peroxides in the mineral oil.
6. *Oxidation of Thiodiglycol (HOCH$_2$CH$_2$)$_2$S+H$_2$O$_2$ → Sulfone Explosion).*[184] The chemist had undertaken to oxidize thiodiglycol with an excess of 30% hydrogen peroxide using acetone as a solvent. At the conclusion of the reaction, the acetone and the excess hydrogen peroxide were removed under vacuum on a steam bath. After about 15 min heating on the steam bath, a violent explosion occurred.
7. *Allyl ether peroxide.*[185] A distillation of allyl ether had been started by treat-

[174]*Chem. Eng. News* **43**, (37), 62 (1965).
[175]*N.S.C. Newsletter, Chem. Sect.*, Oct. 1959.
[176]*Chem Weekblad* **38**, 85 (1941).
[177]Hoffman and Zedtwitz, *Ann. Chem.* **42**, 2031 (1909).
[178]*J. Organometal. Chem.* **5** (3), 292 (1966).
[179]*C.A.* **59**, 2585c; D. C. Noller and D. J. Bolton (Wallace & Tiernan), *Anal. Chem* **35**, 887–893 (1963).
[180]*C.A.* **54**, 898i; A. Rieche and M. Schulz (Deut. Akad. Wiss., Berlin-Aldershof), *Chem. Tech. (Berlin)* **11**, 264 (1959).
[181]*C.A.* **51**, 718f; A. LeRoux, *Mem. poudres* **37**, 49–58 (1955).
[182]M.C.A. Case History No. 128.
[183]M.C.A. Case History No. 202.
[184]M.C.A. Case History No. 223.
[185]M.C.A. Case History No. 412.

ing the ether with caustic. After a two-week delay, the ether distillation was re-started and had progressed to the point where about 500 ml of ether remained in the flask when a violent explosion occurred.

8. *Isobutyryl peroxide (2(CH_3)$_2$CH·CO·Cl + Na$_2$O$_2$ → [(CH_3)$_2$CH·CO] $_2$O$_2$ + 2NaCl).*[186] Ten grams of isobutyryl peroxide were being prepared from isobutyryl chloride and sodium peroxide according to the procedure of Khatasch, Kane, and Brown [*JACS* **63,** 526 (1941)]. The reaction was completed and reaction mixture worked up. On nearing completion of the ether evaporation step, a detonation occurred.

9. *Cumene hydroperoxide (C_6H_5CH(CH$_3$)$_2$ + O$_2$ → C_6H_5(CH$_3$)$_2$OOH).*[187] Cumene was oxidized with pure oxygen to cumene hydroperoxide. The resulting 15% cumene hydroperoxide was concentrated to 35%. When this material had stood for 5 hours at 109°C it exploded.

10. *Methyl ethyl isobutyl hydroperoxide (CH$_3$C$_2$H$_5$C$_4$H$_9$COH + 90% H$_2$O$_2$ → CH$_3$C$_2$H$_5$C$_4$H$_9$COOH).*[188] Methyl ethyl isobutyl carbinol was treated with 90% H$_2$O$_2$ and a trace of H$_2$SO$_4$ at 0°C. While warming up to room temperature overnight, it exploded violently.

11. *Peroxide.*[189] A chemist had concentrated an aqueous mother liquor solution containing some steroid material which had previously undergone a peroxide oxidation and irradiation. The solution had been tested for peroxides with starch KI paper. The delayed slight color change was attributed to air oxidation. When the solution reached dryness, steam and vacuum were removed, and the flask detonated.

12. *The explosive decomposition of benzoyl peroxide by dimethylaniline [Bz$_2$O$_2$ + C_6H_5N(CH$_3$)$_2$].*[190] Finely ground Bz$_2$O$_2$ (12.1 g) is allowed to react by breaking an ampoule with 0.5 g PhNMe$_2$ in an autoclave giving 3.6 g CO$_2$.

13. *Accidents with peroxides.*[191] Vacuum distillation of tertiary butyl perbenzoate above 100°C resulted in an explosion. Ethylene ozonide is highly sensitive to shock; pouring from one vessel to another has led to violent explosion.

14. *t-Butyl perbenzoate.*[192] A liter flask was half-filled with *t*-butyl perbenzoate which was being distilled under high vacuum. The heating mantle temperature was 300°C and the vapor temperature was 50°C. After about 10% of the product had been distilled the distillation rate suddenly increased, the flask contents darkened in color, and the flask exploded with extreme violence.

15. *Reduction of benzoyl peroxide (Bz$_2$O$_2$ + LiAlH$_4$ →).*[193] An attempted reduction of benzoyl peroxide with LiAlH$_4$ resulted in an explosion.

16. *Acetyl peroxide [(CH$_3$CO)$_2$O$_2$].*[194] A 5-g portion of acetyl peroxide crystals in a bottle detonated while being carried from a storage refrigerator to the lab. The force of the explosion was sufficient to tear off both hands of the worker.

[186]M.C.A. Case History No. 579.
[187]M.C.A. Case History No. 906.
[188]Private communication.
[189]Merck Safety Report, April 1954.
[190]C.A. **48,** 10671g; L. Horner and C. Betzel (Univ. Frankfort/Main, Germany), *Chem. Ber.* **86,** 1071–72 (1953).
[191]C.A. **47,** 11737a; R. Criegee (Tech. Hochschule Korlsruke, Germany), *Angew. Chem.* **65,** 398–399 (1953).
[192]Private communications, The Upjohn Co.
[193]C.A. **45,** 8985; D. A. Sutton (Nat'l. Chem. Research Lab., Pretoria, S, Africa), *Chem, Ind.* **1951,** 272.
[194]C.A. **43,** 1949g; L. P. Kuhn, *Chem. Eng. News* **26,** 3197 (1948).

17. *Urea peroxide.*[195]A bottle of urea peroxide erupted, showering the laboratory with urea peroxide. The bottle had been standing in a cabinet for approximately four years.

18. *Interaction between ethyl peroxide and sulfur.*[196] If ethyl peroxide (Et_2O_2) and S are present in ether, the mixture becomes violently explosive.

19. *Benzoyl peroxide.*[197] A bottle containing 100 g Bz_2O_2 exploded spontaneously with total destruction of the bottle and contents.

20. *o-Azidobenzoyl peroxide detonated.*[198] A 2-g sample of recrystallized crystalline o-azidobenzoyl peroxide on a sintered glass funnel detonated "with extreme violence" when touched with a metal spatula. The crystalline peroxide is particularly treacherous because numerous experiments of a similar nature had been done before without incident.

21. *Ozonized products.*[199] After ozonization of about 1 g of isoprene in 50 ml of n-heptane at dry ice-acetone temperature, a violent explosion occurred. Use of dry ice temperatures probably permitted buildup of concentration of peroxides and ozonides.

22. *Benzoyl peroxide and ethylene.*[200] A reaction of ethylene, carbon tetrachloride, and benzoyl peroxide caused an explosion.

23. *Oxidation of tetralin by H_2O_2 in acetone.*[201] The preparation of 2-tetralone according to Treihs, et al., (*C.A.* **49**, 8216a) was accompanied by a violent explosion.

24. *1-Phenyl-2-methyl-2-propyl hydroperoxide.*[202] A violent explosion occurred when concentrated sulfuric acid was added to a mixture of dimethylbenzylcarbinol and 90% hydrogen peroxide.

20. Explosions Involving Phenols

1. *Friedel-Crafts mixture,* see 16–2.
2. *Lead trinitroresorcinolate, $(Pb(NO_2)_3C_6HO_2)$.*[203] 3 kg Pb trinitroresorcinolate detonated from unknown cause in the anteroom of a dry-house. Wet material in two adjacent drying rooms did not explode.
3. *Sodium dinitrophenate,* see 16–16.
4. *Picric acid dust,* see 16–17.
5. *Picric acid crystals,* see 16–18.

21. Explosions Involving Sulfonyl Chlorides

1. *Benzenesulfonyl chloride.*[204] Benzenesulfonyl chloride stood undisturbed for months in a glass container at room temperature, then exploded.
2. *Nitrosyl chloride and acetone in the presence of platinum.*[205] Nitrosyl chloride

[195]M.C.A. Case History No. 719, June 1961.
[196]*C.A.* **31**, 6009; H. F. Taylor, *Mem. Proc. Manchester Lit. and Phil. Soc.* **81**, 15–18 (1936–1937).
[197]*C.A.* **25**, 4127[8]; S.S. Nametkin and L. S. Kichkina, *J. Russ. Phys. Chem. Soc.* **62**, 2193–2194 (1930).
[198]*Chem. Eng. News,* Dec. 2, 1963.
[199]*Chem. Eng. News,* Jan. 16, 1956, p. 292.
[200]*Chem. Eng. News* **25**, 1866 (1947).
[201]*Angew. Chem.* **76** (16), 716 (1964).
[202]*Chem. Eng. News,* Oct. 9, 1967, p. 72.
[203]*C.A.* **26**, 5210[3]; *Anon, Z. ges. Schiess—Sprengstoffu.* **27**, 205 (1932).
[204]*C.A.* **52**, 8559h; Anon, *Chem. Eng. News* **35** (47), 57 (1957).
[205]*C.A.* **52**, 8559g; G. B. Kauffman (Fresno State Coll., Fresno, Calif.), *Chem. Eng. News* **35** (43), 60 (1957).

was distilled at room temperature and condensed into a heavy walled borosilicate glass tube containing a few g of Pt. The tube was sealed and the glass annealed. An explosion shattered the tube.

3. *2,4-Dinitrobenzenesulfonyl chloride.*[206] A detonation occurred when the solvent, *sym*-tetrachloroethane, had been nearly removed from a reaction mixture in which the chlorinolysis of 15 g of 2,4-dinitrophenyl disulfide had been effected.

4. *Acetylenic p-toluenesulfonate,* see 22–5.

5. *Allyl benzenesulfonate.*[207] A benzene extract of allyl benzenesulfonate was prepared from allyl alcohol and benzenesufonyl chloride in the presence of aqueous sodium hydroxide. Under vacuum distillation two fractions came off, then the temperature rose to 135°C, when the residue darkened and exploded.

22. Explosions Involving Unsaturated Compounds

1. *The addition of uncatalyzed trichlorobromomethane to ethylene,* see 11–1.

2. *Propargyl chloride.*[208] An explosion occurred during an attempt to carry out the reaction $HC{\equiv}CCH_2Cl(I) + 2NH_3 \rightarrow HC{\equiv}CCH_2NH_2 + NH_4Cl$ in a 1–liter steel vessel with an initial NH_3 pressure of 8 atm and 150 g of I.

3. *1'-Pentol fractionation ($HC{\equiv}C{-}C(CH_3){=}CH{-}CH_3 + NaOH \rightarrow$ explosion).*[209] During the fractionation of 1'-pentol in a 50-gal stainless steel still under high vacuum, the material spontaneously decomposed with explosive violence. The detonation instantly killed three men. It is believed that a small amount of caustic which had been used occasionally to clean the equipment may have been retained in a line filter and accidentally drawn into the still along with the pentol.

4. *Trichloroethylene,* see 11–19.

5. *Acetylenic p-toluenesulfonate ($HC{\equiv}CCH_2CH_2OH + p{-}CH_3C_6H_4SO_2Cl \rightarrow p{-}CH_3{-}C_6H_4SO_2OCH_2CH_2C{\equiv}CH$).*[210] $HC{\equiv}CCH_2CH_2OH$ (14 g) in 20.5 g C_5H_5N, added slowly (30 min) to 42.6 g p-$CH_3C_6H_4SO_2Cl$ (temperature below 25°C) and kept 18 hr at 20°C gives 84% 3-butyn-1-yl p-toluenesulfonate; attempted distillation at 0.5 mm resulted in explosive decomposition.

6. *Diphenyltetraacetylene*[211] Diphenyltetraacetylene was stable for at least 13 mo. at room temperature in the dark. When placed on a metallic plate, it decomposed explosively with much soot.

7. *Butadiene-crotonaldehyde ($CH_2{=}CHCH{=}CH_2 + CH_3CH{=}CHCHO \rightarrow$).*[212] The Diels-Alder reaction between butadiene and crotonaldehyde under pressure is a logical approach to the preparation of numerous cyclic aldehydes, alcohols, and hydrocarbons. A destructive explosion, including a secondary gas explosion, occurred in carrying out this reaction.

8. *Ethene (C_2H_4).*[213] Workers have observed that C_2H_4 under certain conditions and especially in the presence of $AlCl_3$ as catalyst and finely divided Ni as pro-

[206]Private communication.
[207]*Chem. Eng. News* **28**, 3452 (1950).
[208]*C.A.* **50**, 14229i; E. Banik, *Chem. Ind. (Dusseldorf)* **8**, 45–46 (1956).
[209]M.C.A. Case History No. 363.
[210]*C.A.* **45**, 7053d; G. Eglinton and M. C. Whiting, *J. Chem. Soc.* **1950**, 3650–3656.
[211]*C.A.* **45**, 7082a: Masazumi Nakagawa (Osaka Univ.), *Kagaku No Ryoiki (J. Japan. Chem.)* **4**, 564–565.
[212]*C.A.* **42**, 6537d: K. W. Greenlee (Ohio State Univ., Columbus), *Chem. Eng. News* **26**, 1985 (1948).
[213]*C.A.* **41**, 6721h; H. I. Waterman, W. J. Hessels, and J. van Steenis (Tech. Univ., Delft, Holland), *J. Inst. Petroleum* **33**, 254–255 (1947).

moter may decompose vigorously or even explode into its elements. It has now been found that some solvents for $AlCl_3$ promote polymerization of C_2H_4. C_5H_{12}, CH_3Cl, and $MeNO_2$ have been involved in explosions of C_2H_4.

9. *Uncatalyzed addition of $CBrCl_3$ to ethylene*, see 11–28.

10. *Methyl methacrylate polymerization.*[214] The polymerization of methyl methacrylate ($CH_2=C(CH_3)\ CO_2CH_3$) under the influence of oxygen or peroxides, proceeds with a steady increase in velocity until finally, after an induction period of variable length, the monomer reacts violently.

11. *3-Chlorocyclopentene*, see 11–36.

12. *Chloroethynylation reactions.*[215] During the preparation of a cloroethynyl compound employing lithium chloroacetylide and liquid ammonia, some ammonia was lost by evaporation and a solid crust formed above the surface of the liquid. A violent explosion suddenly shattered the flask, injuring the chemist's face.

13. *4-Bromocyclopentene.*[216] During the preparation of 4-bromocyclopentene by the hydride reduction of cyclopentadiene dibromide, two explosions occurred.

14. *Acetylenic Grignard reagent.*[217] The reagent was prepared in 2-g samples by stirring a refluxing mixture of ethoxyacetylene and EtMgI in Et_2O. An attempt to prepare a larger sample failed, and as the agitator was stopped an explosion occurred.

15. *1-Iodo-2-ethoxy-3-butene.*[218] In the course of preparing this material via the method of Petrov [*C.A.* **44**, 1003d (1950)], the mixture exploded while ethanol was being distilled off under slight vacuum at 35°C.

16. *Ethylene sulfate.*[219] An ethylene sulfate mixture was being vacuum distilled, and when it reached the boiling point, the flask exploded.

23. Explosions due to Miscellaneous Causes

1. *Dimethyl sulfate.*[220] Concentrated NH_4OH and $(MeO)_2SO_2$ exploded on mixing.

2. *Synthesis of sodium aluminum hydride.*[221] $NaAlH_4$ is prepared from its elements by "seeding" it in tetrahydrofuran with heating and agitation. The reaction ends when hydrogen absorption stops, and at this point an explosion occurred.

3. *Phosphorous oxychloride.*[222] $POCl_3$, recovered from the preparation of diacid chloride of ferrrocene-l,l-dicarboxylic acid, exploded just after it was poured into a bottle and capped.

4. *Danger of explosion of Tollens' reagent.*[223] A compilation of references and discussion of the danger in allowing Tollens' reagent to stand.

[214]*C.A.* **36**, 4012': G. V. Schulz and F. Blaschke, *Z. Elektrochem.* **47**, 749–761 (1941).
[215]Quarterly Safety Summary of the British Chem. Industry Safety Council **36**, (142) (1965).
[216]Tetrahedron Letters **1964** (45–46), 3327–3328 (Eng.). *Nachr. Chem. Tech.* **12** (24), 488 (1964) (Ger.).
[217]*Chem. Eng. News* **44** (8), 40 (1966).
[218]*Chem. Eng. News*, Oct. 17, 1966, p. 6.
[219]Merck Safety Report, Jan. 1966.
[220]*C.A.* **58**, 12155c; H. Lindlar (F. Hoffman-LaRoche Co. A–G., Basel, Switz.), *Angew. Chem.* **75**, 297–298 (1963).
[221]*C.A.* **56**, 5010c; Anon., *Chem. Eng. News* **39** (40), 57 (1961).
[222]*C.A.* **56**, 14521a; Anon., *Chem. Eng. News* **40** (3), 55 (1962).
[223]*C.A.* **54**, 6129g; H. Waldmann (F. Hoffman-LaRoche and Co. Akt.-Ges., Basel, Switz.), *Chimia (Switz.)* **13**, 297–298 (1959).

5. *Thermal decomposition of some phosphorothioate insecticides.*[224] A pilot-plant batch of the methyl homolog of parathion exploded as a result of accidental overheating while the solvent was being removed by distillation.

6. *Boron triazide, lithium boroazide, and silicon tetrazide.*[225] The above three compounds and some of their intermediates are extremely sensitive and explosive.

7. *Methyl azide distillation ($2NaN_3 + (CH_3)_2SO_4 \rightarrow 2CH_3N_3 + Na_2SO_4$).*[226] The explosion occurred while the methyl azide was being distilled, seriously injuring the chemist.

8. *Sodium hydrosulfite ($Na_2S_2O_4$).*[227] Water leaked into a broken drum of sodium hydrosulfite causing an explosion.

9. *Platinum catalyst.*[228] Dry, used Adams-type catalyst was being screened to remove coarse particles prior to shipment for reclamation when the first bottle began to smoke. While adding a second bottle to the contents of the first bottle, a shattering explosion occurred.

10. *Tollens reagent.*[229] An unrinsed test tube which had contained Tollens reagent was picked up to admire the silvery deposit when it exploded.

11. *Explosion hazards of ammoniacal silver solutions.*[230] An explosive solution was prepared by precipitating Ag_2O from $AgNO_3$ solution with NaOH, washing the Ag_2O, dissolving it in NH_4OH, adding a few drops of $AgNO_3$ solution until a permanent precipitate was formed, and then centrifuging. On two occasions, this solution exploded after 10–14 days, once quite violently.

12. *Methyl azide.*[231] A serious explosion occurred in the condensation of methyl azide with dimethyl malonate in presence of Na methylate.

13. *Diallyl phosphite (($C_3H_5)_2POH$).*[232] To 117 g allyl alcohol was added 91.1 g freshly distilled PCl_3 over 2 hr in a CO_2 stream with ice cooling and stirring, the vessel evacuated and carefully warmed over 3–5 hr to expel HCl and RCl, and the product cautiously distilled in vacuo in a CO_2 stream. After some two-thirds of the material is distilled, explosions usually take place.

14. *Ethylene oxide.*[233] The explosion of an ethylene oxide tank was attributed to the accidental pumping of ammonia into the tank. The accident caused one death and extensive damage.

15. *Cyanates and thiocyanates of chromium.*[234] $CrO_2(NCO)_2$ in CCl_4 was concentrated to dryness under atmospheric pressure when an explosion occurred. $CrO_2+(SCN)_2$ in CCl_4 decomposes in the cold, explosively at higher temperatures.

[224]*C.A.* **50**, 4446d; J. B. McPherson, Jr. and G. A. Johnson (American Cyanamid Co., Stamford, Conn.), *J. Agr. Food Chem.* **4**, 42–49 (1956).
[225]*C.A.* **49**, 767g; E. Wiberg and H. Michaud (Univ. Munich, Germ.), *Z. Naturforsch* **9b**, 497–500 (1954).
[226]M.C.A. Case History No. 887.
[227]Merck Safety Report, Nov. 1900.
[228]Merck Safety Report, Dec. 1954.
[229]Merck Safety Report, May 1954.
[230]*C.A.* **47**, 4083f,; H. Vasbinder (Rotterdam, Netherlands), *Pharm. Weekblad* **87**, 861–865 (1952).
[231]*C.A.* **45**, 355e; C. Grundmann and H. Haldenwanger, *Angew. Chem.* **62A**, 410 (1950).
[232]*C.A.* **45**, 8970a; E. I. Shugurova and Gil'm Kamai (S. M. Kirov. Chem. Tech. Inst. Kazan), *Zhur. Obshchei Khim. (J. Gen. Chem.)* **21**, 658–662 (1951).
[233]*Chem. Eng. News* April 23, 1962, p. 21.
[234]*C.A.* **38**, 694[1]; G. S. Forbes and H. H. Anderson, *J. Am. Chem. Soc.* **65**, 2271–2274 (1943).

16. *Tollens' reagent.*[235] After testing samples with Tollens' reagent, the unemptied test tubes were allowed to sit over the weekend. Monday morning, upon picking up the test tubes to dispose of the contents, an explosion occurred. While working with ammoniacal silver ion solutions it is difficult *not* to form fulminating silver. Dry fulminating silver is extremely sensitive and is instantly and violently decomposed by the slightest disturbance.

17. *Ammoniacal silver solution.*[236] A violent explosion occurred on picking up a beaker containing a spent silvering bath, made from an ammoniacal silver solution plus sugar. The explosion is attributed to friction on crystals of fulminating silver, believed to be $AgNH_2$.

18. *Ammoniacal silver solution.*[237] An alkaline ammoniacal silver solution prepared exactly as described by B. Tollens [*Ber.* **15,** 1635–1639 (1882)] exploded violently within 1 hr when the containing vessel was moved, although no precipitate and the merest trace of film were present.

19. *Ammoniacal silver solution.*[238] In preparing ammoniacal silver solution grave explosion hazards are caused by attempts to dissolve precipitated AgOH in NH_4OH. The $AgNO_3$ would be dissolved in NH_4OH then added.

20. Explosions with benzazide ($C_6H_5COOC_2H_5 \rightarrow C_6H_5CONHNH_2 \rightarrow C_6H_5CO \cdot N_3 \rightarrow C_5H_5N:CO$).[239] Benzazide was prepared by the Gattermann-Wieland method from ethyl benzoate via the hydrazide and the phenyl cyanate from benzazide. Two explosions occurred causing severe eye injuries. The crude azide from crude hydrazide, heated in a thin-walled test tube exploded between 120 and 165°C.

21. *Mercuric azide.*[240]

$$NaN_3 + H_2SO_4 \rightarrow HN_3$$
$$HN_3 + HgO \rightarrow Hg(N_3)_2$$

The author considers the preparation of $Hg(N_3)_2$ as one of the most dangerous and treacherous of chemical operations. It presents one of the few examples of crystal tension. The $Hg(N_3)_2$ was prepared in hot water and stirred in order to get small crystals rather than long thin needle crystals which detonate at the slightest touch. When filtering off H_2O from the hot water containing $Hg(N_3)_2$, needle crystals formed at the bottom of the funnel. These were washed off with a gentle stream of hot water but they detonated when they came in contact with the bottom of the beaker.

22. *Ammoniacal silver oxide solutions.*[241] Three explosions are described in which concentrated ammonium hydroxide was added to Ag_2O. *Concentrated* ammonium hydroxide added to Ag_2O forms silver fulminate which is highly explosive. *Dilute* ammonia will not form explosive derivatives if all materials are discarded immediately after use.

23. *Liquid oxygen.*[242] The lecturer was demonstrating the ignition of powdered aluminum mixed with liquid oxygen when the mixture exploded. Seventeen persons were injured and one boy lost an eye. This experiment is described in a number of places as a lecture demonstration. It has been carried out suc-

[235]M.C.A. Case History No. 714, May 1961.
[236]*C.A.* **27,** 1176[8]; A. Caille and A. Masselin, *Bull. Soc. Ind. Ronen* **60,** 133–134 (1932).
[237]*C.A.* **27,** 3823[2]; W. Coltof, *Chem. Weekblad.* **29,** 737 (1932).
[238]*C.A.* **26,** 6140[1]; W.L.A. Warnier, *Chem. Weekblad* **29,** 249 (1932).
[239]*C.A.* **21,** 3463[9].
[240]*C.A.* **14,** 3531[7]; A. Stettbacher, *Schweiz. Chem. Ztg.* **27,** 273–274 (1920).
[241]*C.A.* **13,** 2449[3]; E. J. Witzemann, *J. Ind. Eng. Chem.* **11,** 893 (1919).
[242]*Chem. Eng. News,* June 17, 1957, p. 90.

cessfully hundreds of times but there have been a few explosions when the conditions were just right.

24. *Silver azide.*[243] A pinch of solid being scraped from a glass sintered-disk filter crucible exploded violently in Texaco's Beacon, New York, Laboratories. The explosion may have been caused by silver azide contaminating an otherwise innocent precipitate, according to John J. Mitchell.

25. *Potassium.*[244] A piece of potassium metal, about 2 cc in size, had been dried and then sliced with a stainless steel blade. After several cuts had been made, the piece of potassium suddenly exploded and inflamed. Potassium can form the peroxide, K_2O_3, or the superoxide, K_2O_4, even under mineral oil. These oxides may explode violently when handled or cut.

26. *Sodium aluminum hydride.*[245] A violent explosion occurred during the preparation of sodium aluminum hydride from sodium and aluminum in a medium of tetrahydrofuran.

27. *Methyl azide.*[246] During the preparation of methyl azide from reaction of dimethyl sulfate and sodium azide, a violent explosion occurred.

28. *Vinyl azide.*[247] A sample of vinyl azide in a distilling flask with a ground-glass joint detonated when the joint was rotated.

29. *Benzotriazole.*[248] A powerfully destructive explosion occurred during an attempt to distill benzotriazole at 160°C and 2mm Hg.

30. *Triallyl phosphate ($(CH_2{:}CHCH_2)_3PO_4$).*[249] An explosion occurred on distilling triallyl phosphate prepared from phosphorus oxychloride, allyl alcohol, and pyridine.

31. *Chromium trioxide-pyridine complex.*[250] A chemist was preparing a chromium trioxide-pyridine complex when it exploded.

32. *Pyridine perchromate.*[251] About 5 g of this material which had been standing around for some time exploded violently when touched with a stainless steel spatula.

33. *Sodium hydride-dimethyl sulfoxide (DMSO) mixture.*[252] 4.5 moles NaH was added in five portions to 18.4 moles DMSO at 70°C was stirred. As solution was complete, the temperature rose sharply and an explosion occurred. A similar explosion occurred at 50°C.

34. *2,6-Lutidine-N-oxide and POCl₃.*[253] A laboratory explosion resulted from an attempt to prepare a pyridine derivative of 2,6-lutidine-*N*-oxide using phosphorus oxychloride.

35. *Pyridine-N-oxide and phosphorus oxychloride.*[254] Phosphorus oxychloride was being added dropwise to pyridine-*N*-oxide with stirring but without external heating. The temperature rose steadily at 60–65°C, then rose rapidly and an explosion occurred.

36. *Ammonia pressure explosion.* Ammonia gas built up enough pressure to shatter

[243]*Chem. Eng. News,* Sept. 24, 1956, p. 4704.
[244]Quarterly Saftey Summary of the British Council **36,** (141) (1965).
[245]*Chem. Eng. News* **39** (40), 57 (1961).
[246]M.C.A. Case History No. 887, 1963.
[247]*J. Org. Chem.* **22,** 995 (1957).
[248]*Chem. Eng. News* **34,** 2450 (1956).
[249]*Chem. Eng. News* **28,** 3452 (1950).
[250]M.C.A. Case History No. 1284, May 1967.
[251]*J. Chem. Educ.* **43,** (2), 94 (1966).
[252]*Chem. Eng. News* **44,** (15), 45 (1966).
[253]*Chem. Eng. News,* Nov. 22, 1965, p. 40.
[254]Merck Safety Report, Sept. 1965.

a thickwalled glass pressure tube. The effect of the blast caused glass fragments to crack the safety-glass hood sash panel and the glass panel of the overhead exhaust hood lighting. Another pressure tube also exploded, this one containing liquid ammonia at room temperature.

37. *Tetrahydrofuran.*[255] Explosion occurred during distillation of tetrahydrofuran.

38. *t-Butyl hypochlorite.*[256] A hydrogen peroxide oxidation on a half molar scale of a cyclic 1,2-diketone was carried out in alkaline acetone solution. After standing at room temperature for 48 hr, the solution was diluted with water and extracted with diethyl ether. The aqueous phase was acidified to pH 4.5 with 3N HCl and extracted with ethyl acetate. The ethyl acetate extract was backwashed with a 4% solution of NaCl dried over anhydrous Na_2SO_4 and evaporated. When the original 8.0 liters of extract was evaporated under vacuum to about 1.0 liter, the evaporation was discontinued due to gassing. Platinum catalyst was added to the concentrate to decompose residual hydrogen peroxide. Catalyst was added once in the morning and again in the afternoon and the solution allowed to stand overnight at $RT°$. The following morning the catalyst was removed by filtration and the filtrate further evaporated. When there was about 200–300 cc of concentrate remaining there was evidence of vigorous gassing; the flask was removed from the vacuum still and set in the hood. Within a matter of seconds the contents of the flask exploded. Although three people were in close proximity of the hood at the time of the explosion, only a minor laceration was sustained by a doctor and a mild case of shock by all three individuals.

39. *t-Butyl hypochlorite.* A sealed glass ampoule containing 10 g t-butyl hypochlorite exploded violently after several minutes exposure to fluorescent and north-window light. The ambient temperature was not above 25°C.

[255]J. Schurz, and H. Stubchen, (U. of Graz.) *Angew. Chem.* **68,** (5), 182 (March 7, 1956) (in German).
[256]*Chem. Eng. News,* Oct. 22, 1962, pp. 62–63.

APPENDIX 4

HAZARD CHARTS AND WASTE DISPOSAL PROCEDURES

CONTENTS

SECTION I. Introduction

This manual is published as a service of the Safety and Fire Protection Committee of the Manufacturing Chemists Association. Its purpose is to assist laboratory personnel in developing an awareness of chemical hazards and in fulfilling their responsibilities to dispose of chemical wastes without personal injury, without hazardous adulteration of drains, and without excessive contamination of ground, air or water.

Some 1200 chemicals are specifically listed in the manual. Therefore, because of the vast number not included, this publication should be considered preliminary in nature. However, categorization of disposal procedures by chemical classes (See VI) offers a means by which information can be obtained on the disposal of chemicals not specifically listed.

In addition to the normal fire and explosion hazards of flammable liquids and their vapors and the expected hazards of toxic materials, certain other hazards must be anticipated in waste disposal handling. Some materials are corrosive to drainage piping. Some react violently with water or with other chemicals. And others, though possibly relatively nonhazardous in themselves, adversely affect sewage disposal systems.

Chemicals poured down the drain should be non-toxic or in concentrations below the threshold limit. The concentrations which may be transferred to the drain in any given location are controlled by water pollution standards. Copies of such standards can be obtained from the water pollution control agency in the state or interstate region.

In this manual recommendations for treatment of hazardous chemical wastes, prior to disposal, are presented in basic outline. Detail is neither possible nor desirable, but all recommendations are directed towards the elimination or adequate reduction of the hazard potential. Some organic wastes can be converted to the non-toxic naturally-occurring forms. For example, a high valence chromium compound (chromate) can be reduced to a harmless trivalent hydroxide. Certain other inorganic elements and their compounds are highly toxic in any soluble form. Therefore, it is recommended that these be recovered for re-use. Contaminated mercury is an example of this class of material. Recovered material can be shipped to the supplier for reprocessing.

Toxic cyanides can be converted to less toxic cyanates or into complex iron cyanides. Very active chemicals, such as $SnCl_4$ can be made less active by mixing with inert materials such as sand, kaolin or vermiculite, and then converted to less harmful forms which can be safely discharged to the drain.

Certain materials can be safely destroyed by burning. The burning process for slow-burning compounds may be speeded up by dissolving in flammable solvents. Some incinerators are equipped to reduce carbon and carbon monoxide by means of afterburners, and to remove by-products such as SO_2 and NO with scrubbers.

The manual stresses safe procedures for on-site waste disposal from small laboratories, especially those in small communities not possessing sophisticated equipment.

SECTION II.

How to Use the Manual

To obtain information on the hazards of a specific chemical substance and recommendations for its disposal as a laboratory waste, proceed as follows:

(a) Find the chemical substance in the alphabetical listing in Section V.

(If the substance is not listed, refer to the chemical class listings under Section VI in the TABLE OF CONTENTS. The substance will fall into one of these categories and the associated disposal procedure can be used.)

(b) Note the information given in Section V on the health, fire and reactivity hazards of the substance. Note other pertinent physical properties and consult the recommended references for more detailed information.

(c) A waste disposal procedure **number** is given in Section V. Refer to this procedure in Section VI and proceed accordingly.

NOTE: For emergency handling of spills and as a matter of good operational policy it is recommended that disposal equipment and materials be readily available and that all who may have to use them know where they are stored.

Section VII lists essential equipment and materials and suggests minimum quantities required.

SECTION III.

CHART HEADINGS DEFINED

Substance/Formula Substances are listed alphabetically by "common" names. Substances which are cross-referenced (e.g. Acetal, see 1,1-Diethoxyethane) are not included in the sequence numbering.

Waste Disposal Procedure (See VI) The number contained in this column indicates the recommended waste disposal procedure as correspondingly numbered in Section VI.

TLV (ACGIH) PPM (mg/M³) This column lists Threshold Limit Values as recommended by the American Conference of Governmental Industrial Hygienists. TLV's are given in parts of vapor or gas per million parts of air by volume at 25° C. and 760 mm. Hg pressure or (in parentheses) in milligrams of particulate per cubic meter of air.

A threshold limit value refers to air-borne concentrations of substances and represents conditions under which it is believed that nearly all workers may be repeatedly exposed, day after day, without adverse effect. The ACGIH advises that threshold limits should be used as guides in the control of health hazards and should not be regarded as fine lines between safe and dangerous concentrations. For further information contact ACGIH at 1014 Broadway, Cincinnati, Ohio 45202.

NFPA, 704M System: A numerical system for the Identification of the Fire Hazards of Materials developed by the National Fire Protection Association. The numbers given in the three columns have been taken from NFPA publications and other sources. For full definitions of the various degrees of hazard (0 to 4 in each category), see NFPA No. 704M-1969. Abbreviated definitions are as follows:

Health

4 Can cause death or major injury despite medical treatment.
3 Can cause serious injury despite medical treatment.
2 Can cause injury. Requires prompt treatment.
1 Can cause irritation if not treated.
0 No hazard.

Fire

4 Very flammable gases or very volatile flammable liquids.
3 Can be ignited at all normal temperatures.
2 Ignites if moderately heated.
1 Ignites after considerable preheating.
0 Will not burn.

Reactivity (Stability)

4 Readily detonates or explodes.
3 Can detonate or explode but requires strong initiating force or heating under confinement.
2 Normally unstable but will not detonate.
1 Normally stable. Unstable at high temp. and pressure. Reacts with water.
0 Normally stable. Not reactive with water.

Sp. Gr. (Specific Gravity). Specific gravity is expressed in grams per milliliter, or density relative to water.

CHART HEADINGS DEFINED—*Continued*

Vap. Dens. (Vapor density). Vapor density is the relative density of a vapor or gas compared with air expressed as 1.0.

Fl. Pt. (Flash Point). The flash point of a liquid is the temperature at which it gives off vapor sufficient to form an ignitible mixture with the air near the surface of the liquid. The flash point figures in the chart represent closed cup tests except where the open cup flash point is designated by the initials "oc" following the figure.

Ignit. Temp. (Ignition Temperature). The ignition temperature of a substance, whether solid, liquid, or gaseous, is the minimum temperature required to initiate or cause self-sustained combustion independent of the source of heat.

Flam. Limits % (Flammable or Explosive Limits in air). In the case of gases or vapors from flammable mixtures with air or oxygen, there is a minimum concentration of vapor in air or oxygen below which propagation of flame does not occur on contact with a source of ignition, and also a maximum concentration above which propagation of flame does not occur. The concentrations (expressed in % by volume) between which propagation *can* occur are known as the lower and upper flammable (or explosive) limits.

B.P. (Boiling Point). The boiling point of a liquid is the temperature of the liquid at which its vapor pressure equals the atmospheric pressure.

M.P. (Melting Point). Expressed as °C and (°F).

Sol. in H_2O g/100g (Solubility of the substance in grams when dissolved in 100 grams water at room temperature).

The statement "insol." is used for zero solubility, e.g. cuprous cyanide and calcium carbide.

"sl. sol." means less than 5 grams soluble in 100 grams water, e.g., bromine and benzoic acid.

"sol" means 5 to 50 grams soluble in 100 grams water, e.g., barium nitrate.

"v. sol." means over 50 grams soluble in 100 grams water, e.g., potassium carbonate.

"∞" means soluble in all proportions, e.g., sulfuric acid and acetone.

Superscripts indicate the centigrade temperature at which the amount stated (in grams) is soluble in 100 grams of water. For example, .52[15] indicates that .52 grams of this substance will dissolve in 100 grams of water at 15°C.

Misc. Ref. See bibliography (Section VIII).

Abbreviations

Note: The "Misc. Ref." columns in the charts (see Section V) list sources of information. For explanations of these abbreviations and names of publications, see Section VIII.

Other abbreviations used in charts

ace.	—	acetone
alc.	—	ethyl alcohol
atm.	—	atmospheres of pressure
bz.	—	benzene
C	—	Ceiling limit (not to be exceeded)
CCl_4	—	carbon tetrachloride
CAR	—	carcinogenic—Only chemicals listed in 1968 as carcinogenic by the Department of Health of the State of Pennsylvania are marked CAR.
chl.	—	chloroform
CS_2	—	carbon disulfide
dec.	—	decomposes
exp.	—	explodes
h	—	hot
insol.	—	insoluble

iso.	—	isomeric
m.	—	meta
MeOH	—	methyl alcohol
Misc. Ref.	—	Miscellaneous References
mm	—	millimeters of mercury
n	—	normal
o	—	ortho
org. sol.	—	organic solvents
p	—	para
sec.	—	secondary
Skin	—	includes skin absorption (avoid skin contact)
sl. sol.	—	slightly soluble
sol.	—	soluble
subl.	—	sublimes
tert.	—	tertiary
v. sol.	—	very soluble
∞	—	soluble in all proportions
$<$	—	less than
$>$	—	greater than
α	—	alpha
β	—	beta
γ	—	gamma

SECTION V.

REFERENCE CHARTS

Alphabetical Index of Chemical Substances

NOTE: Absence of an entry in any column does not necessarily imply that no information exists.

SUBSTANCE/FORMULA	Waste Disposal Procedure (See VI)	TLV (ACGIH) PPM [mg/M³]	NFPA 704M System			Sp. Gr.	Vap. Dens. (Air=1)	Fl. Pt. °C (°F)	Ignit. Temp. °C (°F)	Flam. Limits %	B.P. °C (°F)	M.P. °C (°F)	Sol. in H₂O g/100g	Other Solvents	Misc. Ref.
			Health	Fire	React.										
1. Acenaphthene $C_{12}H_{10}$	18	-	-	1	-	1.02	-	-	-	-	278 (532)	96 (205)	insol.	bz. chl	
Acetal, see 1,1-Diethoxyethane															
2. Acetaldehyde CH_3CHO	2	200	2	4	2	.78	1.52	-38 (-36)	185 (365)	4-57	21 (69)	-124 (-191)	∞	alc., bz.	BDH, MCA
2a. Acetamide CH_3CONH_2	20	-	-	-	-	1.154	-	-	-	-	221	81	sol.	-	
3. Acetic Acid CH_3COOH	24a	10	2	2	1	1.05	2.1	43 (109)	426 (800)	4-16	118 (244)	17 (61)	∞	org. solv.	MCA, AIA, NSC, NTTC, BDH
4. Acetic Anhydride $(CH_3CO)_2O$	24a	5	2	2	1	1.08	3.5	54 (129)	380 (716)	3-10	140 (284)	-73 (-99)	∞	alc., chl.	MCA, BDH
5. Acetone CH_3COCH_3	18	1000	1	3	0	.79	2.0	-18 (0)	538 (1000)	3-13	56 (133)	-94 (-137)	∞	alc., bz.	MCA, AIA, NSC, BDH, NTTC
6. Acetone Cyanohydrin $(CH_3)_2C(OH)CN$	14	-	4	1	2	.93	2.9	74 (165)	688 (1270)	-	82 (180)	-19 (-2.2)	v. sol.	alc.	BDH, MCB, MCA
7. Acetonitrile CH_3CN	14	40	3	3	3	.79	1.4	6 (43)	524 (975)	4-16	80 (176)	-41 (-42)	∞	alc., chl.	BDH, MCB, MCA
Acetyl Acetone, see 2,4-Pentanedione															
8. Acetyl Bromide CH_3COBr	1a	-	3	3	2	1.6	-	-	-	-	77 (170)	-96 (-141)	dec.	bz., chl.	BDH, MCB
9. Acetyl Chloride CH_3COCl	1a	-	3	3	2	1.1	2.7	4 (40)	390 (734)	-	51 (124)	-112 (-170)	dec.	ace., bz. chl.	BDH, MCB
10. Acetylene C_2H_2	18	-	1	4	3	.91	.9	-	335 (635)	3-82	-83 (-118)	subl.	sl. sol.	ace., bz.	CGA, MGB, NSC
Acetylene Dichloride, see 1,2-Dichloroethylene															

SUBSTANCE/FORMULA	Waste Disposal Procedure (See VI)	TLV (ACGIH) PPM (mg/M³)	NFPA 704M System			Sp. Gr.	Vap. Dens. (Air=1)	Fl. Pt. °C (°F)	Ignit. Temp. °C (°F)	Flam. Limits %	B.P. °C (°F)	M.P. °C (°F)	Sol. in H₂O g/100g	Other Solvents	Misc. Ref.
			Health	Fire	React.										
Acetylene Tetrabromide, see Tetrabromoethane															
Acetylene Tetrachloride, see Tetrachloroethane															
Acetylene Trichloride, see Trichloroethylene															
11. **Acetyl Peroxide** $(CH_3CO)_2O_2$	22b	-	1	2	4	1.2	4.07	113oc	-	-	63 exp. (145)	30 (86)	sl. sol.	CCl₄, alc.	MCB
Acid Chromate Solution, see Cleaning Solution															
12. **Acridine** $C_{13}H_9N$	5	-	2		-	1.005	-	-	-	-	345 (653)	110 (230)	sl. sol.	alc., bz., CS₂	
13. **Acrolein** $CH_2=CHCHO$	2	.1	3	3	3	.84	1.9	-26 (-15)	278 (532)	3-31	53 (127)	-87 (-125)	v. sol.	alc.	MCA, BDH, AIA, NSC
14. **Acrolein Dimer** $(CH_2=CHCHO)_2$	2	-	1	2	1	1.1	-	48 (118)	-	-	151 (304)	-	sol.	-	
15. **Acrylic Acid** $CH_2=CHCOOH$	24a	-	3	2	2	1.05	2.5	52oc (126)	429 (804)	-	142 (288)	12 (54)	∞	alc.	BDH
16. **Acrylonitrile** $CH_2=CHCN$	14	20	4	3	2	.81	1.8	0oc (32)	481 (898)	3-17	77 (171)	-83 (-117)	sol.	alc.	MCA, BDH, AIA, MCB, NTTC
17. **Adipic Acid** $COOH\ (CH_2)_4COOH$	24a	-	1	1	-	1.4	-	191 (376)	422 (792)	-	267 (513)	153 (307)	sl. sol.	alc.	AIA
18. **Adiponitrile** $CN(CH_2)_4CN$	14	-	3	2	-	.97	-	93 (199)	-	-	295 (563)	2,3 (36)	sl. sol.	alc., chl.	

Aerozine 50, see Hydrazine

| SUBSTANCE/FORMULA | Waste Disposal Procedure (See VI) | TLV (ACGIH) PPM (mg/M³) | NFPA 704M System | | | Sp. Gr. | Vap. Dens. (Air=1) | Fl. Pt. °C (°F) | Ignit. Temp. °C (°F) | Flam. Limits % | B.P. °C (°F) | M.P. °C (°F) | Sol. in H₂O g/100g | Other Solvents | Misc. Ref. |
			Health	Fire	React.										
19. Aldrin $C_{12}H_8Cl_6$	4b	(.25) Skin	3	1	0	-	-	-	-	-	-	104 (219)	insol.	alc., bz., ace.	
20. Alizarin $C_{14}H_8O_4$	18	-	1	-	-	-	-	-	-	-	430 (806)	289 (552)	sl. sol.	bz., CS_2, alc.	
21. Allene $CH_2 = C = CH_2$	18	-	-	4	-	1.8	1.4	-	-	-	−34 (−30)	−136 (−213)	-	-	MGB
22. Allyl Acetate $CH_3CO_2CH_2CHCH_2$	18	-	3	1	-	.93	3.4	21oc. (70)	374 (705)	-	103 (217)	-	sl. sol.	alc.	BDH
23. Allyl Alcohol $CH_2=CHCH_2OH$	18	2 Skin	3	3	1	.85	2.0	21 (70)	378 (713)	3-18	97 (207)	−129 (−200)	∞	alc.	BDH, MCA
24. Allyl Amine $CH_2CHCH_2NH_2$	7a	-	3	3	1	.76	2.0	−29 (−20)	374 (705)	2-22	58 (136)	-	∞	alc., chl.	MCB
25. Allyl Bromide CH_2CHCH_2Br	4b	-	3	3	1	1.4	4.2	−1 (30)	295 (563)	4-7	70 (158)	−119 (−182)	insol.	alc., chl. CS_2, CCl_4	BDH, MCB
26. Allyl Chloride CH_2CHCH_2Cl	4b	1	3	3	1	.94	2.6	−32 (−26)	392 (738)	3-11	45 (113)	−136 (−213)	insol.	alc., ace., bz.	BDH, AIA, MCB, MCA
27. Allyl Chloroformate $CH_2CHCH_2CO_2Cl$	4b	-	3	3	1	1.1	4.2	31 (88)	-	-	110 (230)	-	insol.	-	MCB
Allylene, see propyne															
28. Allyl Glycidyl Ether $C_6H_{10}O_2$	15	10C	3	-	-	.97	3.4	57 (135)	-	-	154 (309)	−100 (−148)	sol.	ace.	
29. Allyl Iodide CH_2CHCH_2I	4b	-	3	2	-	1.8	5.8	-	-	-	102 (216)	−99 (−146)	insol.	alc., chl.	MCB

SUBSTANCE/FORMULA	Waste Dis-posal Pro-cedure (See VI)	TLV (ACGIH) PPM (mg/M³)	NFPA 704M System			Sp. Gr.	Vap. Dens. (Air=1)	Fl. Pt. °C (°F)	Ignit. Temp. °C (°F)	Flam. Limits %	B.P. °C (°F)	M.P. °C (°F)	Sol. in H₂O g/100g	Other Solvents	Misc. Ref.
			Health	Fire	React.										
30. Allyl Propyl Disulfide C₃H₅S₂C₃H₇	13	2	3	2	-	-	-	-	-	-	-	-	-	-	MCB
31. Alum KAl(SO₄)₂ 12H₂O	11	-	1	-	-	1.7	-	-	-	-	200 (392)	92.5 (198)	11.420		NTTC
32. Aluminum Al	27a	-	0	1	1	2.7	-	-	-	-	2056 (3733)	660 (1220)	insol.	-	
33. Aluminum Alkyls RALX	3	-	3	4	2	.8	-	<0 (<32)	-	-	-	-	-	-	AIA
34. Aluminum Borohydride Al(BH₄)₃	17	-	3	4	2	-	-	-	-	-	45 (113)	−65 (−85)	-	-	AIA, NSC
35. Aluminum Bromide Al Br₃	1b	-	-	-	2	3.0	-	-	-	-	263 (505)	98 (208)	dec.	alc., ace., CS₂	BDH, MCB
36. Aluminum Carbide Al₄C₃	25	-	-	-	-	2.4	-	-	-	-	dec.	1400 (2552)	dec.		
37. Aluminum Chloride AlCl₃	1b	-	3	0	2	2.4	-	-	-	-	183 (361)	194 (381) at 5 atm.	69.915	alc., CCl₄	MCA, BDH, NSC
37a. Aluminum Chloride, hydrate	11	-	-	-	-	-	-	-	-	-	-	-	-	sl. sol.	
38. Aluminum Ethoxide (C₂H₅)₃AlO₃	3	-	-	-	-	-	-	-	-	-	205 (401)	134 (273)	dec.	-	
Aluminum Lithium Hydride, see Lithium Aluminum Hydride															
39. Aluminum Nitrate Al(NO₃)₃·9H₂O	11	-	2	2	-	-	-	-	-	-	150dec. (302)	70 (158)	63.725	CS₂ alc., ace.	
40. 2-Amino-Diphenylene Oxide C₁₂H₉NO	5	-	CAR	-	-	-	-	-	-	-	-	94 (201)	-	-	

SUBSTANCE/FORMULA	Waste Disposal Procedure (See VI)	TLV (ACGIH) PPM (mg/M³)	NFPA 704M System			Sp. Gr.	Vap. Dens. (Air=1)	Fl. Pt. °C (°F)	Ignit. Temp. °C (°F)	Flam. Limits %	B.P. °C (°F)	M.P. °C (°F)	Sol. in H₂O g/100g	Other Solvents	Misc. Ref.
			Health	Fire	React.										
2-Aminoethanol, see Ethanolamine															
41. Aminoethylethanol Amine $NH_2(CH_2)_2NH(CH_2)_2OH$	7a	-	2	1	0	1.03	3.6	129 (265)	368 (695)	-	244 (471)	-	v. sol.	alc.	NTTC
42. 2-Aminopyridine $C_5H_6N_2$	5	5	3	-	-	-	-	-	-	-	204 (399)	58 (136)	sol.	org. solv.	
43. Ammonia, Anhydrous NH_3	10	50	3	1	0	.77	.59	-	651 (1204)	16-25	-33 (-28)	-78 (-108)	89.9⁰	org. solv.	MCA, CGA, NSC, AIA, NTTC, MGB
44. Ammonia, Aqua NH_4OH	10	50	2	1	-	2.2	1.2	-	-	-	-	-72 (-98)	sol.	-	MCA, BDH,
45. Ammonium Dichromate $(NH_4)_2Cr_2O_7$	12a	-	3	2	-	-	-	-	-	-	-	180dec. (356)	15 31	alc.	MCA
46. Ammonium Fluoride NH_4F	11	-	3	-	-	1.3	-	-	-	-	-	subl.	100⁰	alc.	
Ammonium Hydroxide, see Ammonia, Aqua															
47. Ammonium Nitrate NH_4NO_3	11	-	2	1	3	1.7	-	-	-	exp.	210dec. (410)	169 (336)	118⁰	alc., ace., MeOH	NSC, NTTC
48. Ammonium Perchlorate NH_4ClO_4	12a	-	2	1	4	1.9	-	-	-	-	-	dec.	10.7⁰	ace.	MCB
49. Ammonium Persulfate $(NH_4)_2S_2O_8$	12a	-	1	2	-	19	-	-	-	-	-	120dec. (248)	58.2⁰	dec. H₂O 100	
50. Ammonium Sulfamate $NH_4OSO_2NH_2$	19	(15)	1	1	-	-	-	-	-	-	160dec. (320)	125 (257)	1661⁰		

SUBSTANCE/FORMULA	Waste Disposal Procedure (See VI)	TLV (ACGIH) PPM (mg/M³)	NFPA 704M System Health	NFPA 704M System Fire	NFPA 704M System React.	Sp. Gr.	Vap. Dens. (Air=1)	Fl. Pt. °C (°F)	Ignit. Temp. °C (°F)	Flam. Limits %	B.P. °C (°F)	M.P. °C (°F)	Sol. in H₂O g/100g	Other Solvents	Misc. Ref.
51. Ammonium Sulfide $(NH_4)_2S$	23	-	-	2	-	1.2	-	-	-	-	-	dec.	v. sol.	alc., dec. H_2O^{100}	BDH
52. Ammonium Thiocyanate NH_4CNS	11	-	1	-	-	1.3	-	-	-	-	170dec. (338)	150 (302)	128^0	alc., ace	
53. Ammonium Vanadate NH_4VO_3	27i	-	-	-	-	2.3	-	-	-	-	200dec. (392)	-	$.52^{15}$	dec. H_2O^{100}	
54. n-Amyl Acetate $CH_3COO(CH_2)_4CH_3$	18	100	1	3	0	.88	4.5	25 (77)	379 (714)	1-7.5	148 (298)	-79 (-110)	sl. sol.	alc.	BDH, NSC
55. iso-Amyl Acetate $CH_3COO(CH_2)_2CH(CH_3)_2$	18	100	1	3	0	.88	4.5	23 (73)	380 (715)	1-7.5 @212°F	142 (288)	-	-	alc.	BDH, NSC
56. sec-Amyl Acetate $CH_3CO_2CH_2CHCH_3C_2H_5$	18	125	1	3	0	.86	4.5	32 (89)	-	1-7.5	121 (250)	-	-	alc.	NSC
57. n-Amyl Alcohol $CH_3(CH_2)_4OH$	18	-	1	3	0	.82	3.0	33 (91)	300 (572)	1-10	137 (279)	-79 (-110)	sl. sol.	alc.	BDH, NSC
58. iso-Amyl Alcohol $(CH_3)_2CH(CH_2)_2OH$	18	-	1	2	0	.81	3.0	43 (109)	347 (657)	1-9 @212°F	132 (270)	-	sl.sol.	alc.	BDH
59. tert. Amyl Alcohol $CH_3CH_2C(CH_3)OHCH_3$	18	-	1	3	0	.81	3.0	19 (66)	437 (819)	1-9	102 (216)	-12 (10)	sl. sol.	alc.	BDH
60. Amylamine $C_5H_{11}NH_2$	7a	-	3	3	0	.8	3.0	7oc. (45)	-	-	103 (217)	-55 (-67)	∞	alc.	
61. Amyl Bromide $C_5H_{11}Br$	4b	-	1	3	0	1.22	5.2	32 (90)	-	-	-54 (130)	-95 (-139)	insol.	alc.	
62. Amylene C_5H_{10}	18	-	1	4	0	.66	2.4	-2 (28)	273 (523)	1.5-9	30 (86)	-165 (-265)	insol.	alc.	BDH

SUBSTANCE/FORMULA	Waste Disposal Procedure (See VI)	TLV (ACGIH) PPM (mg/M³)	NFPA 704M System Health	Fire	React.	Sp. Gr.	Vap. Dens. (Air=1)	Fl. Pt. °C (°F)	Ignit. Temp. °C (°F)	Flam. Limits %	B.P. °C (°F)	M.P. °C (°F)	Sol. in H₂O g/100g	Other Solvents	Misc. Ref.
63. n-Amyl Ether $C_{10}H_{22}O$	15	-	1	2	0	.74	5.46	57 (135)	171 (340)	-	190 (374)	-70 (-94)	insol.	alc.	
64. iso-Amyl Formate $HCOOC_5H_{11}$	18	-	1	3	0	.89	4.0	26 (79)	-	-	131 (267)	-74 (-101)	sl. sol.	alc.	BDH
65. Amyl Mercaptan $CH_3(CH_2)_4SH$	13	20	2	3	0	.84	1.1	18 (65)	-	-	127 (261)	-76 (-105)	insol.	alc.	
66. iso-Amyl Nitrate $C_5H_{11}NO_3$	4a	-	1	2	2	.99	-	52 (125)	-	-	152-7 (306-15)	-	sl. sol	alc.	BDH
67. Amyl Nitrite $CH_3(CH_2)_4NO_2$	4a	-	1	2	2	.85	-	10 (50)	209 (408)	-	104 (220)	-	sl. sol	alc.	BDH
68. Aniline $C_6H_5-NH_2$	5	5 Skin	3	2	0	1.02	3.22	70oc (158)	770 (1418)	1.3	184 (363)	-6 (21)	sol.	alc., bz	NTTC, MCA, BDH
69. o-Anisaldehyde $C_8H_8O_2$	2	-	2	1	0	1.12	-	118 (244)oc	-	-	250 (482)	2.5 (36)	insol.	alc.	
70. o-Anisidine $CH_3OC_6H_4NH_2$	5	(.5) Skin	2	-	-	-	-	-	-	-	224 (435)	-	sl. sol	alc.	BDH
71. p-Anisidine $CH_3OC_6H_4NH_2$	5	(.5) Skin	2	-	-	-	-	-	-	-	243 (469)	-	sl. sol.	alc.	BDH
72. Anisole $C_6H_5OCH_3$	15	-	1	2	0	.995	3.72	52oc. (126)	-	-	154 (309)	-37 (-35)	insol.	alc.	
73. Anthracene $C_6H_4(CH)_2C_6H_4$	18	-	1	2	-	1.25	6.15	121 (250)	540 (1004)	.6—	340 (644)	217 (423)	insol.	bz	
74. 2-Anthramine $C_{14}H_{11}N$	5	-	CAR	-	-	-	-	-	-	-	-	236 (457)	insol.	alc.	

SUBSTANCE/FORMULA	Waste Disposal Procedure (See VI)	TLV (ACGIH) PPM (mg/M³)	NFPA 704M System			Sp. Gr.	Vap. Dens. (Air=1)	Fl. Pt. °C (°F)	Ignit. Temp. °C (°F)	Flam. Limits %	B.P. °C (°F)	M.P. °C (°F)	Sol. in H₂O g/100g	Other Solvents	Misc. Ref.
			Health	Fire	React.										
75. Anthraquinone $C_6H_4(CO)_2C_6H_4$	18	-	1	1	-	1.44	7.16	185 (365)	-	-	380 (716)	286 (547)	insol.	alc. (sl. sol.)	
76. Antimony Sb	27d	(.5)	3	2	-	6.68	-	-	-	-	1380 (2516)	630 (1166)	insol.		
77. Antimony Hydride, see Stibine															
77. Antimony Pentachloride $SbCl_5$	27d	(.5)	3	-	-	2.34	-	-	-	-	140 (284)	2.8 (37)	dec.	$CHCl_3$	BDH
78. Antimony Pentasulfide Sb_2S_5	27d	(.5)	3	1	1	-	-	-	-	-	-	75dec. (167)	insol.		
79. Antimony Trichloride $SbCl_3$	27d	(.5)	3	-	-	3.14	-	-	-	-	223 (433)	73 (163)	60^{20}	alc., bz., ace., chl.	BDH, MCA
80. Antimony Trioxide Sb_2O_3	27d	(.5)	3	-	-	5.2	-	-	-	-	1550 (2822)	656 (1213)	sl. sol.		
81. Argon Ar	26	-	-	-	-	1.8	-	-	-	-	-186 (-303)	-189 (-308)	4^0		MGB, CGA
82. Arsenic As_4	27d	(.5)	3	2	-	5.7	-	-	-	-	-	615subl. (1139)	insol.		NSC
83. σ-Arsenic Acid $H_3AsO_4 \cdot \frac{1}{2}H_2O$	27d	(.5)	3	-	-	~2	-	-	-	-	160dec. (320)	35.5 (95.7)	16.7	alc.	BDH
84. Arsenic Trichloride $AsCl_3$	27d	(.5)	3	-	-	2.16	6.25	-	-	-	63 (145)	-9 (15.8)	dec.	alc.	BDH, NSC
85. Arsenic Trioxide As_2O_3	27d	(.5)	3	-	-	3.74	-	-	-	-	-	193subl. (379)	1.2^2	alc.	BDH, MCA, NSC

SUBSTANCE/FORMULA	Waste Dis-posal Pro-cedure (See VI)	TLV (ACGIH) PPM (mg/M³)	NFPA 704M System			Sp. Gr.	Vap. Dens. (Air=1)	Fl. Pt. °C (°F)	Ignit. Temp. °C (°F)	Flam. Limits %	B.P. °C (°F)	M.P. °C (°F)	Sol. in H₂O g/100g	Other Solvents	Misc. Ref.
			Health	Fire	React.										
Arylam, see Sevin															
87. **Asphalt**	26	-	0	1	0	1.1	-	204+ (400)	485 (905)	-	>371 >700	-			NTTC
88. **Auramine** $[(CH_3)_2NC_6H_4]C{:}NHCl$	5	-	-	-	-	-		-	-	-	-	267 (513)	sl. sol.	alc., chl.	
89. **Aziridine** $(CH_2)_2NH$	5	-	3	3	3	.83	1.5	-11 (12)	322 (612)	3.6-46	56 (133)	-72 (-98)	∞	org. solv.	
90. **2,2'-Azonaphthalene** $C_{10}H_7N{:}NC_{10}H_7$	8	-	-	-	-	-		-	-	-	-	208subl. (408)	insol.	hot bz.	
91. **Azoxybenzene** $C_6H_5(NON)C_6H_5$	8	-	-	1	-	1.16		-	-	-	87 (189)	36 (97)	insol.	alc.	
92. **Barium** Ba	27h	(.5)	2	2	-	3.5		-	-	-	1140 (2084)	725 (1337)	dec.	alc.	
93. **Barium Chlorate** $Ba(ClO_3)_2{\cdot}H_2O$	12a	(.5)	1	0	1	-		-	-	-	-	120 (248)	27.4⁴⁰	ace., alc.	
94. **Barium Chloride** $BaCl_2$	27h	(.5)	2	0	-	2.8		-	-	-	1560 (2840)	925 (1697)	37.5²⁶		
95. **Barium Hydroxide** $Ba(OH)_2{\cdot}8H_2O$	27h	(.5)	2	-	-	2.18		-	-	-	780 (1436)	78 (172)	5.6¹⁵	alc.	
96. **Barium Nitrate** $Ba(NO_3)_2$	27h	(.5)	1	0	1	3.2		-	-	-	-	592dec. (1098)	8.7²⁰		
97. **Barium Peroxide** BaO_2	27h	(.5)	1	0	1	4.96		-	-	-	800 (1472)	450 (842)	sl. sol.		

SUBSTANCE/FORMULA	Waste Disposal Procedure (See VI)	TLV (ACGIH) PPM (mg/M³)	NFPA 704M System			Sp. Gr.	Vap. Dens. (Air=1)	Fl. Pt. °C (°F)	Ignit. Temp. °C (°F)	Flam. Limits %	B.P. °C (°F)	M.P. °C (°F)	Sol. in H₂O g/100g	Other Solvents	Misc. Ref.
			Health	Fire	React.										
98. Batteries, Dry Cells	26	-	-	-	-	-	-	-	-	-	-	-	-	-	-
99. Batteries, Wet Cells	27a	-	-	-	-	-	-	-	-	-	-	-	-	-	NSC
100. Benzal Chloride $C_6H_5CHCl_2$	4b	-	-	2	-	1.26	-	-	-	-	205 (401)	−16 (3.2)	insol.	alc.	BDH
101. Benzaldehyde C_6H_5CHO	2	-	2	2	0	1.04	3.7	64 (147)	192 (378)	-	178 (352)	−56 (−69)	sl. sol.	alc.	BDH
102. 1,2-Benzanthracene $C_{18}H_{14}$	18	CAR	-	-	-	-	-	-	-	-	435 (815)	162 (324)	insol.	bz.	
103. Benzene C_6H_6	18	25C	2	3	0	.88	2.8	−11 (12)	562 (1044)	1.4-8	80 (176)	5.4 (41)	sl. sol.	alc., ace.	AIA, BDH, NTTC, MCA
Benzene Monochloride, see Chlorobenzene															
104. Benzene Sulfonic Acid $C_6H_5SO_3H$	4c	-	3	-	-	-	-	-	-	-	-	525dec. (977)	sol.	alc.	BDH
105. Benzene Sulfonyl Chloride $C_6H_5SO_2Cl$	1a	-	-	-	-	-	-	-	-	-	252 (485)	14.5 (58)	insol.	hot alc.	BDH
106. Benzidine $(C_6H_4)_2(NH_2)_2$	5	CAR	3	-	-	1.25	-	-	-	-	402 (756)	128 (262)	sl. sol.	alc.	
107. Benzoic Acid C_6H_5COOH	24a	-	1	1	-	1.32	4.2	121 (250)	574 (1065)	-	249 (480)	122 (252)	sl. sol.	alc., bz.	
108. Benzonitrile C_6H_5CN	14	-	3	-	-	1.01	-	-	-	-	exp. (376)	−13 (8.6)	al. sol.	alc.	BDH
2,3-Benzophenanthrene, see 1,2-Benzanthracene															

SUBSTANCE/FORMULA	Waste Disposal Procedure (See VI)	TLV (ACGIH) PPM (mg/M³)	NFPA 704M System Health	NFPA 704M System Fire	NFPA 704M System React.	Sp. Gr.	Vap. Dens. (Air=1)	Fl. Pt. °C (°F)	Ignit. Temp. °C (°F)	Flam. Limits %	B.P. °C (°F)	M.P. °C (°F)	Sol. in H₂O g/100g	Other Solvents	Misc. Ref.
p-Benzoquinone, see Quinone															
109. Benzotrichloride $C_6H_5CCl_3$	4b	-	2	-	-	1.38	6.8	-	-	-	214 (417)	-5 (23)	insol.	alc., bz.	BDH
110. Benzotrifluoride $C_6H_5CF_3$	4b	-	4	3	0	1.19	5.04	12 (54)	-	-	101 (214)	-29 (-20)	insol.	alc.	BDH
111. Benzoyl Chloride C_6H_5COCl	1a	-	3	2	1	1.22	4.88	72 (162)	-	-	197 (387)	-.5 (31)	dec.	bz., CS_2	BDH
112. Benzoyl Peroxide $(C_6H_5CO)_2O_2$	22b	(5)	1	4	4	1.33	-	-	80 (176)	-	ex.	106 (223)	sl. sol.	bz., ace.	BDH, MCA, NFPA
3,4-Benzophenanthrene see 1,2-Benzanthracene															
113. Benzyl Acetate $CH_3CO_2CH_2C_6H_5$	18	-	1	1	0	1.06	5.1	102 (216)	461 (862)	-	214 (417)	-51.5 (-61)	sl. sol.	alc.	
114. Benzyl Alcohol $C_6H_5CH_2OH$	18	-	2	1	0	˚.04	3.7	101 (213)	436 (817)	-	206 (403)	-15.3 (4)	4	alc., ace., chl.	BDH
115. Benzyl Amine $C_6H_5CH_2NH_2$	5	-	3	-	-	.98	-	63 (145)	-	-	185 (365)		∞	alc.	
116. Benzyl Benzoate $C_6H_5CO_2CH_2C_6H_5$	18	-	1	1	0	1.11	7.3	148 (298)	481 (898)	-	323 (614)	21 (70)	insol.	alc., ace., bz., chl.	
117. Benzyl Bromide $C_6H_5CH_2Br$	4b	-	3	-	-	1.44	5.8	-	-	-	198 (388)	-4 (25)	insol.	alc.	BDH
118. Benzyl Chloride $C_6H_5CH_2Cl$	4b	1	2	2	1	1.10	4.36	67 (153)	585 (1085)	1.1-	179 (354)	-39 (-38)	insol.	alc., chl.	BDH, MCA

SUBSTANCE/FORMULA	Waste Disposal Procedure (See VI)	T L V (ACGIH) PPM (mg/M³)	NFPA 704M System			Sp. Gr.	Vap. Dens. (Air=1)	Fl. Pt. °C (°F)	Ignit. Temp. °C (°F)	Flam. Limits %	B.P. °C (°F)	M.P. °C (°F)	Sol. in H₂O g/100g	Other Solvents	Misc. Ref.
			Health	Fire	React.										
119. **Benzyl Chloroformate** $ClCO_2CH_2C_6H_5$	4b	-	2	-	-			-	-	-	103 (217)	-	dec.	dec., alc.	BDH
120. **Benzyl Cyanide** $C_6H_5CH_2CN$	14	-	3	-	-	1.02	-	-	-	-	234 (453)	−24 (−31)	insol.	alc.	BDH
Benzylidene Chloride see Benzal Chloride															
121. **Benzyl Mercaptan** $C_6H_5CH_2SH$	13	-	-	2	-	1.06	4.3	70 (158)	-	-	194 (381)	-	insol.	alc., CS₂	
122. **Beryllium** Be	27g	(.002)	4	1	1	1.86	-	-	-	-	2970 (5378)	1278 (2332)	insol.		AIA, NSC
123. **Beryllium Salts**	27g	(.002)	4	1	-	-	-	-	-	-	-	-	-	-	AIA, BDH
124. **Biphenyl** $(C_6H_5)_2$	18	.2	2	1	0	1.2	5.3	113 (235)	540 (1004)	.6-5.8 @232°F	256 (493)	70 (158)	insol.	alc.	
125. **N-4-Biphenylacetohydroxamic Acid**	5											171 (340)			
126. **2-Biphenylamine** $C_6H_5C_6H_4NH_2$	5	-	2	1	0	1.16	5.8	-	452 (846)	-	299 (570)	49 (120)	insol.	alc., bz.	
Biphenyl, 4,4'Diamino, see Benzidine															
127. **Bismuth Chloride** $BiCl_3$	27d	-	2	-	-	4.75	-	-	-	-	441 (826)	230 (446)	dec.	alc., ace.	
128. **Borax, Anhydrous** $Na_2B_4O_7$	11	-	2	3	-	1.72	-	-	-	-	1575 (2867)	741 (1366)	1.06⁰		AIA
129. **Boric Acid**	24b	-	2	-	-	2.46	-	-	-	-	1860 (3380)	169 (336)	6.3³⁰	alc., MeOH	AIA

SUBSTANCE/FORMULA	Waste Disposal Procedure (See VI)	TLV (ACGIH) PPM (mg/M³)	NFPA 704M System Health	NFPA 704M System Fire	NFPA 704M System React.	Sp. Gr.	Vap. Dens. (Air=1)	Fl. Pt. °C (°F)	Ignit. Temp. °C (°F)	Flam. Limits %	B.P. °C (°F)	M.P. °C (°F)	Sol. in H₂O g/100g	Other Solvents	Misc. Ref.
130. Borneol $C_{10}H_{17}OH$	18	-	2	2	0	1.01	5.31	66 (151)	-	-	-	212subl (413)	insol.	alc., ace.	
131. Boron B	26	-	2	2	-	3.33	-	-	-	-	2550 (4622)	2300 (4172)	insol.		AIA
132. Boron Fluoride-Ethyl Ether Complex $C_4H_{10}BF_3O$	21	1	3	2	0	1.1	-	64oc. (147)	-	-	-110 (-166)	-128 (-198)	sol.		AIA
Boron Hydrides, see Di-,Penta-, or Deca-Boranes															
133. Boron Oxide B_2O_3	24b	15	2	-	-	2.46	-	-	-	-	1860 (3380)	460 (860)	sol.		AIA
134. Boron Tribromide BBr_3	21	-	2	-	-	2.65	-	-	-	-	92 (196)	-45 (-49)	dec.	alc., CCl4	AIA
135. Boron Trichloride BCl_3	21	-	2	-	-	1.35	4.03	-	-	-	12.5 (54)	-107 (-161)	dec.	alc.	AIA, BDH, MGB
136. Boron Trifluoride BF_3	21	1C	3	0	1	2.99	-	-	-	-	-95 (-146)	-127 (-197)	106	alc.	AIA, MGB
137. Boron Trifluoride Complexes with Acetic Acid and Mentanol	21	-	3	-	-	-	-	-	-	-	-	-	-	-	AIA, BDH
137a. Brass	27a														
138. Bromic Acid $HBrO_3$	12a	-	3	2	-	3.19	-	-	-	-	-	100dec (212)	v. sol.		BDH
139. Bromine Br_2	12a	0.1	4	0	1	3.1	5.5	-	-	-	59 (138)	-7 (19)	4.20	alc., CS2 chl.	AIA, BDH, NTTC, MCA, NSC
140. Bromine Pentafluoride BrF_5	21	-	3	-	4	2.48	6.05	-	-	-	40 (104)	-61 (-78)	dec.	-	MGB

SUBSTANCE/FORMULA	Waste Dis-posal Pro-cedure (See VI)	TLV (ACGIH) PPM (mg/M³)	NFPA 704M System Health	Fire	React.	Sp. Gr.	Vap. Dens. (Air=1)	Fl. Pt. °C (°F)	Ignit. Temp. °C (°F)	Flam. Limits %	B.P. °C (°F)	M.P. °C (°F)	Sol. in H₂O g/100g	Other Solvents	Misc. Ref.
141. Bromine Trifluoride BrF₃	21	-	3	-	4	2.84	-	-	-	-	127 (261)	8.8 (48)	dec.		MGB
142. Bromo-Acetic Acid CH₃BrCOOH	4c	-	2	-	-	1.93	-	-	-	-	208 (406)	49 (120)	∞	alc.	BDH
143. Bromobenzene C₆H₅Br	4b	-	2	2	0	1.52	5.4	51 (124)	566 (1051)	-	155 (311)	-31 (-24)	insol.	alc., bz., chl., CCl₄	BDH
1-Bromobutane, see Butyl Bromide															
144. Bromochloromethane CH₂BrCl	26	-	2	-	-	1.99	4.46	-	-	-	69 (156)	-88 (-126)	insol.	org. sol.	
145. Bromoethane CH₃CH₂Br	4b	-	2	3	0	1.46	3.76	-	511 (952)	6.7-11.3	38 (100)	-119 (-182)	sl. sol.	alc.	BDH
146. Bromoethylene CH₂CHBr	4b	-	2	3	-	1.49	-	-	-	-	15.8 (60)	-138 (-216)	insol.	alc.	
147. Bromoform CHBr₃	27j	.5 Skin	2	-	-	2.89	-	-	-	-	150 (302)	-6 (21)	sl. sol.	alc., bz., chl.	BDH
Bromomethane, see Methyl Bromide															
Bromopentane, see Amyl Bromide															
Bromopropane, see Propyl Bromide															
3-Bromopropene, see Allyl Bromide															

SUBSTANCE/FORMULA	Disposal Procedure (See VI)	TLV (ACGIH) PPM (mg/M³)	NFPA 704M System			Sp. Gr.	Vap. Dens. (Air=1)	Fl. Pt. °C (°F)	Ignit. Temp. °C (°F)	Flam. Limits %	B.P. °C (°F)	M.P. °C (°F)	Sol. in H₂O g/100g	Other Solvents	Misc. Ref.
			Health	Fire	React.										
3-Bromopyne, see Propargyl Bromide															
148. O-Bromotoluene $CH_3C_6H_4Br$	4b	-	2	2	0	1.42	5.9	79 (174)	-	-	181 (358)	-27 (-17)	insol.	alc., bz.	
149. Bromotrifluoromethane $CBrF_3$	26	-	-	-	-			-	-	-	-59 (-74)	-168 (-270)	sl. sol.	chl.	
150. Bronze Alloy	27a	-	0	0	1										
151. Brucine $C_{23}H_{26}N_2O_4$	6	-	3	-	-	-	-	-	-	-		178 (352)	sl. sol.	alc., chl.	
152. 1,3-Butadiene $CH_2=(CH)_2=CH_2$	18	1000	2	4	2	1.88	1.97	<-7 (<20)	429 (804)	2-11.5	-4.7 (23)	-109 (-164)	insol.	org. sol.	MCA
153. n-Butane C_4H_{10}	18	-	1	4	0	.60	2.04	-60 (-76)	405 (761)	1.9-8.5	-.5 (31)	-138 (-216)	v. sol.	alc., chl.	NTTC
154. iso-Butane $CH(CH_3)_3$	18	-	1	4	0	.56	2.01	-	462 (864)	1.8-8.4	-12 (10.4)	-160 (-256)	sol.	alc.	
n-Butanol, see n-Butyl Alcohol															
Butanone, see Methyl Ethyl Ketone															
155. 1-Butene $CH_3CH_2CH=CH_2$	18	-	1	4	0	-	1.9	-	384 (723)	1.6-9.3	-6.1 (21)	-	insol.	alc.	
156. 2-Butene $CH_3CH=CHCH_3$	18	-	1	4	0	-	1.9	-	324 (615)	1.8-9.7	1.1 (34)	-	insol.	alc.	
2-Butoxyethanol, see Butyl Cellosolve															

SUBSTANCE/FORMULA	Waste Disposal Procedure (See VI)	TLV (ACGIH) PPM (mg/M³)	NFPA 704M System Health	Fire	React.	Sp. Gr.	Vap. Dens. (Air=1)	Fl. Pt. °C (°F)	Ignit. Temp. °C (°F)	Flam. Limits %	B.P. °C (°F)	M.P. °C (°F)	Sol. in H₂O g/100g	Other Solvents	Misc. Ref.
157. n-Butyl Acetate $CH_3COOC_4H_9$	18	150	1	3	0	.88	4.0	27 (81)	399 (750)	1.4-7.6	125 (257)	−76 (−105)	sl. sol.	alc.	BDH
158. iso-Butyl Acetate $CH_3COOCH_2CH(CH_3)CH_3$	18	150	1	3	0	.87	4.0	18 (64)	423 (793)	1.3-7.5	117 (243)	−99 (146)	sl. sol.	alc.	BDH
159. sec-Butyl Acetate $CH_3COOCH(CH_3)CH_2CH_3$	18	200	1	3	0	.86	4.0	31 (88)oc	-	1.7-	112 (234)	-	insol.	alc.	
160. tert-Butyl Acetate $CH_3COOC(CH_3)_3$	18	200	-	-	-	.87	-	-	-	-	95 (203)	-	insol.	alc.	MCA
161. n-Butyl Alcohol $CH_3CH_2CH_2CH_2OH$	18	100	1	3	0	.81	2.55	29 (84)	365 (689)	1.4-11	118 (244)	−89 (−128)	sol.	alc.	BDH, MCA
162. iso-Butyl Alcohol $CH_3CH(CH_3)CH_2OH$	18	100	1	3	0	.81	2.55	28 (82)	427 (800)	1.7-10.9 @212 F	107 (225)	−108 (−162)	10¹⁵	alc.	BDH
163. sec-Butyl Alcohol $CH_3CH_2CH(OH)CH_3$	18	150	1	3	0	.81	2.55	24 (75)	406 (763)	1.7-9.8 @212 F	99.5 (211)	−115 (−175)	12.5	alc.	BDH, MCA
164. tert-Butyl Alcohol $(CH_3)_3COH$	18	100	1	3	0	.78	2.55	10 (50)	478 (892)	2.4-8	83 (181)	25 (77)	∞	alc.	BDH, MCA
165. Butylamine $C_4H_9NH_2$	7a	5C	3	3	0	.76	2.5	−12 (10)	312 (594)	1.7-9.8	78 (172)	−50 (−58)	∞	alc.	BDH
166. tert-Butylamine $(CH_3)_3CNH_2$	7a	5C	3	4	0	.70	2.5	-	-	1.7-8.9 @212 F	45 (113)	−67 (−89)	∞	alc.	BDH
167. iso-Butylamine $(CH_3)_2CHCH_2NH_2$	7a	-	3	3	0	.73	2.5	−9 (16)	378 (712)	-	66 (151)	−86 (−123)	∞	alc.	BDH
168. n-Butyl Bromide $CH_3(CH_2)_3Br$	4b	-	2	3	0	1.28	4.7	18 (64)	265 (509)	2-6.6. @212 F	101 (214)	−112 (−170)	.06	alc.	BDH

SUBSTANCE/FORMULA	Waste Disposal Procedure (See VI)	TLV (ACGIH) PPM (mg/M³)	NFPA 704M System Health	NFPA 704M System Fire	NFPA 704M System React.	Sp. Gr.	Vap. Dens. (Air=1)	Fl. Pt. °C (°F)	Ignit. Temp. °C (°F)	Flam. Limits %	B.P. °C (°F)	M.P. °C (°F)	Sol. in H₂O g/100g	Other Solvents	Misc. Ref.
169. Butyl Cellosolve $C_4H_9OCH_2CH_2OH$	18	50 Skin	2	2	-	.91	4.1	61 (142)	-	1.1-12.7	171 (340)	-40 (-40)	∞	alc.	
170. n-Butyl Chloride $CH_3(CH_2)_3Cl$	4b	-	2	3	0	.38	3.2	7 (45)	471 (880)	1.9-10.1	78 (172)	-123 (-189)	.07	alc.	BDH
171. tert-Butyl Chloride $(CH_3)_3CCl$	4b	-	2	3	0	.85	3.2	<21 (<70)	-	-	51 (124)	-27 (-17)	sl. sol.	alc.	BDH
172. tert-Butyl Chromate $C_8H_{18}CrO_4$	12a	(.1)C	3			-	-	-			-	-	-	-	
Butyl Ether, see Dibutyl Ether															
173. n-Butyl Formate $HCOOC_4H_9$	18	-	2	3	0	.91	3.5	18 (64)	322 (612)	1.7-8	107 (225)	-90 (-130)	sl. sol.	alc.	
174. iso-Butyl Formate $HCOOCH_2CH(CH_3)_2$	18	-	-	-	-	.89	-	<21 (<70)	-	-	98 (208)	-95 (-139)	1.122	alc.	BDH
175. n-Butyl Glycidyl Ether $C_4H_9-O-CH_2-CH-O-CH_2$	15	50	-	-	-	.91	-	-	-	-	164 (327)	-	-	-	
176. Butyl Hydroperoxide $(CH_3)_3COOH$	22b	-	1	4	4	.86	2.1	27 (80)	-	-	dec.	6 (43)	sol.	alc., chl.	
177. n-Butyllithium in Hydrocarbon solvents 177a,b,c. $CH_3(CH_2)_3Li$	3	-	-	-	-	-	solvent	-	-	-	-	-	-	-	MCA
177a. In Heptane	3	500	-	-	-	.68	-	-4 (25)	223 (433)	1.2-6.7	98 (209)	-	-	org. solv.	MCA
177b. In Hexane	3	500	-	-	-	.69	-	-22 (-7)	234 (453)	1.2-7.5	69 (156)	-	-	org. solv.	MCA

SUBSTANCE/FORMULA	Waste Dis-posal Pro-cedure (See VI)	TLV (ACGIH) PPM (mg/M³)	NFPA 704M System Health	Fire	React.	Sp. Gr.	Vap. Dens. (Air=1)	Fl. Pt. °C (°F)	Ignit. Temp. °C (°F)	Flam. Limits %	B.P. °C (°F)	M.P. °C (°F)	Sol. in H₂O g/100g	Other Solvents	Misc. Ref.
177c. In Pentane	3	1000	-	-	-	.70	-	−40 (−40)	309 (588)	1.5-7.8	36 (97)	-	-	org. solv.	MCA
178. Butyl Mercaptan CH₃(CH₂)₃SH	13	10	2	3	0	.84	3.1	2 (36)	-	-	98 (208)	−116 (−117)	sl. sol.	alc.	
179. Butyl Methacrylate CH₂C(CH₃)COOC₄H₉	18	-	2	2	0	.89	4.8	52 (126)oc	-	-	163 (325)	-	insol.	alc.	NTTC
180. Isobutyl Methyl Ketone CH₃COC₄H₉	18	100	2	3	0		3.5	17 (63)	460 (860)	1.2-8	126 (259)	−57 (−71)	-	alc.	BDH
181. tert-Butyl Peracetate CH₃CO₃C(CH₃)₃	22b	25	2	3	1	.93	-	27 (80)	-	-					
182. tert-Butyl Perbenzoate C₆H₅CO₃C(CH₃)₃	22b	-	1	3	4	1.0	-	107 (225)	-	-	113dec. (235)				
183. Butyl Peroxypivalate (in 75% sol. of Mineral Spirits) (CH₃)₃CO₂COC(CH₃)₃	22b	-	0	3	4		-	>68 (>155)		-		−19 (−2)			
Butyl Peroxytrimethyl Acetate, see Butyl Peroxypiualate															
184. di-Butyl Phosphate (C₄H₉O)₂PO(OH)	7b	-												alc., ace.	
185. tri-Butyl Phosphate See 1054															
186. p-tert-Butyl Toluene CH₃C₆H₄C₄H₉	18	10	3	2	-	.85	-	-			193 (379)	−52 (−62)	insol.	ace., bz., chl.	
187. Butyl Vinyl Ether C₄H₉OCH=CH₂	15	-	2	3	2	.77	3.4	−9 (16)oc		-	94 (201)	−92 (−134)	insol.	org. sol.	

SUBSTANCE/FORMULA	Waste Disposal Procedure (See VI)	TLV (ACGIH) PPM (mg/M³)	NFPA 704M System			Sp. Gr.	Vap. Dens. (Air=1)	Fl. Pt. °C (°F)	Ignit. Temp. °C (°F)	Flam. Limits %	B.P. °C (°F)	M.P. °C (°F)	Sol. in H₂O g/100g	Other Solvents	Misc. Ref.
			Health	Fire	React.										
188. 1-Butyne $CHCCH_2CH_3$	18	-	-	2	-	.67	-	-	-	-	8.1 (46)	-127 (-197)	insol.	alc.	
189. n-Butyraldehyde $CH_3CH_2CH_2CHO$	2	-	2	3	1	.82	2.5	-6.7 (20)	230 (446)	2.5-	76 (169)	-89 (-146)	4	alc.	BDH, MCA
190. iso-Butyraldehyde $(CH_3)_2CHCHO$	2	-	2	3	1	.79	2.5	-40 (-40)	254 (490)	1.6-10.6	64 (147)	-66 (-87)	4	alc.	BDH, MCA
191. n-Butyric Acid $CH_3(CH_2)_2COOH$	24a	-	2	2	0	.96	3.0	66 (151)	452 (846)	2-10	164 (327)	-7.9 (17.5)	∞	alc.	BDH
192. iso-Butyric Acid $(CH_3)_2CHCOOH$	24a	-	1	2	-	.95	3.0	62 (144)	502 (935)	-	154 (309)	-47 (-53)	20^{20}	alc., chl.	BDH
193. n-Butyric Anhydride $[CH_3(CH_2)_2CO]_2O$	24a	-	3	2	1	.97	5.4	88 (190)	307 (585)	-	198 (388)	-73 (-99)	dec.	dec., alc.	
194. 2-Butyrolactone $CH_2CH_2CH_2COO$	18	-	1	2	-	1.05	3.0	98 (209)	-	-	206 (403)	-44 (-47)	∞	alc.	BDH
195. n-Butyronitrile $CH_3(CH_2)_3CN$	14	-			-	.8	-	26 (79)	-	-	117 (243)	-112 (-170)	sl. sol.	alc., bz.	BDH
196. Butyryl Chloride C_3H_7COCl	1a	-	3	2	-	1.03	3.7	<21 (<70)	-	-	107 (225)	-89 (-128)	dec.	alc.	BDH
197. Cadmium Cd	27a	0.2	3	-	-	8.6	-	-	-	-	767 (1413)	321 (610)	insol.	-	
198. Cadmium Chloride $CdCl_2$	27f	(.2)	3	-	-	4.0	-	-	-	-	960 (1054)	568 (760)	140^{20}	sl. sol., alc.	
199. Cadmium Oxide CdO	27f	(.1)	3	-	-	8.2	-	-	-	-	1559 (2838)	900 (1652)	insol.	-	

SUBSTANCE/FORMULA	Waste Disposal Procedure (See VI)	TLV (ACGIH) PPM (mg/M³)	NFPA 704M System Health	Fire	React.	Sp. Gr.	Vap. Dens. (Air=1)	Fl. Pt. °C (°F)	Ignit. Temp. °C (°F)	Flam. Limits %	B.P. °C (°F)	M.P. °C (°F)	Sol. in H₂O g/100g	Other Solvents	Misc. Ref.
200. Calcium Ca	3	(5)	1	1	2	1.5	-	-	-	-	1240 (1548)	842 (2264)	dec.	sl. sol., alc.	BDH
201. Calcium Arsenate $Ca(AsO_4)_2$	27d	(1)	3	-	-	3.6	-	-	-	-	-	1.45 (34.5)	.0132^{25}	-	
202. Calcium Carbide CaC_2	25	-	1	4	2	2.2	-	-	-	-	2300 (4172)	447 (837)	dec.	-	BDH, MCA
203. Calcium Carbonate $CaCO_3$	26	-	1	-	-	2.7	-	-	-	-	899 (1650)	825 (1517)	sl. sol.	-	
204. Calcium Chlorate $Ca(ClO_3)_2 \cdot 2H_2O$	12a	-	2	2	-	2.71	-	-	-	-	-	100 (212)	178^{8}	alc., ace.	AIA
205. Calcium Chloride $CaCl_2$	11	-	-	-	-	2.15	-	-	-	-	<1600 (<2912)	772 (1422)	75^{20}	alc., ace.	NTTC
206. Calcium Cyanide $Ca(CN)_2$	14	10	2	0	0	-	-	-	-	-	-	350dec. (662)	dec.	-	BDH, MCA, MCB
207. Calcium Hydride CaH_2	17	-	3	-	-	1.8	-	-	-	-	-	816 (1501)	dec.	dec., alc.	BDH
208. Calcium Hydroxide $Ca(OH)_2$	10	-	2	0	-	2.5	-	-	-	-	dec.	580 (1076)	.18^{0}	-	
209. Calcium Hypochlorite $Ca(ClO)_2$	12a	-	2	1	2	2.35	-	-	-	-	-	100dec. (212)	dec.	-	
210. Calcium Oxide CaO	26 or 10	(5)	1	0	1	3.25	-	-	-	-	2850 (5162)	2580 (4676)	.13^{10}	-	BDH
210a. Calcium Sulfide CaS	23														
211. Camphor $C_{10}H_{16}O$	18	(2)	2	2	0	.99	5.24	66 (151)	466 (871)	.6-3.5	209 (408)	175 (347)	sl. sol.	ace., alc., bz., chl.	

SUBSTANCE/FORMULA	Waste Disposal Procedure (See VI)	TLV (ACGIH) PPM (mg/M³)	NFPA 704M System Health	Fire	React.	Sp. Gr.	Vap. Dens. (Air=1)	Fl. Pt. °C (°F)	Ignit. Temp. °C (°F)	Flam. Limits %	B.P. °C (°F)	M.P. °C (°F)	Sol. in H₂O g/100g	Other Solvents	Misc. Ref.
212. Caproic Acid C$_5$H$_{11}$COOH	24a	-	2	1	0	.93	4.0	102 (215)oc	-	-	205 (401)	−5.4 (22)	1.120	alc.	-
Capryl Alcohol, see 2-Octanol															
213. Caprylaldehyde CH$_3$(CH$_2$)$_6$CHO	2	-	2	2	0	.8	4.5	52 (125)	-	-	168 (335)	-	-	org. solv.	
Caprylic Acid, see Octanoic Acid															
Caprylic Alcohol, see 1-Octanol															
214. Carbazole C$_6$H$_4$NHC$_6$H$_4$	6	-	1	1	-	1.10	-	-	-	-	355 (671)	245 (473)	insol.	alc., hot bz.	
Carbitol, see Diethylene Glycol Monoethyl Ether															
Carbolic Acid, see Phenol															
215. Carbon, Black C	26	(3.5)	1	1	-	~2	-	-	-	-	~4200 (~7592)	~3680 (~6656)	insol.	-	
216. Carbon Disulfide CS$_2$	9	20 Skin	2	3	0	1.26	2.6	−30 (−22)	100 (212)	1-44	46 (115)	−112 (−169)	.2⁰	alc., chl.	AIA, BDH, MCA, NSC
217. Carbon Monoxide CO	18	50	2	4	0	.81	.97	-	609 (1128)	12.5-74	−192 (−314)	−207 (−341)	.004⁰	alc., bz.	
Carbon Oxysulfide, see Carbonyl Sulfide															

SUBSTANCE/FORMULA	Waste Disposal Procedure (See VI)	TLV (ACGIH) PPM (mg/M³)	NFPA 704M System			Sp. Gr.	Vap. Dens. (Air=1)	Fl. Pt. °C (°F)	Ignit. Temp. °C (°F)	Flam. Limits %	B.P. °C (°F)	M.P. °C (°F)	Sol. in H₂O g/100g	Other Solvents	Misc. Ref.
			Health	Fire	React.										
218. Carbon Tetrabromide CBr_4	27j	-	3	0	1	3.4	-	-	-	-	190 (374)	90 (194)	insol.	alc., chl.	
219. Carbon Tetrachloride CCl_4	27j	10 Skin	3	0	0	1.58	5.3	none	-	none	77 (171)	-23 (-9.4)	insol.	alc., bz., chl.	AIA, BDH, MCA, NTTC, NSC
220. Carbon Tetrafluoride CF_4	21	-	2	-	-	3.42	-	-	-	-	-127 (-197)	-184 (-299)	sl. sol.	-	
Carbonyl Chloride see Phosgene															
221. Carbonyl Fluoride COF_2	21	1	2	3	-	1.14	-	-	-	-	-83 (-117)	-114 (-173)	dec.	dec., alc.	MGB
222. Carbonyl Sulfide COS	13	-	3	4	1	1.07	2.1	-	-	12-29	-50 (-58)	-138 (-216)	80[14]	alc.	MGB
Carvene, see Dipentene															
223. Caustic Potash KOH	10	-	3	0	1	2.04	-	none	-	none	1320 (2408)	360 (680)	107[15]	alc.	BDH, MCA
224. Caustic Soda $NaOH$	10	(2)	3	0	1	2.13	-	none	-	none	1390 (2534)	315 (599)	42[0]	alc.	BDH, NTTC, MCA
225. Cellosolve $CH_2OHCH_2OC_2H_5$	18	-	2	2	-	.93	3.10	94 (202)	-	1.7-15.6	135 (275)	-	∞	ace., alc., chl.	BDH
226. Cellosolve Acetate $CH_3COOCH_2CH_2OC_2H_5$	18	100 Skin	2	2	-	.97	4.7	55 (131)	382 (720)	1.2-12.7	156 (313)	-62 (-80)	v. sol.	org. solv.	
227. Cellulose Nitrate $C_{12}H_{17}(ONO_2)_3O_7$	28	-	2	3	3	1.66	-	13 (55)	-	-	-	-	insol.	ace., MeOH	

SUBSTANCE/FORMULA	Waste Disposal Procedure (See VI)	TLV (ACGIH) PPM (mg/M³)	NFPA 704M System			Sp. Gr.	Vap. Dens. (Air=1)	Fl. Pt. °C (°F)	Ignit. Temp. °C (°F)	Flam. Limits %	B.P. °C (°F)	M.P. °C (°F)	Sol. in H₂O g/100g	Other Solvents	Misc. Ref.
			Health	Fire	React.										
228. Celluloid, Lower Nitrates of Cellulose and Camphor	28	-	-	2	-	-	-	-	-	-	-	-	-	-	
229. Chloral CCl_3CHO	2	-	3	-	-	1.51	5.1	none	-	-	98 (208)	-58 (-72)	v. sol.	hot alc.	
230. Chloral Hydrate $CCl_3CH(OH)_2$	2	-	2	1	-	1.64	-	-	-	-	96 (205)	52 (126)	47417	alc, chl, CS_2	
231. Chlordane $C_{10}H_6Cl_8$	4b	(.5)	3	-	-	~1.6	-	56 (133)	-	-	175 (347)	-	-	-	
232. Chloric Acid $HClO_3 \cdot 7H_2O$	12a	-	3	3	3	1.28	-	-	-	-	40dec. (104)	-20 (-4)	v. sol.	-	BDH
233. Chlorine Cl_2	12a	1C	3	0	1	2.48	2.49	-	-	-	-34 (-29)	-101 (-150)	1.4^0	-	Cl
234. Chlorine Dioxide ClO_2	12a	.1	3	3	-	3.09	2.3	-	100 (212)	-	9.9 (50)	-59 (-74)	dec.	expl. with compounds of carbon.	
235. Chlorine Trifluoride ClF_3	21	.1C	3	3	3	1.77	3.1	-	-	-	12 (54)	-83 (-117)	dec.	-	MCA, MGB, NASA
236. Chloroacetaldehyde CH_2ClCHO	2	1C	3	2	-	1.19	2.7	88 (190)	-	-	85 (185)	-16 (3.2)	-	-	
237. Chloroacetamide $CH_2ClCONH_2$	20	-	-	-	-	-	-	-	-	-	225 (437)	120 (248)	10^{24}	alc.	
238. Chloroacetic Acid $CH_2ClCOOH$	4c	-	4	2	-	1.40	3.3	-	-	-	189 (372)	63 (145)	v. sol.	alc., bz., chl., CS_2	BDH
239. Chloroacetonitrile CH_2ClCN	14	-	3	-	-	1.19	-	-	-	-	123 (253)	-	-	-	

SUBSTANCE/FORMULA	Waste Disposal Procedure (See VI)	TLV (ACGIH) PPM (mg/M³)	NFPA 704M System Health	Fire	React.	Sp. Gr.	Vap. Dens. (Air=1)	Fl. Pt. °C (°F)	Ignit. Temp. °C (°F)	Flam. Limits %	B.P. °C (°F)	M.P. °C (°F)	Sol. in H₂O g/100g	Other Solvents	Misc. Ref.
240. Chloroacetophenone $C_6H_5COCH_2Cl$	4b	.05	3	-	-	1.19	5.2	-	-	-	247 (477)	59 (138)	insol.	ace., alc., bz., CS₂	
241. Chloroacetyl Chloride $CH_2ClCOCl$	1a	-	3	-	-	1.5	-	-	-	-	108 (226)	−23 (−9)	dec.	dec., alc.	BDH
242. m-Chloroaniline $C_6H_4NH_2Cl$	5	-	3	-	-	1.22	-	-	-	-	229 (444)	−10 (14)	insol.	org. solv.	BDH
243. o-Chloroaniline $C_6H_4NH_2Cl$	5	-	3	-	-	1.21	-	-	-	-	209 (408)	−14 (6.8)	insol.	org. solv.	BDH
244. p-Chloroaniline $C_6H_4NH_2Cl$	5	-	3	-	-	1.43					231 (448)	70 (158)	sol. (hot)	org. solv.	BDH
Chlorobenzaldehyde, see Benzoyl Chloride															
245. Chlorobenzene C_6H_5Cl	4b	75	2	3	0	1.11	3.9	29 (84)	638 (1180)	1.3-7.1	132 (270)	−56 (−69)	insol.	alc., bz., chl., CS₂	AIA, BDH, NTTC
246. o-Chlorobenzoyl Chloride ClC_6H_4COCl	1a	-	2	2	-	-	-	-	-	-	225 (437)	−10 (14)	dec.	dec. alc.	
247. Chlorobromomethane CH_2ClBr	26	200	2	-	-	1.93	4.5	-	-	-	68 (154)	−88 (−126)	insol.	org. solv.	
Chlorobutadiene, see Chloroprene															
248. p-Chloro-m-cresol $C_6H_3CH_3OHCl$	4b	-	3	-	-	-	-	<−1 (<30)	-	-	235 (455)	66 (151)	sl. sol.	alc.	BDH
249.															

SUBSTANCE/FORMULA	Waste Dis-posal Pro-cedure (See VI)	TLV (ACGIH) PPM (mg/M³)	NFPA 704M System			Sp. Gr.	Vap. Dens. (Air=1)	Fl. Pt. °C (°F)	Ignit. Temp. °C (°F)	Flam. Limits %	B.P. °C (°F)	M.P. °C (°F)	Sol. in H₂O g/100g	Other Solvents	Misc. Ref.
			Health	Fire	React.										
250. **1-Chloro-1,1-Difluoroethane** $C_2H_3ClF_2$	26	-	2	0	-	1.12	-	-	632 (1170)	9-14.8	-9.2 (15)	-	-	-	
251. **1-Chloro-2,4-Dinitrobenzene** $C_6H_3Cl(NO_2)_2$	6	-	3	1	4	1.7	-	194 (381)	432 (810)	2-22	315 (599)	43 (109)	insol.	bz., CS_2	BDH
252. **Chlorodiphenyls** $C_{12}H_xCl_y$	4b	(.05)	3	1	-	~1.4	-	176-80 (394-50) oc	-	-	340-75 (644-707)	-	-	-	
1-Chloro-2,3-Epoxypropane, see Epichlorohydrin															
Chloroethane, see Ethyl Chloride															
Chloroethanol, see Ethylene Chlorohydrin															
253. **Chloroform** $CHCl_3$	27j	50C	3	1	0	1.49	4.1	none	-	none	61 (142)	-64 (-83)	sl. sol.	alc., bz.	AIA, BDH, NTTC, MCA
Chloromethane, see Methyl Chloride															
254. **Chloromethyl Ether** $ClCH_2OCH_3$	15	-	-	-	-	1.07	-	-	-	-	59 (138)	-103 (-153)	dec.	alc.	
255. **2-Chloro-2-Methyl Propene** $ClCH_2-C(CH_3)CH_2$	4b	-	-	-	-	.93	-	-	-	-	71 (160)	-	-	alc.	
256. **Chloronaphthalene** $C_{10}H_7Cl$	4b	-	3	-	-	1.19	5.6	132 (270)	558 (1036)	-	256 (493)	-20 (-4)	insol.	CS_2, chl., alc., bz.	BDH

SUBSTANCE/FORMULA	Waste Disposal Procedure (See VI)	TLV (ACGIH) PPM (mg/M³)	NFPA 704M System			Sp. Gr.	Vap. Dens. (Air=1)	Fl. Pt. °C (°F)	Ignit. Temp. °C (°F)	Flam. Limits %	B.P. °C (°F)	M.P. °C (°F)	Sol. in H₂O g/100g	Other Solvents	Misc. Ref.
			Health	Fire	React.										
257. Chloronaphthylamine C₁₀H₈ClN	6	-	CAR	-	-	-	-	-	-	-	-	60 (140)	-	alc.	
258. Chloronitroanilines NO₂ClC₆H₃NH₂	6	-	3	1	-	-	-	-	-	-	-	108-16 (226-41)	-	alc., ace.	BDH
259. Chloronitrobenzenes C₆H₄ClNO₂	6	-	3	1	1	1.37	-	127 (261)	-	-	242 (468)	32-46 (90-115)	insol.	alc., bz.	BDH
260. 1-Chloro-1-Nitropropane C₂H₅CHClNO₂	4a	20	3	2	3	1.21	4.3	62 (144)	-	-	142 (288)	-	sl. sol.	alc.	
261. Chloropentafluoroethane C₂ClF₅	26	1000	-	0	0	1.26	-	-	-	-	-39 (-38)	-106 (-159)	insol.	alc.	
262. m-Chlorophenol C₆H₄ClOH	6	-	3	2	-	1.24	-	-	-	-	214 (417)	32.5 (90.5)	sol.(hot)	alc., bz.	BDH
263. o-Chlorophenol C₆H₄ClOH	6	-	3	2	-	1.24	-	85 (185)	-	-	175 (347)	7 (45)	v. sol.	bz.	BDH
264. p-Chlorophenol C₆H₄ClOH	6	-	3	2	-	1.24	-	121 (250)	-	-	220 (428)	43 (109)	sol.	alc., bz.	BDH
265. o-Chlorophenyl Diphenyl Phosphate ClC₆H₄O(C₆H₅O)₂PO	7b	-	-	1	-	-	12.5	215 (419)	-	-	240-55 (464-91)	<0 (<32)	-	-	
266. Chloropicrin CCl₃NO₂	4a	.1	4	0	1	1.66	5.7	-	-	-	112 (234)	-64 (-83)	sl. sol.	alc., bz.	BDH
267. Chloroprene CH₂CHCClCH₂	4b	25	3	-	-	.95	3.0	-20 (-4)	-	4-20	59 (138)	-	sl. sol.	org. sol.	
Chloropropane, see n-Propyl Chloride															

SUBSTANCE/FORMULA	Waste Disposal Procedure (See VI)	TLV (ACGIH) PPM (mg/M³)	NFPA 704M System			Sp. Gr.	Vap. Dens. (Air=1)	Fl. Pt. °C (°F)	Ignit. Temp. °C (°F)	Flam. Limits %	B.P. °C (°F)	M.P. °C (°F)	Sol. in H₂O g/100g	Other Solvents	Misc. Ref.
			Health	Fire	React.										
268. 3-Chloro-Propionly Chloride ClCH₂CH₂COCl	1a	-	-	-	-	1.33	-	-	-	-	143 (289)	-	sl. sol.	alc., chl.	BDH
269. Chlorosulfonic Acid HOSO₂Cl	1b	-	3	0	2	1.79	4.0	none	-	none	152 (306)	−80 (−112)	dec.	dec. alc.	BDH, MCA, NSC
270. Chlorotrifluoroethylene FCCl=CF₂	26	-	3	4	2	1.31	-	-	-	8.4—39	−28 (−18)	−158 (−252)	-	-	
271. Chromates (CrO$_H^=$) and Dichromates (Cr₂O₇$^=$)	12a	(.1)	-	-	-	-	-	-	-	-	-	-	-	-	
272. Chromic Salts (soluble) Cr+++	11	(.5)	-	-	-	-	-	-	-	-	-	-	sol.	-	
273. Chromium Cr	26	(.5)	0	2	-	7.2	-	-	-	-	2480 (4496)	1890 (3434)	insol.	-	
274. Chromium Trioxide (Chromic Acid) CrO₃	12a	(.1)	1	0	1	2.70	-	-	-	-	dec.	196 (385)	62^0	alc.	BDH, MCA
275. Chromous Salts (soluble) Cr++	12b	(.5)	3	-	-	-	-	-	-	-	-	-	sol.	-	
276. Chromyl Chloride (Chromium Oxycloride) CrO₂Cl₂	12a	-	3	-	-	1.91	-	-	-	-	116 (241)	−96 (−141)	dec.	org. sol.	BDH
277. Cinnamaldehyde C₆H₅CH=CHCHO	2	-	-	-	-	1.05	-	49 (120)	-	-	253 (487)	−7.5 (18)	sl. sol.	alc., chl.	
278. Citraconic Anhydride C₅H₄O₃	24a	-	-	-	-	1.25	-	-	-	-	213 (416)	7-8 (45—6)	dec.	alc.	
Citrene, see Dipentene															
279. Citric Acid C₃H₄(OH)(COOH)₃	24a	-	1	1	-	1.54	-	-	-	-	dec.	153 (307)	v. sol.	alc.	

SUBSTANCE/FORMULA	Waste Disposal Procedure (See VI)	TLV (ACGIH) PPM (mg/M³)	NFPA 704M System Health	Fire	React.	Sp. Gr.	Vap. Dens. (Air=1)	Fl. Pt. °C (°F)	Ignit. Temp. °C (°F)	Flam. Limits %	B.P. °C (°F)	M.P. °C (°F)	Sol. in H₂O g/100g	Other Solvents	Misc. Ref.
280. Cleaning Solution	12a	-	-	-	-	-	-	-	-	-	-	-	-	-	
281. Cobalt Co	27a	.1	2	-	-	8.9	-	-	-	-	2900 (5252)	1495 (2723)	insol.	-	
282. Cobaltous Nitrate Co(NO₃)₂·6H₂O	11	-	1	0	1	1.87	-	-	-	-	-	55dec. (131)	134⁰	alc., ace.	
283. Collodion C₁₂H₁₆O₆(NO₃)₄ C₁₃H₁₇O₇(NO₃)₃	28	-	1	4	0	-	-	$\lesssim$-18 (<0)	-	-	-	-	-	-	
284. Copper Cu	27a	.1	2	-	-	8.9	-	-	-	-	2595 (4703)	1083 (1981)	insol.	-	
285. Copper Nitrate Cu(NO₃)₂·6H₂O	11	-	1	0	1	2.07	-	-	-	-	198dec. (388)	26 (79)	240⁰	alc.	
286. Crag 974 C₅H₁₀N₂S₂	13	(15)	2	-	-	-	5.6	-	-	-	-	106 (223)	-	-	
287. m-Cresol CH₃C₆H₄OH	18	5	2	1	0	1.03	3.7	94 (202)	559 (1038)	1.06-1.35 @302°F	203 (397)	12 (54)	sol.	org. sol.	BDH, MCA
288. o-Cresol CH₃C₆H₄OH	18	5	2	2	0	1.05	3.7	81 (178)	599 (1110)	1.35-	191 (376)	31 (88)	sol. (hot)	org. sol.	BDH, MCA
289. p-Cresol CH₃C₆H₄OH	18	5	2	1	0	1.04	3.7	94 (202)	559 (1038)	1.06-1.4 @302°F	202 (396)	35 (95)	sol. (hot)	org. sol.	BDH, MCA
290. Creosote (Mixed Phenols)	18	-	2	2	0	1.07	-	74-82 (165-180)	336 (637)	-	200-250 (392-482)	-	-	-	NTTC
Cresylic Acid, see o-Cresol															

SUBSTANCE/FORMULA	Waste Disposal Procedure (See VI)	TLV (ACGIH) PPM (mg/M³)	NFPA 704M System			Sp. Gr.	Vap. Dens. (Air=1)	Fl. Pt. °C (°F)	Ignit. Temp. °C (°F)	Flam. Limits %	B.P. °C (°F)	M.P. °C (°F)	Sol. in H2O g/100g	Other Solvents	Misc. Ref.
			Health	Fire	React.										
291. Crotonaldehyde $CH_3CHCHCHO$	2	2	3	3	2	.87	2.4	13 (55)	207 (405)	2.1-15.5	104 (219)	-76 (-105)	v. sol.	alc., bz.	BDH
292. Croton Oil	18	-	3	1	-	.94	-	-	-	-	-	-	-	-	
293. Crotonitrile $CH_3CH=CHCN$	14	-	-	-	-	.83	2.3	<100 (<212)	-	-	122 (252)	-52 (-62)	-	-	
294. Crotonyl Chloride $CH_3CH=CHCH_2Cl$	1a	-	1	-	-	1.09	-	-	-	-	124 (255)	-	dec.	-	
295. Crude Oil (Petroleum)	18	-	2	2	-	.78-.97	-	(-7)-(+32) (20-90)	-	-	-	<-46 (<-50)	-	-	NTTC
296. Cumene $C_6H_5CH(CH_3)_2$	18	50	0	2	0	.86	4.1	44 (111)	424 (795)	.9-6.5	152 (306)	-96 (-141)	insol.	alc., bz.	BDH
297. Cumene Hydroperoxide $C_6H_5C(CH_3)_2OOH$	22b	-	1	2	4	1.05	-	79 (174)	-	-	153 (307)	-	-	-	NFPA
298. Cuprous Chloride $CuCl$	11	-	3	-	-	3.53	-	-	-	-	1490 (2714)	422 (792)	sl. sol.	-	
299. Cuprous Cyanide $CuCN$	14	-	3	-	-	2.92	-	-	-	-	dec.	473 (883)	insol.	-	
300. Cyanamide H_2NCN	14	-	2	1	-	1.07	1.45	141 (286)	-	-	140 (284) @19mm.	42 (108)	v. sol.	alc., bz.	
301. Cyanides $-CN^-$	14	(5)	3	2	-	-	-	-	-	-	-	-	-	-	AIA, BDH
302. Cyanoacetamide $CNCH_2CONH_2$	14	-	3	2	-	-	-	-	-	-	dec.	118 (244)	15	sl. sol. alc.	

SUBSTANCE/FORMULA	Waste Disposal Procedure (See VI)	TLV (ACGIH) PPM (mg/M³)	NFPA 704M System			Sp. Gr.	Vap. Dens. (Air=1)	Fl. Pt. °C (°F)	Ignit. Temp. °C (°F)	Flam. Limits %	B.P. °C (°F)	M.P. °C (°F)	Sol. in H₂O g/100g	Other Solvents	Misc. Ref.
			Health	Fire	React.										
303. Cyanogen (CN)₂	14	10	4	4	2	.95	1.8	-	-	6-32	−21 (−5.8)	−34 (−29)	45020	alc.	MGB
304. Cyanogen Chloride CNCl	14	.5	3	-	-	1.2	1.9	-	-	-	13 (55)	−6.5 (20)	sl. sol.	alc.	
305. Cycloheptanone C₆H₁₂CO	18	-	2	-	-	.95	-	-	-	-	178 (352)	-	insol.	alc.	
306. Cyclohexane C₆H₁₂	18	300	1	3	0	.77	2.91	−20 (−4)	260 (500)	1.3-8.4	80 (176)	4.1 (39)	insol.	alc., ace., bz.	BDH, MCA
307. Cyclohexanol C₅H₁₀CHOH	18	50	1	2	0	.96	3.45	68 (154)	300 (572)	-	161 (322)	25 (77)	3.6	CS₂, bz.	
308. Cyclohexanone C₅H₁₀CO	18	50	1	2	0	.99	3.4	44 (111)	420 (788)	1.1-8.1	156 (313)	−16 (3.2)	sl. sol.	alc.	BDH
309. Cyclohexene C₆H₁₀	18	300	1	3	0	.81	2.8	−6 (21)	-	-	83 (181)	−104 (−155)	insol.	alc., ace., bz.	BDH
310. Cyclohexylamine C₆H₁₁NH₂	7a	-	2	3	0	.87	3.4	32 (90)	293 (560)	-	135 (275)	−18 (−4)	sol.	org. sol.	BDH
311. Cyclohexylbenzene C₆H₁₁C₆H₅	18	-	2	1	0	.95	-	99 (210) oc	104 (220)	-	238 (460)	7.5 (45)	insol.	alc.	
312. Cyclonite-RDX C₃H₆N₆O₆	4a	-	2	3	4	1.8	-	-	-	-	-	204 (399)	-	ace.	
313. Cyclopentadiene C₅H₆	18	75	-	2	-	.80	-	-	-	-	42 (108)	−97 (−143)	insol.	alc., bz.	
314. Cyclopentane C₅H₁₀	18	-	1	3	0	.75	2.42	−7 (<20)	-	-	49 (120)	−94 (−137)	insol.	alc.	

SUBSTANCE/FORMULA	Waste Disposal Procedure (See VI)	TLV (ACGIH) PPM (mg/M³)	NFPA 704M System Health	Fire	React.	Sp. Gr.	Vap. Dens. (Air=1)	Fl. Pt. °C (°F)	Ignit. Temp. °C (°F)	Flam. Limits %	B.P. °C (°F)	M.P. °C (°F)	Sol. in H₂O g/100g	Other Solvents	Misc. Ref.
315. Cyclopentanone C_4H_8CO	18	-	2	3	0	.95	2.3	26 (79)	-	-	131 (268)	-58 (-72)	insol.	alc., MeOH	
316. Cyclopropane C_3H_6	18	-	1	4	0	.72	1.45	-	498 (928)	2.4-10.4	-33 (-27)	-127 (-197)	sl. sol.	alc.	BDH
317. p-cymene $CH_3C_6H_4CH(CH_3)_2$	18	-	2	2	0	.86	4.62	47 (117)	436 (817)	.7-5.6	177 (351)	-68 (-90)	insol.	alc., chl.	BDH
318. Decaborane $B_{10}H_{14}$	17	.05 Skin	3	2	1	.94	-	80 (176)	149 (300)	-	213 (415)	100 (212)	sl. sol.	alc., bz., CS₂	AIA, NFPA, NSC
319. Decahydronaphthalene $C_{10}H_{18}$	18	-	2	2	0	.87	4.76	57 (135)	250 (482)	.7-4.9	186 (367)	-45 (-49)	insol.	alc., chl.	BDH
320. iso-Decaldehyde $C_9H_{19}CHO$	2	-	-	-	-	.8	5.4	85 (186)	-	-	197 (387)	-	-	-	
Decalin, see Decahydronaphthalene															
321. n-Decane $CH_3(CH_2)_8CH_3$	18	-	0	2	0	.73	4.90	46 (115)	208 (406)	.8-5.4	174 (345)	-30 (-22)	insol.	alc.	
322. m-Decyl Alcohol $CH_3(CH_2)_8CH_2OH$	18	-	-	2	-	.83	5.3	82oc (180)	-	-	229 (444)	7 (45)	insol.	alc., ace., bz., chl.	
Dematon, see Systox															
323. Deuterium D_2	27k	-	0	4	0	.18	-	-	-	5-75	-250 (-418)	-255 (-427) @121mm	-	-	
324. Diacetone Alcohol $CH_3COCH_2C(OH)(CH_3)_2$	18	50	1	2	0	-	4.0	64 (147)	603 (1117)	1.8-6.9	168 (334)	-50 (-58)	∞	alc.	BDH
Diaminoethane, see Ethylene Diamine															

SUBSTANCE/FORMULA	Waste Disposal Procedure (See VI)	TLV (ACGIH) PPM (mg/M³)	NFPA 704M System Health	Fire	React.	Sp. Gr.	Vap. Dens. (Air=1)	Fl. Pt. °C (°F)	Ignit. Temp. °C (°F)	Flam. Limits %	B.P. °C (°F)	M.P. °C (°F)	Sol. in H₂O g/100g	Other Solvents	Misc. Ref.
Diatomaceous Earth, see Silica															
325. Diazomethane CH_2N_2	8	.2	3	-	-	1.45	-	-	100exp. (212)	-	−23 (−9.4)	−145 (−229)	dec.	hot alc.	
326. 1,2,5,6-Dibenzacridine $C_{21}H_{15}N$	5	-	-	-	-	-	-	-	-	-	-	-	-	-	
327. 1,2,7,8-Dibenzacridine $C_{21}H_{15}N$	5	-	-	-	-	-	-	-	-	-	-	-	-	-	
328. 1,2,5,6-Dibenzanthracene $C_{22}H_{14}$	18	-	-	-	-	-	-	-	-	-	-	268 (514)	insol.	ace., bz., CS₂	
329. 1,2,5,6-Dibenzcarbazole $C_{20}H_{17}N$	5	-	-	-	-	-	-	-	-	-	-	-	-	-	
330. 3,4,5,6-Dibenzcarbazole $C_{20}H_{17}N$	5	-	-	-	-	-	-	-	-	-	-	-	-	-	
331. 1,2,5,6-Dibenzofluorene $C_{21}H_{18}$	18	-	-	-	-	-	-	-	-	-	280 (536) @ .3mm	174 (345)	-	bz.	
332. 1,2,3,4-Dibenzophenanthrene $C_{22}H_{18}$	18	-	-	-	-	-	-	-	-	-	-	-	-	-	
Dibenzopyrrole, see Carbazole															
333. 3,4,8,9-Dibenzpyrene $C_{24}H_{17}$	18	-	-	-	-	-	-	-	-	-	-	-	-	-	
334. Dibenzylamine $HN(CH_2C_6H_5)_2$	5	-	3	1	-	1.03	-	-	-	-	300dec. (572)	−26 (−15)	sol.	alc.	

SUBSTANCE/FORMULA	Waste Disposal Procedure (See VI)	TLV (ACGIH) PPM (mg/M³)	NFPA 704M System			Sp. Gr.	Vap. Dens. (Air=1)	Fl. Pt. °C (°F)	Ignit. Temp. °C (°F)	Flam. Limits %	B.P. °C (°F)	M.P. °C (°F)	Sol. in H₂O g/100g	Other Solvents	Misc. Ref.
			Health	Fire	React.										
335. Diborane B_2H_6	17	0.1	3	4	3	.46	.96	-90 (-130)	145 (293)	.9-98	-93 (-135)	-165 (-265)	dec.	dec., alc.	AIA, MCA, NSC
Dibrom, see Dimethyl-1,2-Dibromo-2,2-Dichloroethylphos															
336. 2, 6-Dibromo-N-Chloro-p-Benzo-Quinonimine $C_6H_2Br_2ClNO$	6	-	-	-	-	-	-	-	-	-	-	85 (185)	insol.	-	
1,2-Dibromoethane see Ethylene Dibromide															
337. 1,2,-Dibromotetrafluoromethane (Freon-114B2) $C_2Br_2F_4$	26	-	-	0	-	2.72	-	-	-	-	24.5 (76)	-142 (-224)	-	-	
338. Dibutylamine $(C_4H_9)_2NH$	7a	-	3	2	0	.50	4.46	52 (126)oc	-	-	159 (318)	-51 (-60)	sol.	alc.	
339. Dibutyldichlorotin $(C_4H_9)_2 SnCl_2$	4b	-	3	1	-	1.36	10.5	355 (671)	-	-	135 (275) @ 10mm	43 (109)	dec.	alc., bz.	
340. Di-n-Butyl Ether $[CH_3(CH_2)_3]_2O$	15	-	2	3	0	.77	4.5	25 (77)	194 (381)	1.5-7.6	141 (286)	-98 (-144)	insol.	alc.	BDH
341. Dibutyl Oxalate $(C_4H_9)_2C_2O_4$	18	-	3	1	0	1.01	7.0	104 (220)	-	-	246 (475)	-30 (-22)	insol.	alc.	
342. Dibutyl Peroxide $(CH_3)_3COOC(CH_3)_3$	22b	1000	2	3	4	.79	5.0	18 (65)	-	-	111 (232)	-40 (-40)	insol.	ace.	NFPA
343. Dibutyl Phosphite $(C_4H_9O)_2POH$	7b	1	3	2	0	.97	6.7	49 (120)	-	-	115 (239) @ 10mm	-	-	-	

SUBSTANCE/FORMULA	Waste Disposal Procedure (See VII)	TLV (ACGIH) PPM (mg/M³)	NFPA 704M System			Sp. Gr.	Vap. Dens. (Air=1)	Fl. Pt. °C (°F)	Ignit. Temp. °C (°F)	Flam. Limits %	B.P. °C (°F)	M.P. °C (°F)	Sol. in H₂O g/100g	Other Solvents	Misc. Ref.
			Health	Fire	React.										
344. Dibutyl Phthalate $C_6H_4(COO)_2(C_4H_9)_2$	18	(5)	0	1	0	1.04	9.6	157 (315)	403 (757)	-	340 (644)	-35 (-31)	insol.	bz, ace., alc.	-
345. Dibutyl Tin Dilaurate $[CH_3(CH_2)_{10}COO]_2$ $Sn(C_4H_9)_2$	18	-	-	1	-	1.05	21.8	235 (455)	-	-	-	27 (81)	insol.	ace., bz.	
346. Dichloroacetic Acid $CHCl_2COOH$	4c	-	3	-	-	1.56	4.45	-	-	-	192 (378)	9.7 (50)	∞	alc.	BDH
347. Dichloroacetyl Chloride $CHCl_2COCl$	1a	-	3	2	1	1.53	5.1	66 (151)	-	-	107 (225)	-	dec.	dec., alc.	BDH
348. 2,5-Dichloroaniline $Cl_2C_6H_3NH_2$	6	-	3	1	0	-	5.6	166 (331)	-	-	251 (484)	50 (122)	sl. sol.	alc., bz.	BDH
349. o-Dichlorobenzene $C_6H_4Cl_2$	6	50C	2	2	0	1.30	5.07	66 (151)	648 (1198)	2-9	180 (356)	-18 (-4)	insol.	alc.	AIA, BDH, NTTC, MCA
350. 3,3'-Dichlorobenzidine $ClNH_2C_6H_3C_6H_3NH_2Cl$	6	CAR	-	-	-	-	8.7	-	-	-	dec.	132 (270)	insol.	bz, ace., alc.	
351. 1,4-Dichlorobutane $CH_2Cl(CH_2)_3Cl$	4b	-	2	2	0	1.1	4.4	52 (126)	-	-	162 (324)	-39 (-38)	insol.	chl.	
352. 1,3-Dichloro-2-Butene $CH_2ClCHCClCH_3$	4b	-	2	3	0	-	4.3	27 (81)	-	-	123 (253)	-	insol.	org. sol.	
353. Dichlorodifluoromethane CCl_2F_2	26	1000	1	-	-	1.30	4.2	-	-	-	-29 (-20)	-158 (-252)	sol.	alc.	
354. 1,3-Dichloro-5,5-Dimethylhydantoin $C_5H_6Cl_2N_2O_2$	6	(.2)	3	-	-	1.5	6.8	-	-	-	-	132 (270)	-	-	
355. 1,1-Dichloroethane CH_3CHCl_2	4b	100	2	3	0	1.18	3.4	-6 (21)	458 (856)	5.6-11.4	+57 (+135)	-97 (-143)	.70	alc.	BDH

SUBSTANCE/FORMULA	Waste Disposal Procedure (See VI)	TLV (ACGIH) PPM (mg/M³)	NFPA 704M System Health	Fire	React.	Sp. Gr.	Vap. Dens. (Air=1)	Fl. Pt. °C (°F)	Ignit. Temp. °C (°F)	Flam. Limits %	B.P. °C (°F)	M.P. °C (°F)	Sol. in H₂O g/100g	Other Solvents	Misc. Ref.
1,2-Dichloroethane, see Ethylene Dichloride															
1,1-Dichloroethylene, see Vinylidene Chloride															
356. **1,2-Dichloroethylene** ClCH=CHCl	4b	200	2	3	2	1.3	3.3	2–4 (36–9)	-	9.7-12.8	48–60 (118–40)	−80 (−112)	sl. sol.	alc.	AIA, BDH, MCB
357. **2,2'-Dichloroethyl Ether** (Cl C₂H₄)₂O	15	15C Skin	2	2	0	1.22	4.9	55 (131)	369 (696)	-	178 (352)	−24 (−11)	insol.	org. sol.	AIA, BDH MCB
358. **Dichloromethane** CH₂Cl₂	26	500	2	0	0	1.34	2.93	-	662 (1224)	15.5-66 in O₂	40 (104)	−97 (−143)	220	alc.	BDH, NTTC, AIA, MCA
359. **Dichloromonofluoromethane** CH Cl₂F	26	1000	1	-	-	1.48	3.8	-	552 (1026)	-	9 (48)	−135 (−211)	insol.	alc.	
360. **1,1-Dichloro-1-Nitroethane** H₃C(NO₂)CCl₂	4a	10C	3	2	3	1.42	4.97	76 (168)	-	-	124 (255)	-	.520	-	
361. **2,4-Dichlorophenol** C₆H₃OHCl₂	6	-	-	1	0	1.38	5.6	114 (237)	-	-	210 (410)	45 (113)	sl. sol.	alc., bz., chl.	BDH
362. **Dichlorophenylphosphine** C₆H₅PCl₂	7b	-	3	-	-	1.3	6.2	-	-	-	225 (437)	-	dec.	bz., CS₂	
363. **1,2-Dichloropropane** H₂ClCHClCH₃	4b	75	2	3	0	1.16	3.9	16 (61)	557 (1035)	3.4-14.5	97 (207)	−100 (−148)	sl. sol.	alc.	AIA, NTTC
364. **1,3-Dichloropropene** C₃H₄Cl₂	4b	-	3	2	-	1.23	3.8	35 (95)	-	-	104 (219)	-	insol.	chl.	
365. **Dichlorotetrafluoroethane** CCl₂F-CF₃	26	1000	1	0	-	1.44	-	-	-	-	3.6 (38)	−94 (−137)	insol.	alc.	

SUBSTANCE/FORMULA	Waste Dis-posal Pro-cedure (See VI)	TLV (ACGIH) PPM (mg/M³)	NFPA 704M System			Sp. Gr.	Vap. Dens. (Air=1)	Fl. Pt. °C (°F)	Ignit. Temp. °C (°F)	Flam. Limits %	B.P. °C (°F)	M.P. °C (°F)	Sol. in H₂O g/100g	Other Solvents	Misc. Ref.
			Health	Fire	React.										
366. **Dicyclohexylamine** (C₆H₁₁)₂NH	7a	-	-	2	-	.93	6.3	99oc (210)	-	-	256 (493)	-1 (32)	.1628	alc., bz.	
367. **Dicyclopentadiene** C₁₀H₁₂	18	-	1	3	1	.93	4.55	35 (95)	-	-	170 (338)	33 (91)	-	alc.	
368. **Dieldrin** C₁₂H₁₀Cl₆O	4b	(.25)	-	-	-	1.75	13.2	-	-	-	-	175 (347)	insol.	ace., bz.	
369. **Diethanolamine** (HOCH₂CH₂)₂NH	7a	-	1	1	0	1.09	3.6	152 (306)oc	662 (1224)	-	270 (518)	28 (82)	v. sol.	alc.	NTTC
370. **1,1-Diethoxyethane** CH₃CH(OC₂H₅)₂	15	-	2	3	0	.83	4.08	-20 (-5)	230 (446)	1.7-10.4	102 (216)	-100 (-148)	21²⁰	alc.	BDH
371. **Diethyladipate** C₂H₅OOC(CH₂)₄COOC₂H₅	18	-	-	-	-	1.01	-	-	-	-	240-5 (464-73)	-21 (-5.8)	.43³⁰	alc.	
372. **Diethylaluminum Chloride** (C₂H₅)₂AlCl	4b	-	3	3	3	-	-	-	Ignites in air	-	208 (406)	-50 (-58)	-	-	NFPA
373. **Diethylamine** (C₂H₅)₂NH	7a	25	3	3	0	.71	2.5	<-26 (<-15)	312 (594)	1.8-10.1	56 (133)	-48 (-54)	v. sol.	alc.	BDH, MCA
374. **2-Diethyl-Amino-Ethanol** (C₂H₅)₂NCH₂CH₂OH	7a	10 Skin	3	2	0	.88	4.03	60 (140)oc	-	-	163 (325)	-	∞	ace., alc., bz.	BDH, NTTC
375. **N, N-Diethylaniline** C₆H₅N(C₂H₅)₂	5	-	3	2	0	.94	5.15	85 (185)	332 (630)	-	216 (421)	-38 (-36)	1.412	alc.	BDH
376. **Diethylcarbonate** (C₂H₅)₂CO₃	18	-	2	3	1	.98	4.07	25 (77)	-	-	126 (259)	-43 (-45)	insol.	alc.	BDH
377. **Diethylene Glycol** CH₂OHCH₂OCH₂CH₂OH	18	-	1	1	0	1.12	3.66	124 (255)	229 (444)	2-	245 (473)	-8 (+18)	sol.	alc.	

SUBSTANCE/FORMULA	Disposal Procedure (See VI)	TLV (ACGIH) PPM (mg/M³)	NFPA 704M System Health	Fire	React.	Sp. Gr.	Vap. Dens. (Air=1)	Fl. Pt. °C (°F)	Ignit. Temp. °C (°F)	Flam. Limits %	B.P. °C (°F)	M.P. °C (°F)	Sol. in H₂O g/100g	Other Solvents	Misc. Ref.
378. Diethylene Glycol-Monoethyl Ether $C_2H_5O(CH_2)_2O(CH_2)_2OH$	15	-	1	1	0	1.11 -1.23	4.6	96 (205)	204 (400)	1.2-	202 (396)	-10 (-14)	∞	org. sol.	
379. Diethylenetriamine $(NH_2C_2H_4)_2NH$	7a	-	3	1	0	.95	3.5	102 (216)oc	399 (750)	-	207 (405)	-39 (-38)	∞	alc.	NTTC, MCA
Diethyl Ether, see Ethyl Ether															
380. Diethyl Ethyl Phosphate $(C_2H_5)_2PO(OC_2H_5)$	7b	-	4	1	0	1.03	5.7	105 (221)	-	-	83 (181) @11mm	-	sl. sol.	alc.	
381. Diethyl Ketone $(C_2H_5)_2CO$	18	-	1	3	0	.82	2.96	13 (55)oc	452 (846)	-	101 (214)	-42 (-44)	4,720	alc.	
382. Diethyl Malonate $CH_2(CO_2C_2H_5)_2$	18	-	0	1	0	1.06	5.5	93 (200)oc	-	-	199 (390)	-50 (-58)	2,120	alc.	
383. o-Diethyl Phthalate $(C_6H_4)(COOC_2H_5)_2$	18	-	0	1	0	1.23	7.7	117 (243)	-	-	302 (576)	-40 (-40)	insol.	alc., bz.	
384. p-Diethyl Phthalate $(C_6H_4)(COOC_2H_5)_2$	18	-	0	1	0	1.23	7.7	163 (325)	-	-	296 (565)	-5 (23)	insol.	alc., bz.	
385. Diethyl Sulfate $(C_2H_5)_2SO_2$	4b	-	3	1	1	1.18	5.3	104 (219)	436 (817)	-	208 (406)	-25 (-13)	insol.	alc.	BDH
386. Diethyl Zinc $Zn(C_2H_5)_2$	3	-	0	3	3	1.18	-	-	Ignites in air	-	124 (255)	28 (82)	dec.	dec., alc.	
387. Diglycidyl Ether $[(CH_2-CHCH_2)O]_2O$	15	.5C	3	-	-	1.25	-	-	-	-	260 (500)	-	-	-	
Diglycol, see Diethylene Glycol															

SUBSTANCE/FORMULA	Waste Disposal Procedure (See VI)	TLV (ACGIH) PPM (mg/M³)	NFPA 704M System			Sp. Gr.	Vap. Dens. (Air=1)	Fl. Pt. °C (°F)	Ignit. Temp. °C (°F)	Flam. Limits %	B.P. °C (°F)	M.P. °C (°F)	Sol. in H₂O g/100g	Other Solvents	Misc. Ref.
			Health	Fire	React.										
1,2-Dihydroxyanthraquinone, see Alizarin Dyes															
388. 3,4-Dihydro-2H-Pyran C_5H_8O	18	-	-	3	0	.92	2.9	−18 (0)	-	-	86 (187)	-	sol.	alc.	
389. Diisobutyl Ketone $[(CH_3)_2CHCH_2]_2CO$	18	50	1	2	0	.81	4.9	60 (140)	-	.8-6.2 @212°F	168 (334)	-	insol.	alc.	
390. Diisopropyl Amine $[(CH_3)_2CH]_2NH$	7a	5 Skin	3	3	0	.72	3.5	−1 (30)oc	-	−84	83 (181)	−61 (−78)	sl. sol.	-	
391. Diisopropyl Ether $(CH_3)_2CHOCH(CH_3)_2$	15	500	2	3	1	.72	3.5	−28 (−18)	443 (829)	1.4-7.9	69 (155)	−86 (−123)	sl. sol.	alc.	BDH
392. Diisopropyl Fluorophosphate $C_6H_{14}FPO_3$	7b	-	3	-	-	~1.1	5.2	-	-	-	46 (115) @5mm	−82 (−116)	-	-	
393. Diisopropylperoxydicarbonate $CH_3\text{-}CH\text{-}O\text{-}C\text{-}O\text{-}O\text{-}C\text{-}O\text{-}CH\ CH_3$	22b	-	0	4	4	-	-	-	-	-	-	12exp. (53)	-	-	NFPA
394. 3,3'-Dimethoxybenzidine $[NH_2(OCH_3)C_6H_3]_2$	5	-	CAR	-	-	-	8.5	206 (403)	-	-	-	137 (279)	insol.	bz., ace., alc.	
395. Dimethoxyethane $(CH_3O)_2C_2H_4$	15	-	-	2	0	.85	3.1	40 (104)	-	-	65 (149)	−113 (−171)	sol.	alc., chl.	BDH
396. Dimethoxymethane $(CH_3O)_2CH_2$	15	1000	2	3	2	.86	2.63	−18 (0)oc	237 (459)	-	46 (115)	−105 (−157)	33	alc.	
397. Dimethoxypropane $(CH_3O)_2C_3H_6$	15	-	-	2	-	.85	3.6	−7 (19)	-	-	95 (203)	-	-	-	
398. N,N-Dimethyl Acetamide $CH_3CON(CH_3)_2$	20	10 Skin	2	1	-	.94	3.0	77 (171)	354 (669)	1.8-13.8	165 (329)	−20 (−4)	∞	alc., bz.	

SUBSTANCE/FORMULA	Waste Disposal Procedure (See VI)	TLV (ACGIH) PPM (mg/M³)	NFPA 704M System Health	NFPA 704M System Fire	NFPA 704M System React.	Sp. Gr.	Vap. Dens. (Air=1)	Fl. Pt. °C (°F)	Ignit. Temp. °C (°F)	Flam. Limits %	B.P. °C (°F)	M.P. °C (°F)	Sol. in H₂O g/100g	Other Solvents	Misc. Ref.
399. **Dimethylamine** C_2H_7N	7a	10	3	4	0	.68	1.65	−50 (−58)	402 (755)	2.8-14.4	7.4 (45)	−92 (−134)	v. sol.	alc.	BDH, MCA, MGB
400. **Dimethylaminoazobenzene-2-Naphthalene** $C_{18}H_{19}N_3$	8	-	CAR	-	-	1.08	-	-	-	-	174 (345)	-	-	hot bz.	
401. **2-Dimethylaminofluorene** $C_{15}H_{15}N$	6	-	CAR	-	-	-	-	-	-	-	-	-	-	-	
402. **N,N-Dimethylaniline** $C_6H_5N(CH_3)_2$	5	5 Skin	3	2	0	.95	4.17	63 (145)	371 (700)	-	193 (379)	2.5 (36)	sl. sol.	alc., bz.	AIA, BDH
403. **2,3-Dimethylazobenzene** $(CH_3)_2C_6H_3N_2C_6H_5$	8	-	CAR	-	-	-	-	-	-	-	-	-	-	-	
404. **9,10'-Dimethyl-1,2-Benzanthracene** $(CH_3)_2C_{18}H_{10}$	6	-	CAR	-	-	-	-	-	-	-	-	122 (252)	insol.	CS₂, bz.	
Dimethylbenzylhydroperoxide, see Cumene Hydroperoxide															
405. **N,N-Dimethyl-4-Biphenylamine** $(CH_3)_2(C_6H_4)_2NH$	6	-	CAR	-	-	-	-	-	-	-	-	-	-	-	
406. **2,2-Dimethyl Butane** $(CH_3)_3CCH_2CH_3$	18	-	1	3	0	.65	3.00	−48 (−54)	425 (797)	1.2-7.0	50 (122)	−98 (−144)	insol.	alc.	
407. **Dimethyl Carbonate** $OC(OCH_3)_2$	18	-	-	3	1	1.1	3.1	19 (66)oc	-	-	90 (194)	.5 (32)	insol.	alc.	BDH
408. **1,2-Dimethyl Chrysene** $(CH_3)_2C_{18}H_{10}$	18	-	CAR	-	-	-	-	-	-	-	448 (838)	250 (482)	insol.	sl. sol., bz.	
409. **Dimethyl-1,2-Dibromo-2,2-Dichloroethyl Phosphate** $(CH_2Br)_2C_2H_3Cl_2PO_4$	7b	(3)	-	-	-	-	-	-	-	-	~200 (~392)	-	-	-	

SUBSTANCE/FORMULA	Waste Dis-posal Pro-cedure (See VI)	TLV (ACGIH) PPM (mg/M³)	NFPA 704M System Health	NFPA 704M System Fire	NFPA 704M System React.	Sp. Gr.	Vap. Dens. (Air=1)	Fl. Pt. °C (°F)	Ignit. Temp. °C (°F)	Flam. Limits %	B.P. °C (°F)	M.P. °C (°F)	Sol. in H₂O g/100g	Other Solvents	Misc. Ref.
Dimethyl Ether, see Methyl Ether															
410. Dimethylformamide (CH₃)₂NCHO	20	10 Skin	1	2	0	.94	2.51	58 (136)	445 (833)	2.2-15.2	153 (307)	−61 (−78)	∞	alc., bz.	BDH
411. Dimethyl Fumarate (CHCOOCH₃)₂	18	-	-	-	-	-	-	-	-	-	192 (378)	102 (216)	-	chl.	
2,6-Dimethyl-4-Heptanone, see Diisobutyl Ketone															
1,1-Dimethylhydrazine, see Unsymmetrical Dimethyl-Hydrazine															
412. 1,4-Dimethylnaphthalene C₁₀H₆(CH₃)₂	18	-	-	-	-	1.02	-	-	-	-	262 (504)	−18 (0)	insol.	alc.	
Dimethylnitrosoamine, see N-Nitrosodimethylamine															
413. Dimethyl Phthalate C₆H₄(CO₂CH₃)₂	18	(5)	0	1	0	1.19	6.7	146 (295)	556 (1032)	-	288 (550)	-	insol.	alc., bz.	
414. 2,2-Dimethyl Propane (CH₃)₄C	18	-	-	4	0	.61	2.48	<−7 (<20)	450 (842)	1.4-7.5	9.5 (49)	−18 (0)	insol.	alc.	
415. Dimethyl Sulfate (CH₃)₂SO₄	4b	1 Skin	4	2	0	1.33	4.35	83 (181)	188 (370)	-	189dec. (372)	−31 (−24)	sol.	alc., bz.	AIA, BDH, NTTC, MCA
416. Dimethyl Sulfide (CH₃)₂S	13	-	4	4	0	.8	2.1	<−18 (<0)	206 (403)	2.2-19.7	37 (99)	−83 (−117)	insol.	alc., MeOH	
417. Dimethyl Sulfoxide (CH₃)₂SO	4b	-	1	1	0	1.01	-	95 (203)oc	215 (419)	2.6-28.5	189 (372)	18 (65)	sol.	alc., ace.	

SUBSTANCE/FORMULA	Waste Disposal Procedure (See VI)	TLV (ACGIH) PPM (mg/M³)	NFPA 704M System			Sp. Gr.	Vap. Dens. (Air=1)	Fl. Pt. °C (°F)	Ignit. Temp. °C (°F)	Flam. Limits %	B.P. °C (°F)	M.P. °C (°F)	Sol. in H₂O g/100g	Other Solvents	Misc. Ref.
			Health	Fire	React.										
418. 2,4-Dinitroaniline $(NO_2)_2C_6H_3NH_2$	6	-	3	1	3	1.62	6.3	244 (471)	-	-	-	188 (370)	insol.	sl. sol., alc.	
419. m-Dinitrobenzene $C_6H_4(NO_2)_2$	6	(1) Skin	4	-	4	1.58	-	150 (302)	-	-	291 (556)	90 (194)	.39⁹⁹	bz.	
420. o-Dinitrobenzene $C_6H_4(NO_2)_2$	6	(1) Skin	3	1	4	1.31	5.8	150 (302)	-	-	319 (606)	118 (244)	sl. sol.	alc.	
421. p-Dinitrobenzene $C_6H_4(NO_2)_2$	6	(1) Skin	3	1	4	1.63	-	150 (302)	-	-	299 (570)	172 (342)	.18³⁸	chl.	
422. 4,6-Dinitro-o-Cresol $(NO_2)_2C_6H_2CH_3OH$	6	(.2) Skin	3	2	1	-	6.8	-	-	-	-	85 (185)	sl. sol.	alc., ace.	BDH
423. 2,7-Dinitrofluorene $C_{13}H_7(NO_2)_2$	6	-	CAR	-	-	-	-	-	-	-	-	160 (320)	-	-	
424. 2,4-Dinitrophenol $(NO_2)_2C_6H_3OH$	6	-	3	2	3	1.68	6.4	-	-	-	-	114 (237)	sl. sol.	alc., bz.	BDH
425. 1,4-Dinitrosopiperazino $C_4H_8N_2(NO_2)_2$	6	-	CAR	-	-	-	4.97	-	-	-	-	158 (316)	sl. sol.	alc.	
426. 2,4-Dinitrotoluene $C_6H_3CH_3(NO_2)_2$	6	(1.5) Skin	3	1	3	1.28	6.3	207 (404)	-	-	300 (572)	70 (158)	.03²⁰	alc., bz.	AIA, MCA
427. Di-m-Octyl Phthalate $C_{26}H_{42}O_2$	18	-	0	1	0	.97	14.4	219 (426)oc	-	-	385 (726)	−30 (−22)	-	-	
428. Di-sec-Octyl Phthalate $C_6H_4(COOC_8H_{17})_2$	18	-	0	1	0	.99	-	218 (425)oc	410 (770)	-	358 (676)	-	-	-	
429. 1,4-Dioxane $C_4H_8O_2$	18	100	2	3	1	1.04	3.0	12 (54)	180 (356)	2-22.2	101 (214)	10 (50)	∞	org. sol.	BDH

SUBSTANCE/FORMULA	Waste Dis-posal Pro-cedure (See VI)	TLV (ACGIH) PPM (mg/M³)	NFPA 704M System			Sp. Gr.	Vap. Dens. (Air=1)	Fl. Pt. °C (°F)	Ignit. Temp. °C (°F)	Flam. Limits %	B.P. °C (°F)	M.P. °C (°F)	Sol. in H₂O g/100g	Other Solvents	Misc. Ref.
			Health	Fire	React.										
430. **Dipentene** $C_{10}H_{16}$	18	-	2	2	0	.85	4.66	45 (113)	237 (458)	.7-6.1, @302°F	178 (352)	-97 (-143)	insol.	alc.	BDH
Dipentene Dioxide, see p-Mentha-1,8-Diene															
431. **Dipentene Monoxide** $C_{10}H_{14}O$	18	-	-	-	-	.93	4.45	67 (152)	-	-	75 (167)	<-6 (21)	-	-	
432. **Diphenylamine** $(C_6H_5)_2NH$	5	-	3	1	0	1.16	5.8	153 (307)	634 (1173)	-	302 (576)	53 (127)	insol.	alc., bz., CS₂	
433. **Diphenylmethane** $(C_6H_5)_2CH_2$	18	-	1	1	0	1.01	5.8	130 (266)	436 (907)	-	266 (511)	26 (79)	insol.	alc., chl.	
434. **Diphenylsulfide** $(C_6H_5)_2S$	13	-	-	-	-	1.12	-	-	-	-	296 (565)	<-40 (<-40)	insol.	bz., CS₂	
435. **Dipropylene Glycol Methyl Ether** $CH_3OC_3H_6OC_3H_6OH$	15	100	2	2	-	.95	5.1	85 (185)oc	-	-	189 (372)	-	-	-	
436. **m-Dodecane** $C_{12}H_{26}$	18	-	0	2	0	.75	5.96	74 (165)	204 (399)	.6-	216 (421)	-12 (10)	insol.	alc., ace., CS₂	
437. **Dodecyl Sodium Sulfate** $C_{12}H_{25}NaSO_4$	4b	-	1	-	-	-	-	-	-	-	-	-	-	-	
438. **Elon** $(HOC_6H_4NHCH_3)_2H_2SO_4$	6	-	-	-	-	-	-	-	-	-	-	260dec (500)	425	alc.	
439. **Endrin** $C_{12}H_8OCl_6$	6	.1 Skin	3	1	0	-	-	27 (80)	-	-	-	200 (392)	-	-	
440. **Epichlorohydrin** C_3H_5ClO	4b	5 Skin	3	3	2	1.18	3.29	41 (106)oc	-	-	117 (243)	-48 (-54)	<5	alc., bz.	BDH, MCA

SUBSTANCE/FORMULA	Waste Disposal Procedure (See VI)	TLV (ACGIH) PPM (mg/M³)	NFPA 704M System			Sp. Gr.	Vap. Dens. (Air=1)	Fl. Pt. °C (°F)	Ignit. Temp. °C (°F)	Flam. Limits %	B.P. °C (°F)	M.P. °C (°F)	Sol. in H₂O g/100g	Other Solvents	Misc. Ref.
			Health	Fire	React.										
441. EPN $C_{14}H_{14}O_5NPS$	4a	(.5) Skin	3	-	-	1.27	-	-	-	-	-	36 (97)	insol.	alc., bz.	
442. 1,2-Epoxy-3-Phenoxypropane $H_2COCHCH_2OC_6H_5$	15	10	-	-	-	1.11	4.4	-	-	-	245 (473)	3.5 (38)	-	-	BDH
443. 1,2-Epoxy-Propane $H_3C\text{-}CHOCH_2$	15	100	2	4	2	.83	2.0	-37 (-35)	-	2.1-21.5	34 (93)	-104 (-155)	v. sol.	alc.	
444. Epoxy Resin Systems	26	-	-	-	-	-	-	-	-	-	-	-	-	-	AIA, NSC
445. Erbium Er	27k	-	-	2	-	9.05	-	-	-	-	2900 (5252)	1497 (2727)	insol.	-	
446. Ethane C_2H_6	18	-	1	4	0	.57	1.04	-	515 (959)	3-12.5	-89 (-128)	-183 (-297)	insol.	alc.	
Ethanediol, see Ethylene Glycol															
447. Ethanethiol C_2H_5SH	13	10C	2	4	0	.84	2.14	<27 (<81)	299 (570)	2.8-18	37 (99)	-144 (-227)	1.5	alc.	BDH
Ethanoic Acid, see Acetic Acid															
448. Ethanol CH_3CH_2OH	18	1000	0	3	0	.79	1.59	12 (54)	423 (793)	3.3-19	79 (174)	-114 (-173)	∞	ace.	BDH, NTTC, MCA
449. Ethanolamine $NH_2CH_2CH_2OH$	7a	3	2	2	0	1.02	2.11	85 (185)	-	-	170 (338)	11 (52)	∞	alc.	NTTC, MCB
450. Ethoxy Acetylene C_2H_5OCCH	15	-	-	-	-	.79	-	<-7 (<19)	100exp. (212)	-	50 (122)	-	insol.	alc.	
2-Ethoxy Ethanol, see Cellosolve															

SUBSTANCE/FORMULA	Waste Disposal Procedure (See VI)	TLV (ACGIH) PPM (mg/M³)	NFPA 704M System Health	Fire	React.	Sp. Gr.	Vap. Dens. (Air=1)	Fl. Pt. °C (°F)	Ignit. Temp. °C (°F)	Flam. Limits %	B.P. °C (°F)	M.P. °C (°F)	Sol. in H₂O g/100g	Other Solvents	Misc. Ref.
2-Ethoxy Ethylacetate, see Cellosolve Acetate															
451. 4-Ethoxy-2-Nitro-Aniline $C_2H_5OC_6H_3NO_2NH_2$	6	-	-	-	-	-	-	-	-	-	-	113 (235)	-	chl.	
452. Ethyl Acetanilide $CH_3CON(C_2H_5)(C_6H_5)$	20	-	0	2	0	.94	5.62	52 (126)	-	-	258 (496)	54 (129)	insol.	alc.	BDH, NTTC, MCA
453. Ethyl Acetate $CH_3COOC_2H_5$	18	400	1	3	0	.90	3.04	-4.4 (24)	427 (800)	2.18-9	77 (171)	-84 (-119)	7.5²⁰	alc.	BDH
454. Ethyl Acetoacetate $CH_3COCH_2COOC_2H_5$	18	-	2	2	0	1.03	4.48	84 (183)oc	-	-	181 (358)	<-80 (<-112)	13¹⁷	alc., bz.	BDH, NTTC, MCA
455. Ethyl Acrylate $CH_2CHCOOC_2H_5$	18	25 Skin	2	3	2	.92	3.5	16 (60)	273 (524)	1.8-	100 (212)	<-72 (<-98)	sol.	alc.	BDH, NTTC, MCA
Ethyl Alcohol, see Ethanol															
Ethyl Aldehyde, see Acetaldehyde															
456. Ethylamine $C_2H_5NH_2$	7a	10	3	4	0	.80	1.56	<-18 (0)	384 (723)	3.5-14	17 (63)	-81 (-114)	∞	alc.	BDH, NTTC, MGB
457. Ethyl sec-Amyl Ketone $CH_3CH_2CHCH_3CH_2COCH_2CH_3$	18	25	-	-	-	.85	-	57 (135)oc	-	-	161 (322)	-	insol.	alc.	
458. N-Ethyl Aniline $C_2H_5NHl(C_6H_5)$	5	-	3	2	0	.96	4.18	85 (185)oc	-	-	205 (401)	-64 (-83)	insol.	alc.	BDH
459. Ethyl Benzene $C_6H_5C_2H_5$	18	100	2	3	0	.9	3.7	15 (59)	432 (810)	1-6.7	136 (277)	-95 (-139)	.0115	alc.	BDH

SUBSTANCE/FORMULA	Waste Disposal Procedure (See VI)	TLV (ACGIH) PPM (mg/M³)	NFPA 704M System Health	Fire	React.	Sp. Gr.	Vap. Dens. (Air=1)	Fl. Pt. °C (°F)	Ignit. Temp. °C (°F)	Flam. Limits %	B.P. °C (°F)	M.P. °C (°F)	Sol. in H₂O g/100g	Other Solvents	Misc. Ref.
460. Ethyl Benzoate $C_6H_5COOC_2H_5$	18	-	1	2	0	1.15	5.17	96 (205)	-	-	213 (415)	-35 (-31)	insol.	alc.	
Ethyl Bromide, see Bromoethane															
461. Ethyl Bromoacetate $CH_2BrCO_2C_2H_5$	4b	-	3	-	-	1.51	5.8	48 (118)	-	-	159 (318)	<-20 (<-4)	insol.	alc.	
462. Ethyl Butyl Ketone $C_2H_5COC_4H_9$	18	-	1	2	0	.82	3.93	46 (115)oc	-	-	148 (298)	-37 (-35)	insol.	alc.	
463. Ethyl Butyrate $C_3H_7CO_2C_2H_5$	18	-	2	3	0	.88	4.0	26 (79)	463 (865)	-	121 (250)	-97 (-143)	.68 25	alc.	BDH
464. Ethyl Chloride CH_3CH_2Cl	4b	-	2	4	0	.92	2.2	-50 (-58)	519 (966)	3.6-15.4	12 (54)	-139 (-218)	.45 0	alc.	BDH, MCA
465. Ethyl Chloroacetate $ClCH_2COOC_2H_5$	4b	-	2	2	0	1.26	4.3	66 (150)	-	-	144 (291)	-27 (-17)	insol.	alc.	BDH
466. Ethyl Chloroformate $ClCO_2C_2H_5$	4b	-	3	3	1	1.36	3.74	16 (61)	-	-	95 (203)	-81 (-114)	dec.	bz., dec. alc.	BDH
467. Ethyl Crotonate $CH_3CHCHCO_2C_2H_5$	18	-	2	3	0	.92	3.93	2 (36)	-	-	143-7 (289-96)	45 (113)	insol.	alc.	
468. Ethyl Cyanoacetate $CNCH_2CO_2C_2H_5$	14	-	3	1	0	1.06	3.9	110 (230)	-	-	206 (403)	-23 (-9.4)	2 25	alc.	BDH
Ethyl Cyanide, see Propionitrile															
469. Ethylene $H_2C=CH_2$	18	-	1	4	2	.001	1.0	-	450 (842)	3.1-32	-104 (-155)	-169 (-272)	26 0	sl. sol., alc.	

SUBSTANCE/FORMULA	Waste Disposal Procedure (See VII)	TLV (ACGIH) PPM (mg/M³)	NFPA 704M System			Sp. Gr.	Vap. Dens. (Air=1)	Fl. Pt. °C (°F)	Ignit. Temp. °C (°F)	Flam. Limits %	B.P. °C (°F)	M.P. °C (°F)	Sol. in H2O g/100g	Other Solvents	Misc. Ref.
			Health	Fire	React.										
470. Ethylene Chlorohydrin CH$_2$OHCH$_2$Cl	4a	5 Skin	3	2	0	1.21	2.78	60 (140)oc	425 (797)	4.9-15.9	128 (262)	-69 (-92)	∞	alc.	AIA, BDH, MCB
471. Ethylene Diamine NH$_2$CH$_2$CH$_2$NH$_2$	7a	10	3	2	0	.90	2.07	43 (110)	-	-	117 (243)	8.5 (47)	v. sol.	alc.	BDH, NTTC
472. Ethylene Dibromide BrCH$_2$CH$_2$Br	4b	25C	3	-	-	2.18	6.48	-	-	-	131 (268)	9.3 (48)	sl. sol.	alc.	BDH
473. Ethylene Dichloride ClCH$_2$CH$_2$Cl	4b	50	2	3	0	1.26	3.4	13 (55)	413 (775)	6.2-15.9	83 (182)	-36 (-32)	sl. sol.	alc.	BDH, AIA, NTTC, MCA
474. Ethylenedinitrile Tetracetic Acid Dipotassium Salt (EDTA) K$_2$C$_{10}$O$_9$N$_2$H$_{19}$	4c	-	-	-	-	-	-	-	-	-	-	-	-	-	CED
475. Ethylene Glycol CH$_2$OHCH$_2$OH	18		1	1	0	1.11	2.14	111 (232)	413 (775)	3.2-	198 (388)	-13 (8.6)	∞	alc., ace.	BDH, NTTC
Ethylene Glycol Monobutyl Ether, see Butyl Cellosolve															
Ethylene Glycol Monoethyl Ether, see Cellosolve															
Ethylene Glycol Monomethyl Ether, see Methyl Cellosolve															
Ethylenimine, see Aziridine															
476. Ethylene Nitrate C$_2$H$_4$(ONO$_2$)$_2$	4a	.2	3	2	-	1.49	5.25	-	-	-	114exp. (237)	-20 (-4)	insol.	alc.	
477. Ethylene Oxide (CH$_2$)$_2$O	15	50	2	4	3	.87	1.49	<-18 (<0)	429 (804)	3-100	11 (51)	-111 (-168)	sol.	alc., bz.	AIA, MCA

SUBSTANCE/FORMULA	Disposal Procedure (See VI)	TLV (ACGIH) PPM (mg/M³)	NFPA 704M System Health	Fire	React.	Sp. Gr.	Vap. Dens. (Air=1)	Fl. Pt. °C (°F)	Ignit. Temp. °C (°F)	Flam. Limits %	B.P. °C (°F)	M.P. °C (°F)	Sol. in H₂O g/100g	Other Solvents	Misc. Ref.
478. Ethyl Ether $C_2H_5OC_2H_5$	15	400	2	4	1	.71	2.55	−45 (−49)	180 (356)	1.85-48	34 (94)	−123 (−189)	7.5^{20}	bz.	AIA, BDH, MCA
Ethyl Ethynyl Ether, see Ethoxyacetylene															
479. Ethyl Fluoride CH_3CH_2F	4b	-	1	-	-	.82	1.66	-	-	-	−38 (−36)	−143 (−225)	198^{14}	alc.	
480. Ethyl Fluoroacetate $FCH_2COOC_2H_5$	4b	-	4	-	-	-	3.7	-	-	-	116 (241)	-	-	-	
481. Ethyl Formate $HCOOC_2H_5$	18	100	2	3	0	.95	2.55	−20 (−4)	455 (851)	2.7-13.5	54 (129)	−79 (−110)	1118	alc.	BDH
482. 2-Ethyl Hexanol $C_4H_9CH(C_2H_5)CH_2OH$	18	-	2	2	0	.83	4.49	84 (184)	-	-	180-5 (356-65)	<−76 (−105)	insol.	alc.	BDH
483. Ethyl Iodide C_2H_5I	4b	-	2	2	-	1.95	5.4	-	-	-	72 (162)	−105 (−157)	4^{20}	alc.	BDH
484. Ethyl Lactate $CH_3CHOHCO_2C_2H_5$	18	-	2	2	0	1.04	4.07	46 (115)	400 (752)	1.5-30 @212°F	154 (309)	-	∞	alc.	BDH
Ethyl Malonate, see Diethyl Malonate															
Ethyl Mercaptan, see Ethanethiol															
485. N-Ethylmorpholine $C_6H_{13}NO$	5	20 Skin	2	3	0	.92	4.00	32 (90)oc	-	-	138 (280)	-	∞	alc.	
486. Ethyl Nitrite C_2H_5ONO	4a	-	2	4	4	.90	2.59	−35 (−30)	90 (194)	4.1-50	17 (63)	-	dec.	alc.	

SUBSTANCE/FORMULA	Waste Disposal Procedure (See VI)	TLV (ACGIH) PPM (mg/M³)	NFPA 704M System Health	Fire	React.	Sp. Gr.	Vap. Dens. (Air=1)	Fl. Pt. °C (°F)	Ignit. Temp. °C (°F)	Flam. Limits %	B.P. °C (°F)	M.P. °C (°F)	Sol. in H₂O g/100g	Other Solvents	Misc. Ref.
487. N-Ethyl-N-Nitroso-N-Butylamine $C_6H_4N_2O$	7a	-	CAR	-	-	-	-	-	-	-	-	-	-	-	
488. N-Ethyl-N-Nitrosovinylamine $C_4H_8N_2O$	7a	-	CAR	-	-	-	-	-	-	-	-	-	-	-	
489. Ethyl Oxalate $(COOC_2H_5)_2$	18	-	3	2	0	1.08	5.04	76 (168)	-	-	186 (367)	-41 (-42)	sl. sol.	alc.	
490. p-Ethyl Phenol $CH_3CH_2C_6H_4OH$	18	-	-	1	0	1.0	-	104 (219)	-	-	219 (426)	45 (113)	sl. sol.	alc., bz., CS_2	
491. 2-Ethyl-3-Propyl Acrolein $CH_3(CH_2)_2CHC(C_2H_5)CHO$	2	-	2	2	1	.85	4.35	68 (155)	-	-	163 (325)	<100 (<212)	insol.	alc.	
492. Ethyl Silicate $(C_2H_5)_4SiO_4$	18	100	2	2	0	.93	7.22	52 (125)	-	-	-	110subl. (230)	dec.	sl. sol., bz.	BDH
493. Ethyl Vinyl Ether $CH_2CHOC_2H_5$	15	-	2	4	2	.75	2.46	<-46 (<-50)	202 (395)	1.7-28	36 (97)	-115 (-175)	sl. sol.	alc.	
494. Excelsior	29	-	-	-	-	-	-	-	-	-	-	-	-	-	
495. Fatty Acids	24a	-	-	-	-	-	-	-	-	-	-	-	-	-	NTTC
496. Ferbam $Fe[CH_3)_2NCS_2]_3$	13	(15)	2	-	-	-	-	-	-	-	-	dec.	-	-	
497. Ferric Chloride (60%) $FeCl_3$	1b	-	1	-	-	2.80	-	-	-	-	-	306dec. (583)	74⁰	alc., MeOH	
498. Ferrosilicon Fe + Si	26	-	-	2	-	5.4	-	-	-	-	-	-	insol.		NSC
499. Ferrous Ammonium Sulfate $FeN_2H_8S_2O_8·6H_2O$	11	-	-	-	-	1.86	-	-	-	-	-	100-10 dec. (212-30)	26.9²⁰		

SUBSTANCE/FORMULA	Waste Disposal Procedure (See VI)	TLV (ACGIH) PPM (mg/M³)	NFPA 704M System Health	Fire	React.	Sp. Gr.	Vap. Dens. (Air=1)	Fl. Pt. °C (°F)	Ignit. Temp. °C (°F)	Flam. Limits %	B.P. °C (°F)	M.P. °C (°F)	Sol. in H₂O g/100g	Other Solvents	Misc. Ref.
500. Ferrous Chloride $FeCl_2$	11	-	1	-	-	3.16	-	-	-	-	1026 (1879)	670 (1238)	64.410	alc.	
501. Ferrous Sulfate $FeSO_4 \cdot H_2O$	11	-	1	-	-	2.97	-	-	-	-	-	-	sl. sol.	-	
502. Ferrovanadium Dust FeV	26	(1)	3	2	-	-	-	-	-	-	-	-	-	-	
503. N-2-Fluorenylacetamide $C_{15}H_{13}NO$	6	-	CAR	-	-	-	-	-	-	-	-	192 (378)	insol.	alc.	
504. Fluorine F_2	12a	.1	4	0	3	1.1	1.7	-	-	-	-188 (-307)	-218 (-360)	dec.	-	AIA, MCA, MGB, NASA
505. Fluoroacetic Acid CH_2FCOOH	4c	-	2	-	-	-	-	-	-	-	165 (329)	33 (91)	sol.h	alc.	
506. Fluoro-Boric Acid HBF_4	24b	-	3	-	-	-	-	-	-	-	130dec. (266)	-	∞	alc.	BDH
507. 4-Fluoro-4-Biphenylamine $C_{12}H_{10}NF$	6	-	CAR											-	
Fluoroethane, see Ethyl Fluoride															
508. Fluoroethylene C_2H_3F	4b	-	2	4	1	-	-	-	-	2.6-21.7	-51 (-60)	-160 (-256)	insol.	alc., bz.	
Fluorohydric Acid, see Hydrofluoric Acid															
509. 2'-Fluoro-4-Phenylacetanilide $C_{14}H_{12}NOF$	6	-	CAR	-										-	

SUBSTANCE/FORMULA	Waste Disposal Procedure (See VI)	TLV (ACGIH) PPM (mg/M³)	NFPA 704M System			Sp. Gr.	Vap. Dens. (Air=1)	Fl. Pt. °C (°F)	Ignit. Temp. °C (°F)	Flam. Limits %	B.P. °C (°F)	M.P. °C (°F)	Sol. in H₂O g/100g	Other Solvents	Misc. Ref.
			Health	Fire	React.										
510. 2'-Fluoro-4'-Phenylacetanilide $C_{14}H_{12}NOF$	6	-	CAR	-	-	-	-	-	-	-	-	-	-	-	
511. 4'''-Fluoro-4'-Phenylacetanilide $C_{14}H_{12}NOF$	6	-	CAR	-	-	-	-	-	-	-	-	-	-	-	
512. Fluoro-Silicic Acid H_2SiF_6	24b	-	3	-	-	-	-	-	-	-	-	dec.	sol.	-	BDH
513. Fluorotrichloromethane CCl_3F	26	1000	1	-	-	1.49	-	-	-	-	24 (75)	-111 (-168)	insol.	alc.	
514. Formaldehyde HCHO	2	5C	2	4	0	.82	1.08	-	430 (806)	7.0-73	-19 (-3)	-92 (-134)	sol.	alc.	BDH, NTTC, MCA, NSC
515. Formalin (39% Formaldehyde Methanol Free)	2	5	2	2	0	.82	-	85 (185)	-	-	101 (214)	-	-	-	
516. Formalin (37% Formaldehyde- 15% Methanol)	2	5	-	2	-	-	-	50 (122)	-	-	101 (214)	-	-	-	
517. Formamide $HCONH_2$	20	-	2	2	-	1.13	-	155 (310)	-	-	211dec. (412)	2.6 (36)	∞	alc.	
Form-Dimethylamide, see Dimethylformamide															
518. Formic Acid HCOOH	24a	5	3	2	0	1.22	1.59	69 (156)	601 (1114)	18-57	101 (213)	8.2 (47)	∞	alc.	BDH
519. Fumaric Acid HOOCCH=CHCOOH	24a	-	1	1	-	1.64	-	-	-	-	-	290subl. (554)	sl. sol.	alc.	
520. Furan C_4H_4O	18	-	1	4	1	.94	2.35	<0 (<32)	-	-	32 (90)	-86 (-123)	insol.	alc.	

SUBSTANCE/FORMULA	Waste Disposal Procedure (See VI)	TLV (ACGIH) PPM (mg/M³)	NFPA 704M System Health	Fire	React.	Sp. Gr.	Vap. Dens. (Air=1)	Fl. Pt. °C (°F)	Ignit. Temp. °C (°F)	Flam. Limits %	B.P. °C (°F)	M.P. °C (°F)	Sol. in H₂O g/100g	Other Solvents	Misc. Ref.
521. Furfural C_4H_3OCHO	2	5 Skin	1	2	1	1.16	3.31	60 (140)	316 (600)	2.1-	162 (324)	−37 (−35)	9.113	alc.	BDH, NTTC
522. Furfuryl Alcohol $C_4H_3OCH_2OH$	18	50	1	2	1	1.13	3.37	75 (167)oc	491 (916)	1.8-16.3	171 (340)	−31 (−24)	∞	alc.	CED
523. Gadolinium Gd	27k	-	-	-	-	7.9	-	-	-	-	∼3000 (∼5432)	1312 (2394)	insol.	-	
524. Gallic Acid $C_6H_2(OH)_3COOH$	24a	-	-	-	-	1.69	-	-	-	-	-	253dec. (487)	1 13	alc., ace.	
525. Gallium Ga	27k	-	1	-	-	5.90	-	-	-	-	2403 (4357)	30 (86)	insol.	-	
526. Garments, Contaminated	29	-	-	-	-	-	-	-	-	-	-	-	-	-	
527. Gasoline C_5H_{12} to C_9H_2O	18	-	1	3	0	.8	3.0-4.0	−43 (−45)	280-456 (536-853)	1.4-7.6	38-204 (100-400)	-	-	-	NTTC
528. Germane $Ge\,H_4$	26	-	2	3	-	3.43	-	-	-	-	−90 (−130)	−165 (−265)	-	-	
529. Germanium Ge	27k	-	2	2	-	5.35	-	-	-	-	2700 (4892)	958 (1756)	insol.	-	AIA
530. Germanium Dioxide GeO_2	27k	-	2	-	-	6.24	-	-	-	-	-	1100 (2012)	insol.	-	AIA
531. Germanium Hydride Ge_2H_6	17	-	-	-	-	6.74	1.5	-	-	-	29 (84)	−109 (−164)	dec.	-	AIA
532. Germanium Tetrachloride $GeCl_4$	1b	-	3	-	-	1.88	-	-	-	-	86 (187)	−50 (−57)	dec.	alc.	AIA

SUBSTANCE/FORMULA	Waste Dis-posal Pro-cedure (See VI)	TLV (ACGIH) PPM (mg/M³)	NFPA 704M System			Sp. Gr.	Vap. Dens. (Air=1)	Fl. Pt. °C (°F)	Ignit. Temp. °C (°F)	Flam. Limits %	B.P. °C (°F)	M.P. °C (°F)	Sol. in H₂O g/100g	Other Solvents	Misc. Ref.
			Health	Fire	React.										
533. Glutaraldehyde OCH(CH₂)₃CHO	2	-	-	-	-	.72	3.4	-	-	-	188dec. (370)	-	∞	alc.	
534. Glutaric Anhydride C₅H₆O₃	24a	-	-	-	-	-	-	-	-	-	303 (577)	-	-	-	
535. Glycerol CH₂OHCHOHCH₂OH	18	-	1	1	0	1.26	3.17	160 (320)	393 (739)	-	290dec. (554)	18 (64)	∞	alc., CS₂	NTTC
536. Glycidol C₃H₆O₂	15	50	3	-	-	1.12	2.15	-	-	-	160dec. (320)	-	∞	alc., bz.	
537. Glycolic Acid HOCH₂COOH	24a	-	2	-	-	-	-	-	-	-	dec.	79 (174)	sol.	alc.	
538. Glyoxal OCHCHO	2	-	2	1	-	1.14	-	-	-	-	50 (122)	15 (59)	v. sol.	alc.	BDH
538a. Gold	27a														
539. Grain Fumigants (Agricultural Insecticides)	7b	-	-	-	-	-	-	122-140 (50-60) oc	-	-	-	-	-	-	NTTC
540. Greases	18	-	-	-	-	-	-	-	-	-	-	-	-	-	NTTC
541. Guaiacol HOC₆H₄OCH₃	15	-	2	2	-	1.13	-	82 (180)	-	-	205 (401)	28 (82)	1.7^{15}	alc.	
542. Hafnium Hf	27k	(.5)	-	-	-	13.31	-	-	20 (68)	-	5400 (9752)	2150 (3902)	insol.	-	
543. Helium He	26	-	1	-	-	.13	-	-	-	-	-269 (-452)	-272 (-458) @26atm	.9	-	CGA
544. Heptachlor C₁₀H₅Cl₇	4b	(.5) Skin	3	-	-	1.58	-	-	-	-	-	95 (203)	insol.	alc.	

SUBSTANCE/FORMULA	Waste Disposal Procedure (See VI)	TLV (ACGIH) PPM (mg/M³)	NFPA 704M System			Sp. Gr.	Vap. Dens. (Air=1)	Fl. Pt. °C (°F)	Ignit. Temp. °C (°F)	Flam. Limits %	B.P. °C (°F)	M.P. °C (°F)	Sol. in H₂O g/100g	Other Solvents	Misc. Ref.
			Health	Fire	React.										
545. Heptane $CH_3(CH_2)_5CH_3$	18	500	1	3	0	.68	3.45	-4 (25)	223 (433)	1.2-6.7	98 (208)	-91 (-132)	insol.	alc.	BDH
2-Heptanone, see Methyl Amyl Ketone															
3-Heptanone, see Ethyl Butyl Ketone															
546. n-Heptylamine $C_7H_{15}NH_2$	7a	-	2	2	0	.77	4.0	54 (129)oc	-	-	158 (316)	-23 (-7.6)	sl. sol.	alc.	
547. Hexachlorobenzene C_6Cl_6	4b	-	1	1	-	1.57	9.8	242 (468)	-	-	322 (612)	230 (446)	insol.	bz.	
Hexachlorocyclohexane, see Lindane															
548. Hexachloroethane Cl_3CCCl_3	26	1	2	-	-	2.09	-	-	-	-	-	186 (367)	insol.	alc.	
549. Hexachloronaphthalene $C_{10}H_2Cl_6$	4b	(.2) Skin	3	-	-	-	-	-	-	-	-	-	-	-	
550. Hexafluoroethane C_2F_6	26	-	-	0	-	1.59	-	-	-	-	-79 (-110)	-94 (-137)	insol.	sl. sol., alc.	
551. Hexamethylene Tetramine $(CH_2)_6N_4$	7a	-	2	1	-	1.33	-	250 (482)	-	-	-	280subl. (536)	v. sol.	alc., ace.	
552. Hexane $CH_3(CH_2)_4CH_3$	18	500	1	3	0	.66	2.97	-22 (-7)	261 (502)	1.1-7.5	68 (154)	-96 (-141)	insol.	alc.	BDH, NTTC
553. 1,6-Hexanediamine $NH_2(CH_2)_6NH_2$	7a	-	2	1	-	-	-	-	-	-	204 (399)	42 (108)	v. sol.	alc., bz.	

| SUBSTANCE/FORMULA | Waste Disposal Procedure (See VI) | TLV (ACGIH) PPM (mg/M³) | NFPA 704M System | | | Sp. Gr. | Vap. Dens. (Air=1) | Fl. Pt. °C (°F) | Ignit. Temp. °C (°F) | Flam. Limits % | B.P. °C (°F) | M.P. °C (°F) | Sol. in H₂O g/100g | Other Solvents | Misc. Ref. |
			Health	Fire	React.										
Hexanedioic Acid, see Adipic Acid															
Hexanoic Acid, see Caproic Acid															
554. **m-Hexanol** CH₃(CH₂)₄CH₂OH	18	-	1	2	0	.81	3.52	60 (140)	293 (559)	-	158 (316)	−45 (−49)	sl. sol.	alc., bz.	BDH
2-Hexanone, see Methyl Butyl Ketone															
555. **1-Hexene** H₂C=CH(CH₂)₃CH₃	18	-	1	3	0	.7	2.97	<−7 (<20)	-	1.2-6.9	63 (145)	−139 (−218)	insol.	alc.	
556. **2-Hexene** CH₃CH=CHCH₂CH₂CH₃	18	-	1	3	0	.68	2.92	<−7 (<20)	-	-	69 (156)	−146 (−231)	insol.	alc.	
Hexone, see iso-Butyl Methyl Ketone															
557. **sec-Hexyl Acetate** (CH₃)₂CHCH₂CH₂CH₂OOCCH₃	18	50	1	2	0	.86	4.97	45 (113)	-	-	141 (286)	−64 (−83)	insol.	alc.	
Hexyl Alcohol, see m-Hexanol															
558. **Hexyl Amine** C₆H₁₃NH₂	7a	-	2	3	0	.76	3.49	29 (84)oc	-	-	129 (264)	−19 (−2.2)	sl. sol.	alc.	
559. **Holmium** Ho	27k	-	-	-	-	8.80	-	-	-	-	2600 (4712)	1461 (2662)	insol.	-	
560. **Hydracrylic Acid-β-Lactone** C₃H₄O₂	24a	CAR	4	2	-	1.15	2.5	74 (165)	-	2.9-	155 (311)	−33 (−27)	dec.	chl.	

SUBSTANCE/FORMULA	Waste Disposal Procedure (See VI)	TLV (ACGIH) PPM (mg/M³)	NFPA 704M System			Sp. Gr.	Vap. Dens. (Air=1)	Fl. Pt. °C (°F)	Ignit. Temp. °C (°F)	Flam. Limits %	B.P. °C (°F)	M.P. °C (°F)	Sol. in H₂O g/100g	Other Solvents	Misc. Ref.
			Health	Fire	React.										
561. Hydrazine H_2NNH_2	16	1 Skin	3	3	3	1.01	1.1	38 (100)	Varies with Surface (74-518°F)	4.7-100	113 (236)	1.4 (34)	v. sol.	alc.	AIA, BDH, MCA, MCB, NASA
562. Hydrazoic Acid HN_3	8	1	3	-	4	1.09	-	-	exp.	-	37 (99)	-80 (-112)	∞	alc.	
563. Hydrazine Salts and Solutions	16	1	3	2	-	-	-	-	-	-	-	-	-	-	BDH, MCA
564. Hydriodic Acid HI	24b	3	3	-	-	5.66	-	-	-	-	-35 (-31) @4 atm	-51 (-60)	42.5^0	alc.	BDH
565. Hydrobromic Acid HBr	24b	3	3	-	-	3.50	-	-	-	··	-67 (-89)	-87 (-125)	221^0	alc.	BDH, MGB
566. Hydrochloric Acid HCl	24b	5C	3	0	0	1.19	-	none	-	-	-85 (-121)	-115 (-175)	82^0	alc., bz.	AIA, BDH, NTTC, MCA, MGB
567. Hydrocyanic Acid HCN	14	10 Skin	4	4	2	.69	.93	-18 (0)	538 (1000)	6-41	26 (79)	-14 (6.8)	∞	alc.	BDH, MCA
568. Hydrofluoric Acid HF	24b	3	4	0	0	1.0	.71	-	-	none	20 (68)	-83 (-117)	∞	-	AIA, BDH, MCA, SDS, MGB, NSC
569. Hydrogen H_2	26	-	0	4	-	.09	.069	-	585 (1085)	4-75	-253 (-423)	-259 (-434)	2.1^0	alc.	CGA, NASA
Hydrogen Bromide, see Hydrobromic Acid															
Hydrogen Chloride, see Hydrochloric Acid															
Hydrogen Cyanide, see Hydrocyanic Acid															

SUBSTANCE/FORMULA	Waste Disposal Procedure (See VI)	TLV (ACGIH) PPM (mg/M³)	NFPA 704M System Health	Fire	React.	Sp. Gr.	Vap. Dens. (Air=1)	Fl. Pt. °C (°F)	Ignit. Temp. °C (°F)	Flam. Limits %	B.P. °C (°F)	M.P. °C (°F)	Sol. in H2O g/100g	Other Solvents	Misc. Ref.
Hydrogen Fluoride, see Hydrofluoric Acid															
Hydrogen Iodide, see Hydriodic Acid															
570. **Hydrogen Peroxide** H_2O_2 (90%)	22a	1	2	0	3	1.39	-	-	-	-	140 (284)	-	∞	-	MCA
571. **Hydrogen Peroxide** (27-52%)	22a	-	2	0	1	1.39	-	-	-	-	107 (225) for 35%	−11 (12)	∞	-	BDH, MCA
572. **Hydrogen Selenide** H_2Se	27e	.05	3	3	-	2.00	2.8	-	-	-	−42 (−44)	−64 (−83)	377^4	CS_2	
573. **Hydrogen Sulfide** H_2S	23	10	3	4	0	1.19	1.2	-	260 (500)	4.3-46	−60 (−76)	−83 (−117)	437^0	alc., CS_2	AIA, MCA, MGB, NSC
574. **p-Hydroquinone** $C_6H_4(OH)_2$	18	(2)	2	1	-	1.33	3.8	165 (329)	-	-	285 (545)	171 (340)	sol.	alc.	
575. **Hydroquinone Monomethyl Ether** $CH_3OC_6H_4OH$	15	-	2	1	-	1.55	-	131 (268)	421 (790)	-	246 (475)	54 (129)	sol.	alc., bz.	
576. **Hydroxydimethylarsine Oxide** $(CH_3)_2AsOOH$	7b	-	3	-	-	-	-	-	-	-	-	200 (392)	v. sol.	alc.	
577. **Hydroxylamine** NH_2OH	7a	-	1	3	3	1.20	-	129exp. (264)	-	-	56 (133)	34 (93)	sol.	alc., MeOH	
578. **Hydroxylamine Hydrochloride** $NH_2OH \cdot HCl$	7a	-	2	3	-	1.67	-	-	-	-	dec.	153 (307)	83^{17}	alc., MeOH	
Hypo, see Sodium Thiosulfate															

SUBSTANCE/FORMULA	Waste Disposal Procedure (See VI)	TLV (ACGIH) PPM (mg/M³)	NFPA 704M System Health	Fire	React.	Sp. Gr.	Vap. Dens. (Air=1)	Fl. Pt. °C (°F)	Ignit. Temp. °C (°F)	Flam. Limits %	B.P. °C (°F)	M.P. °C (°F)	Sol. in H₂O g/100g	Other Solvents	Misc. Ref.
2,2'-Iminodiethanol, see Diethanolamine															
579. 5-Indanol $C_9H_{10}O$	18	-	-	-	-	-	-	-	-	-	255 (491)	55 (131)	sl. sol.	alc.	
580. Indium In	27k	-	3	2	-	7.3	-	-	-	-	2000 (3632)	155 (311)	insol.	-	
581. Industrial Gas (Liquefied Hydrocarbon Gas)	18	-	-	-	-	-	-	−108 (−162)	-	-	-	-	-	-	NTTC
582. Iodic Acid HIO_3	24b	-	3	2	-	4.63	-	-	-	-	-	110dec. (230)	286¹⁸	alc.	
583. Iodine I_2	12a	.1C	3	-	-	4.93	-	-	-	-	184 (363)	113 (235)	.03²⁰	alc., bz., CS_2	BDH, NSC
584. Iodine Chloride ICl	21	-	3	2	-	3.18	-	-	-	-	97 (207)	14-27 (57-81)	dec.	alc.	BDH
585. Iodine Pentafluoride IF_5	21	-	3	-	3	3.75	-	-	-	-	98 (208)	9.6 (49)	dec.	-	MGB
586. Iodine Trichloride ICl_3	21	-	3	1	-	3.12	-	-	-	-	77dec. (171)	101 (214) @16 atm.	dec.	alc., bz.	BDH
587. Iodoacetic Acid $C_2H_3IO_2$	4c	-	1	-	-	-	-	-	-	-	dec.	82 (180)	sol.	alc.	
Iodoethane, see Ethyl Iodide															
Iodomethane, see Methyl Iodide															

SUBSTANCE/FORMULA	Waste Disposal Procedure (See VI)	TLV (ACGIH) PPM (mg/M³)	NFPA 704M System Health	Fire	React.	Sp. Gr.	Vap. Dens. (Air=1)	Fl. Pt. °C (°F)	Ignit. Temp. °C (°F)	Flam. Limits %	B.P. °C (°F)	M.P. °C (°F)	Sol. in H₂O g/100g	Other Solvents	Misc. Ref.
3-Iodopropene, see Allyl Iodide															
588. Isoprene CH₂C(CH₃)CHCH₂	18	-	2	4	1	.68	2.35	-54 (-65)	220 (428)	-	34 (93)	-147 (-233)	insol.	alc.	
Isothiocyanic Acid, see Methyl Isothiocyanate															
589a Jet Fuel JP-1	18	-	0	2	0	-	-	35-63 (95-145)	228 (442)	-	-	<-46 (<-50)	-	-	NTTC
589b Jet Fuel JP-4	18	-	1	3	0	-	-	-23-1 (-10-30)	242 (468)	-	-	<-46 (<-50)	-	-	NTTC
589c Jet Fuel JP-5	18	-	0	2	0	-	-	35-63 (95-145)	246 (475)	-	-	<-46 (<-50)	-	-	NTTC
589d Jet Fuel JP-6	18	-	-	-	-	.8	<1	38 (100)	224 (435)	.6-3.7	121 (250)	<-46 (<-50)	-	-	NTTC
590. Kerosene	18	-	0	2	0	.81	4.5	38-66 (100-150)	229 (444)	.7-5	170-300 (338-572)	<-46 (<-50)	-	-	NTTC
591. Ketene CH₂CO	18	.5	3	-	-	-	1.45	-	-	-	-56 (-69)	-151 (-240)	dec.	dec., alc.	
592. Lactonitrile CH₃CH(OH)CN	14	-	3	2	-	.99	2.45	77 (171)	-	-	182dec. (360)	-40 (-40)	∞	CS₂	
593. Lamps, Fluorescent	26	-	-	-	-	-	-	-	-	-	-	-	-	-	
594. Lanthanum La	27k	-	1	3	-	6.15	-	-	-	-	3470 (6278)	920 (1688)	dec.	-	

SUBSTANCE/FORMULA	Waste Disposal Procedure (See VI)	TLV (ACGIH) PPM (mg/M³)	NFPA 704M System			Sp. Gr.	Vap. Dens. (Air=1)	Fl. Pt. °C (°F)	Ignit. Temp. °C (°F)	Flam. Limits %	B.P. °C (°F)	M.P. °C (°F)	Sol. in H₂O g/100g	Other Solvents	Misc. Ref.
			Health	Fire	React.										
595. Lacquer Diluent	18	-	-	3	0	.7	-	−11 (12)	232-88 (450-550)	1.2-6	88-107 (190-225)	-	-	-	-
596. Latex	26	-	2	-	-	-	-	-	-	-	-	-	-	-	NTTC
597. Lauric Acid CH₃(CH₂)₁₀COOH	24a	-	-	1	-	.87	-	-	-	-	299 (570)	48 (118)	insol.	alc., bz.	-
598. Lauroyl Peroxide (C₁₁H₂₃CO)₂O₂	22b	-	0	2	3	-	-	-	-	-	-	49 (120)	insol.	-	NFPA
599. Lead Pb	27a	(.2)	3	2	-	11.3	-	-	-	-	1744 (3171)	327 (621)	insol.	-	
600. Lead Acetate Pb(C₂H₃O₂)₂.2H₂O	27f	-	3	-	-	3.25	-	-	-	-	-	75 (167)	44^{20}	-	
601. Lead Arsenate PbHAsO₄	27d	(.15)	3	-	-	7.80	-	-	-	-	-	720dec. (1328)	insol.	-	
602. Lead Carbonate (PbCO₃)₂ Pb(OH)₂	27f	-	3	-	-	61.4	-	-	-	-	-	400dec. (752)	insol.	-	
603. Lead Nitrate Pb(NO₃)₂	27f	-	1	0	1	4.53	-	-	-	-	-	470dec. (878)	38^{0}	alc.	
604. Lead Oxide PbO	27f	-	3	1	-	~9.5	-	-	-	-	-	888 (1630)	insol.	-	BDH, MCA
605. Lead Thiocyanate Pb(SCN)₂	27f	-	1	1	1	3.82	-	-	-	-	-	190dec. (374)	insol.	-	
606. Ligroin	18	-	-	-	-	-	-	<22 (<72)	-	-	-	-	-	-	BDH

SUBSTANCE/FORMULA	Waste Disposal Procedure (See VI)	TLV (ACGIH) PPM	TLV (mg/M³)	NFPA 704M Health	NFPA Fire	NFPA React.	Sp. Gr.	Vap. Dens. (Air=1)	Fl. Pt. °C (°F)	Ignit. Temp. °C (°F)	Flam. Limits %	B.P. °C (°F)	M.P. °C (°F)	Sol. in H₂O g/100g	Other Solvents	Misc. Ref.
Limonene, see Dipentene see p-Moutha-1,8-Diene																
607. Lindane $C_6H_6Cl_6$	4b	(.5)		3	-	-	1.87	-	-	-	-	288 (550)	157 (315)	insol.	ace., bz.	
608. Liquefied Petroleum Gas (LPG)	18	1000		-	3	-	-	-	-	-	-	-	<−46 (<−50)	-	"	NTTC
609. Lithium Li	3	-		1	1	2	.53	-	-	-	-	1337 (2437)	179 (354)	dec.	-	AIA, BDH, NSC
610. Lithium Aluminum Hydride $LiAlH_4$	17	-		3	1	2	.92	-	-	-	-	-	125dec. (257)	dec.	-	BDH, NSC
611. Lithium Borohydride $LiBH_4$	17	-		2	-	-	.66	-	-	-	-	-	284 (543)	dec.	-	BDH, NSC
612. Lithium Carbonate Li_2CO_3	11	-		-	-	-	2.11	-	-	-	-	1310dec (2390)	723 (1333)	sl. sol.	-	BDH
613. Lithium Hydride LiH	17	(.025)		1	4	2	.82	-	-	May Ignite Spontaneously	-	-	680 (1256)	dec.	-	BDH
614. Lutecium Lu	27k	-		-	2	-	9.84	-	-	-	-	3327 (6020)	1652 (3006)	-	-	
615. 2,6-Lutidine $(CH_3)_2C_5H_3N$	7a	-		-	-	-	.92	3.7	38-46 (100-15)	-	-	144 (291)	−6.6 (20)	∞	sl. sol., alc.	BDH
616. Magnesium Mg	27a	-		0	1	2	1.74	-	-	-	-	1107 (2025)	651 (1204)	dec.100	-	
617. Magnesium Chlorate $Mg(ClO_3)_2 \cdot 6H_2O$	12a	-		2	3	-	1.80	-	-	-	-	120dec. (248)	35 (95)	129^{10}	alc.	AIA

392

SUBSTANCE/FORMULA	Waste Disposal Procedure (See VI)	TLV (ACGIH) PPM (mg/M³)	NFPA 704M System			Sp. Gr.	Vap. Dens. (Air=1)	Fl. Pt. °C (°F)	Ignit. Temp. °C (°F)	Flam. Limits %	B.P. °C (°F)	M.P. °C (°F)	Sol. in H₂O g/100g	Other Solvents	Misc. Ref.
			Health	Fire	React.										
618. Magnesium Chloride MgCl₂	11	-	1	-	-	2.32	-	-	-	-	1412 (2574)	708 (1306)	54^{20}	alc.	
619. Magnesium Nitrate Mg(NO₃)₂·2H₂O	11	-	1	0	1	2.03	-	-	-	-	330dec. (626)	129 (264)	sol.	alc.	
620. Magnesium Oxide MgO	26	(15)	2	-	-	3.65	-	-	-	-	3600 (6512)	~2600 (4712)	si.sol		
621. Magnesium Perchlorate Mg(ClO₄)₂	12a	-	1	0	1	2.6	-	-	-	-	-	251dec. (484)	50^{25}	alc.	
622. Malathion C₁₀H₁₉O₆PS₂	7b	(15) Skin	3	-	-	1.23	-	-	-	-	156 (313) @.7mm	2.9 (37)	sl. sol.	alc., bz.	
623. Maleic Acid HOOC(CH)₂COOH	24a	-	2	1	-	1.59	4.0	-	-	-	135dec. (275)	-	v. sol.	alc., ace. ether	
624. Maleic Anhydride (COCH)₂O	24a	(.25)	3	1	1	.9	3.4	102 (216)	477 (890)	1.4-7.1	202 (396)	58 (127)	dec.	ace.	AIA, MCA
625. Manganese Mn	27a	(5)C	2	2	-	7.20	-	-	-	-	2030 (3686)	1260 (2300)	dec.	-	
626. Manganese Sulfate MnSO₄	11	-	2	-	-	3.25	-	-	-	-	850dec. (1562)	700 (1292)	52^{5}	alc.	
627. p-Mentha-1,8-Diene C₁₀H₁₆	18	-	-	2	0	3.84	7.4	45 (113)	237 (458)	.7-6.1	170 (338)	<60 (<140)	insol	-	
Mercaptoacetic Acid, see Thioglycolic Acid															
628. 2-Mercaptoethanol HSCH₂CH₂OH	13	-	2	2	-	1.14	2.69	74oc (165)	-	-	157 (315)	-	sol.	alc.	

SUBSTANCE/FORMULA	Waste Disposal Procedure (See VI)	TLV (ACGIH) PPM (mg/M³)	NFPA 704M System Health	Fire	React.	Sp. Gr.	Vap. Dens. (Air=1)	Fl. Pt. °C (°F)	Ignit. Temp. °C (°F)	Flam. Limits %	B.P. °C (°F)	M.P. °C (°F)	Sol. in H₂O g/100g	Other Solvents	Misc. Ref.
629. Mercuric Chloride $HgCl_2$	27b	-	3	-	-	5.44	-	-	-	-	304 (579)	277 (531)	6920	alc.	
630. Mercury Hg	27b	(.1) Skin	3	-	-	13.6	-	-	-	-	357 (674)	−39 (−38)	insol.	-	AIA, BDH NSC
631. Mercury Compounds (organic)	27b	(.01) Skin	3	-	-	-	-	-	-	-	-	-	-	-	BDH
631a. Mercury Compounds (inorganic)	27b	-	-	-	-	-	-	-	-	-	-	-	-	-	
632. Mercury Fulminate $HgC_2N_2O_2$	27b	-	3	3	4	4.42	-	-	-	-	-	exp.	sl. sol.	alc.	NSC
633. Mesityl Oxide $C_6H_{10}O$	18	25	3	3	0	.86	3.5	31 (88)	344 (652)	-	130 (266)	−59 (−74)	sol.	alc.	BDH
633a. Metal Scraps	26														
634. α-Methacrylic Acid $CH_2C(CH_3)COOH$	24a	-	3	2	2	1.02	-	77oc (171)	-	-	158 (316)	16 (61)	sol.	alc.	
635. Methane CH_4	18	-	1	4	0	.42	.6	-	537 (999)	5-15	−161 (−258)	−183 (−297)	sl. sol.	bz.	
636. Methanesulfonic Acid CH_4O_3S	4c	-	-	-	-	1.48	-	-	-	-	167 (333)	20 (68)	v. sol.	alc.	
Methanethiol, see Methyl Mercaptan															
637. Methoxychlor (DMDT) $(CH_3OC_6H_4)_2CHCCl_3$	4b	(15)	1	-	-	-	12	-	-	-	-	78 (172)	insol.	org. sol.	
2-Methoxyphenol, see Guaiacol															
4-Methoxy Phenol, see Hydroquinone Monomethyl Ether															

SUBSTANCE/FORMULA	Waste Disposal Procedure (See VI)	TLV (ACGIH) PPM (mg/M³)	NFPA 704M System Health	Fire	React.	Sp. Gr.	Vap. Dens. (Air=1)	Fl. Pt. °C (°F)	Ignit. Temp. °C (°F)	Flam. Limits %	B.P. °C (°F)	M.P. °C (°F)	Sol. in H₂O g/100g	Other Solvents	Misc. Ref.
3-(3-Methoxypropoxy)-1-Propanol, see Dipropylene Glycol Methyl Ether															
638. Methyl Acetate CH₃CO₂CH₃	18	200	1	3	0	.97	2.55	−9 (15)	502 (935)	3.1-16	57 (135)	−99 (−146)	v. sol.	alc.	BDH
Methyl Acetylene, see Propyne															
639. Methyl Acrylate CH₂CHCOOCH₃	18	10 Skin	2	3	2	.95	3.0	−3oc (27)	-	2.8-25	80 (176)	−75 (−103)	sl. sol.	alc.	CED, MCA
Methylal, see Dimethoxymethane															
640. Methyl Alcohol CH₃OH	18	200	1	3	0	.79	1.11	12 (54)	464 (867)	6-36.5	65 (149)	−98 (−144)	∞	alc., ace.	AIA, BDH, MCA, NSC
p-Methylaminophenol Sulfate, see Elon															
641. Methyl Amyl Alcohol (CH₃)₂CHCH₂CHOHCH₃	18	25 Skin	2	2	0	.80	3.5	41 (106)	-	1-5.5	130 (266)	<−90 (<−130)	sl. sol.	alc.	
642. Methyl-n-Amyl Ketone CH₃(CH₂)₄COCH₃	18	100	1	2	0	.81	3.9	49oc (120)	533 (991)	-	151 (304)	−35 (−31)	sl. sol.	alc.	
N-Methylaniline, see o-Toluidine															
2-Methylaziridine, see Propylenimine															
643. 6-Methyl-1, 2-Benzanthracene C₁₈H₁₁CH₃	18	-	Car	-	-			-	-	-	-	150 (302)	insol.	CS₂	

SUBSTANCE/FORMULA	Waste Disposal Procedure (See VI)	TLV (ACGIH) PPM (mg/M³)	NFPA 704M System			Sp. Gr.	Vap. Dens. (Air=1)	Fl. Pt. °C (°F)	Ignit. Temp. °C (°F)	Flam. Limits %	B.P. °C (°F)	M.P. °C (°F)	Sol. in H₂O g/100g	Other Solvents	Misc. Ref.
			Health	Fire	React.										
644. 10-Methyl-1, 2-Benzanthracene $C_{18}H_{11}CH_3$	18	-	CAR	-	-	-	-	-	-	-	-	141 (286)	insol.	alc., ace., CS_2	
645. Methyl Benzoate $C_6H_5CO_2CH_3$	18	-	0	2	0	1.09	4.69	83 (181)	-	-	200 (392)	-13 (8.6)	insol.	alc., MeOH	
646. α-Methylbenzyl Alcohol $C_6H_5CH(CH_3)OH$	18	-	1	1	0	1.02	4.21	96 (205)	-	-	203 (397)	21 (70)	sl. sol.	-	
647. Methyl Bromide CH_3Br	4b	20C Skin	3	1	0	1.73	3.27	-	537 (999)	10-16	4 (40)	-95 (-139)	insol.	alc., CS_2	AIA, BDH, MCA, MGB
648. 2-Methyl-1-Butene C_5H_{10}	18	-	2	4	0	.66	2.4	<-7 (<-20)	-	-	39 (102)	-134 (-209)	insol.	-	
649. 2-Methyl-2-Butene C_5H_{10}	18	-	2	3	0	.67	2.4	<-7 (<20)	-	-	38 (100)	-123 (-189)	sl. sol.	alc.	
650. 3-Methyl-1-Butene C_5H_{10}	18	-	2	4	0	.67	2.42	<-7 (<20)	365 (689)	1.5-9.1	20 (68)	-68 (-90)	insol.	alc.	
651. N-Methylbutylamine $CH_3(CH_2)_3NHCH_3$	7a	-	3	3	0	.74	3.0	13oc (55)	-	-	91 (196)	-	sol.	-	
652. Methyl Butyl Ketone $CH_3OC(CH_2)_3CH_3$	18	100	2	3	0	.81	3.45	35oc (95)	533 (991)	1.2-8	126 (259)	-57 (-71)	-	alc.	
653. Methyl Butyrate $CH_3COOC_3H_7$	18	-	2	3	0	.90	3.53	14 (57)	-	-	102 (216)	<-97 (<-143)	sl. sol.	alc.	
654. Methyl Cellosolve $H_3C-O-CH_2-CH_2OH$	18	25 Skin	2	2	0	.97	2.62	46 (115)	288 (551)	2.5-14	125 (257)	-87 (-125)	∞	bz.	BDH
655. Methyl Cellosolve Acetate $CH_3COOCH_2CH_2OCH_3$	18	25 Skin	3	2	-	1.01	4.07	56 (132)	394 (740)	1.7-8.2	145 (293)	-70 (-94)	sol.	-	

SUBSTANCE/FORMULA	Waste Disposal Procedure (See VI)	TLV (ACGIH) PPM (mg/M³)	NFPA 704M System			Sp. Gr.	Vap. Dens. (Air=1)	Fl. Pt. °C (°F)	Ignit. Temp. °C (°F)	Flam. Limits %	B.P. °C (°F)	M.P. °C (°F)	Sol. in H₂O g/100g	Other Solvents	Misc. Ref.
			Health	Fire	React.										
656. Methyl Chloride CH_3Cl	4b	100C	2	4	0	0.98	1.8	-	632 (1170)	10.7-17.4	-24 (-11)	-98 (-144)	sl. sol.	alc.	MCA, MGB
Methyl Chloroform, see 1, 1, 1-Trichloroethane															
657. Methyl Chloroformate $ClCOOCH_3$	4b	-	3	-	-	1.24	3.26	12 (54)	504 (940)	-	71 (160)	-	dec.	alc., bz.	BDH
658. 3-Methyl Cholanthrene $C_{21}H_{16}$	18	-	CAR	-	-	1.28	-	-	-	-	280 (536) @80mm	180 (356)	-	-	
Methyl Cyanide, see Acetonitrile															
659. Methyl Cyclohexane $C_6H_{11}CH_3$	18	500	2	3	0	.77	3.39	-4 (25)	285 (545)	1.2-	100 (212)	-126 (-195)	insol.	alc.	BDH
660. σ-Methyl Cyclohexanol $CH_3C_6H_{10}OH$	18	100	3	2	0	.92	3.93	68 (154)	296 (565)	-	165 (329)	-20 (-4)	sl. sol.	alc.	BDH
661. 2-Methyl Cyclohexanone $CH_3C_5H_9CO$	18	100 Skin	3	2	0	.92	3.86	48 (118)	-	-	165 (329)	-	insol.	alc.	BDH
662. 4-Methyl Cyclohexene $C_6H_9CH_3$	18	-	1	3	0	.80	3.34	-1oc (30)	-	-	103 (217)	-116 (-177)	insol.	alc.	
663. 3-Methyl-4-Dimethylaminoazo-Benzene $C_{15}H_{17}N_3$	8	-	CAR	-	-	-	-	-	-	-	-	-	-	-	
664. Methylene-Bi-Phenylisocyanate $OCNC_6H_4CH_2C_6H_4NCO$	6	-	-	-	-	1.19	-	-	-	-	194–9 (381–90) @ 5mm	37 (99)	-	-	
Methylene Chloride, see Dichloromethane															

SUBSTANCE/FORMULA	Waste Disposal Procedure (See VI)	TLV (ACGIH) PPM (mg/M³)	NFPA 704M System			Sp. Gr.	Vap. Dens. (Air=1)	Fl. Pt. °C (°F)	Ignit. Temp. °C (°F)	Flam. Limits %	B.P. °C (°F)	M.P. °C (°F)	Sol. in H₂O g/100g	Other Solvents	Misc. Ref.
			Health	Fire	React.										
Methylene Chlorobromide, see Bromochloromethane															
Methylene Di-p-Phenylene Isocyanate, see Methylene Biphenyl Isocyanate															
665. Methyl Ether CH_3OCH_3	15	-	3	4	0	.66	1.56	-41 (-42)	350 (662)	3.4-18	-24 (-11)	-139 (-218)	sol.	alc.	MCA
666. Methyl Ethyl Ether $CH_3OC_2H_5$	15	-	2	4	1	.73	2.07	-37 (-35)	190 (374)	2-10.1	11 (52)	-	sol.	alc.	
667. Methyl Ethyl Ketone $CH_3COCH_2CH_3$	18	200	1	3	0	.81	2.5	-7 (20)	515 (960)	2-10	80 (175)	-87 (-124)	v. sol.	alc. bz.	AIA, BDH, NTTC, MCA
668. Methyl Ethyl Ketone Peroxide $CH_3CO_2C_2H_5$	22b	-	2	2	4	-	-	52-93 (125-200)	-	-	-	110exp. (230)	sol.	alc.	NFPA
669. Methyl Ethylnitrosocarbamate $C_4H_8N_2O_3$	4a	-	CAR	-	-	-	-	-	-	-	-	-	-	'	
670. Methyl Formate $HCOOCH_3$	18	100	2	4	0	.99	2.1	-19 (-2)	456 (853)	5.9-20	32 (90)	-100 (-148)	v. sol.	alc.	BDH
671. 2-Methyl Furan C_5H_6O	18	-	2	3	1	.92	2.8	-30 (-22)	-	-	63 (145)	-89 (-128)	insol.	alc.	
5-Methyl-3-Heptanone, see Ethyl Amyl Ketone															
672. Methyl Hydrazine CH_3NHNH_2	16	.2C Skin	-	-	-	.9	1.6	<27 (<80)	-	-	87 (189)	<-80 (<-112)	sol.	alc.	MCA
673. Methyl Iodide CH_3I	4b	5 Skin	3	-	-	2.28	4.89	-	-	-	43 (109)	-67 (-89)	sl. sol.	alc.	BDH

SUBSTANCE/FORMULA	Waste Disposal Procedure (See VI)	TLV (ACGIH) PPM (mg/M³)	NFPA 704M System			Sp. Gr.	Vap. Dens. (Air=1)	Fl. Pt. °C (°F)	Ignit. Temp. °C (°F)	Flam. Limits %	B.P. °C (°F)	M.P. °C (°F)	Sol. in H₂O g/100g	Other Solvents	Misc. Ref.
			Health	Fire	React.										
674. Methyl Isobutyl Ketone $CH_3COCH_2CH(CH_3)_2$	18	100	2	3	0	.80	3.5	23 (73)	460 (860)	1.4-7.5	117 (243)	−85 (−121)	sl. sol.	alc., bz.	MCA
675. Methyl Isobutyrate $(CH_3)_2CHCO_2CH_3$	18	-	-	-	-	.86	3.5	13oc (55)	482 (900)	-	92 (198)	−84 (−119)	sl. sol.	alc.	
676. Methyl Isocyanate $CH_3N{=}C{=}O$	18	.02 Skin	-	-	-	.97	-	-	-	-	44 (111)	-	-	-	
677. Methyl Isothiocyanate CH_3NCS	6	-	3	-	-	1.67	-	-	-	-	119 (246)	36 (97)	sl. sol.	alc.	
2-M Ethyllactonitrile, see Acetone Cyanohydrin															
678. Methyl Mercaptan CH_3SH	13	10C	2	4	0	.87	1.66	<−18 (<0)	-	3.9-21.8	7.6 (45)	−123 (−189)	sl. sol.	alc.	MGB
679. Methyl Methacrylate $CH_2C(CH_3)COOCH_3$	18	100	2	3	2	.94	3.6	10 (50)	421 (790)	2.1-12.5	100 (212)	−50 (−58)	sl. sol.	alc.	BDH, MCA
680. 1-Methylnaphthalene $C_{10}H_7CH_3$	18	-	2	2	0	1.03	-	-	528 (982)	-	240-3 (464-9)	−22 (−7.6)	insol.	alc.	
681. 1-Methyl-2-Naphthylamine $CH_3C_{10}H_6NH_2$	5	-	CAR	-	-	-	-	-	-	-	51 (124)	-	-	org. solv.	
682. N-Methyl-N-Nitrosoacetamide $C_3H_6N_2O_2$	6	-	CAR	-	-	-	-	-	-	-	-	-	-	-	
683. N-Methyl-N-Nitrosoallylamine $C_4H_8N_2O$	6	-	CAR	-	-	-	-	-	-	-	-	-	-	-	
684. N-Methyl-N-Nitrosoaniline $C_6H_5N(NO)CH_3$	6	-	CAR	-	-	1.13	-	-	-	-	255dec. (491)	15 (59)	insol.	alc.	

SUBSTANCE/FORMULA	Waste Disposal Procedure (See VI)	TLV (ACGIH) PPM (mg/M³)	NFPA 704M System Health	Fire	React.	Sp. Gr.	Vap. Dens. (Air=1)	Fl. Pt. °C (°F)	Ignit. Temp. °C (°F)	Flam. Limits %	B.P. °C (°F)	M.P. °C (°F)	Sol. in H₂O g/100g	Other Solvents	Misc. Ref.
685. N-Methyl-N-Nitrosobenzyla-mine $C_8H_{10}N_2O$	6	-	CAR	-	-	-	-	-	-	-	-	-	-	-	
686. 1-Methyl-1-Nitrosourea $H_2NCON(NO)CH_3$	6	-	CAR	-	-	-	-	-	-	-	-	123dec. (253)	insol.	alc., bz.	
687. N-Methyl-N-Nitrosovinyla-mine $C_3H_6N_2O$	6	-	CAR	-	-	-	-	-	-	-	-	-	-	-	
688. Methyl Parathion (Xylene solution) $C_{10}H_{10}NO_5PS$	7b	(.1) Skin	4	3	-3	-	-	46 (115)	-	-	-	-			
2-Methyl-2-Propanethiol, see t-Butyl Mercaptan															
689. Methyl-n- Propyl Ketone $CH_3CO(CH_2)_2CH_3$	18	200	2	3	0	.81	3.0	7 (45)	505 (941)	1.6-8.2	102 (216)	-78 (-108)	sl. sol.	alc.	BDH
690. 1-Methylpyrrole $CH_3C_4H_4N$	6	-	-	3	-	.91	2.8	16 (61)	-	-	115 (239)	-	insol.	alc.	
691. Methyl Salicylate $HOC_6H_4COOCH_3$	18	-	1	1	0	1.18	5.24	101 (214)	454 (850)	-	223 (433)	-8.3 (18)	sl. sol.	alc.	
692. a-Methyl Styrene $C_8H_7CH_3$	18	100C	1	2	1	.92	4.08	54 (129)	574 (1065)	1.9-6.1	167-70 (333-8)	-23 (-9.4)	insol.	alc.	
693. m,p-Methyl Styrene $CH_2CHC_6H_4CH_3$	18	100	2	2	0	.89	4.08	57 (134)	494 (921)	.9	170 (338)	-83 (-117)	-	-	NTTC
Methyl Sulfate, see Dimethyl Sulfate															
Methyl Sulfide, see Dimethyl Sulfide															

SUBSTANCE/FORMULA	Waste Disposal Procedure (See VI)	TLV (ACGIH) PPM (mg/M³)	NFPA 704M System			Sp. Gr.	Vap. Dens. (Air=1)	Fl. Pt. °C (°F)	Ignit. Temp. °C (°F)	Flam. Limits %	B.P. °C (°F)	M.P. °C (°F)	Sol. in H₂O g/100g	Other Solvents	Misc. Ref.
			Health	Fire	React.										
Methyl Sulfoxide, see Dimethyl Sulfoxide															
694. N-Methyl-N-2,4,6-Tetranitro-aniline $C_7H_5N_5O_8$	6	-	2	3	-	1.57	-	-	-	-	187exp. (369)	130 (266)	insol.	sl. sol	-
695. Methyl-p-Toluenesulfonate $CH_3C_6H_4SO_3CH_3$	18	-	-	-	-	1.23	6.45	-	-	-	-	27 (81)	insol.	bz., alc.	-
696. Methyl Vinyl Ether $H_2C=CH-O-CH_3$	15	-	2	4	2	.77	2.0	-51 (-60)	-	-	8 (46)	-122 (-188)	sl. sol.	org. solv.	-
697. Mixed Acids $(HNO_3+H_2SO_4)$	24b	-	4	-	-	-	-	None	-	None	Varies	Varies	-	-	MCA
698. Molybdenum Mo	27a	-	-	2	-	10.2	-	-	-	-	5560 (10040)	2622 (4751)	-	-	
699. Molybdenum Compounds (Insoluble)	26	(15)	1	-	-	-	-	-	-	-	-	-	-	-	
700. Molybdenum Compounds (Soluble)	11	(5)	1	-	-	-	-	-	-	-	-	-	-	-	
701. Monochloroamine NH_2Cl	19	-	-	-	-	-	-	-	-	-	-	-66 (-87)	sol.	alc.	
Monochlorobenzene, see Chlorobenzene															
702. Monomethylamine CH_3NH_2	7a	10	3	4	0	-	1.1	-10 (14)	430 (806)	4.9-20.8	-6.3 (21)	-94 (-137)	v. sol.	alc.	BDH, MCA, MGB
703. Morpholine C_4H_9NO	5	20 Skin	2	3	0	.99	3.00	38oc (100)	310 (590)	-	128 (262)	-4.9 (23)	∞	org. solv.	

SUBSTANCE/FORMULA	Waste Disposal Procedure (See VI)	TLV (ACGIH) PPM (mg/M³)	NFPA 704M System Health	Fire	React.	Sp. Gr.	Vap. Dens. (Air=1)	Fl. Pt. °C (°F)	Ignit. Temp. °C (°F)	Flam. Limits %	B.P. °C (°F)	M.P. °C (°F)	Sol. in H₂O g/100g	Other Solvents	Misc. Ref.
Motor Fuel Antiknock Compounds, see Tetraethyl Lead															
Muriatic Acid, see Hydrochloric Acid															
Mylone, see Crag 974															
704. Naphtha (Coal Tar)	18	100	2	2	0	.87	-	42 (107)	277 (531)	-	149-216 (300-421)	-	-	-	NTTC
Naphtha (Petroleum), see Petroleum Ether															
705. Naphtha, Varnish Makers and Painters, 500°Flash	18	-	1	3	0	<1	4.1	10 (50)	232 (450)	.9-6.7	116-43 (241-89)	-	-	-	
706. Naphtha, Varnish Makers and Painters, High Flash	18	-	1	3	0	<1	4.3	29 (85)	232 (450)	1-6	139-77 (282-351)	-	-	-	
707. Naphtha, Varnish Makers and Painters, Regular	18	-	1	3	0	<1	-	-2 (28)	232 (450)	.9-6	100-60 (212-320)	-	-	-	
708. Naphthalene C₁₀H₈	18	10	2	2	0	1.15	4.42	79 (174)	526 (979)	.9-5.9	210 (410)	80 (176)	insol.	CS₂ alc., bz.	MCA
Naphthalene Ethylene, see Acenaphthene															
709. 1-Naphthol C₁₀H₇OH	18	-	2	1	-	1.10	-	-	-	-	288 (550)	96 (205)	insol.	alc., bz.	
710. 2-Naphthol C₁₀H₇OH	18	-	2	1	-	1.22	4.97	153 (307)	-	-	295 (563)	123 (253)	insol.	alc., bz.	

SUBSTANCE/FORMULA	Waste Disposal Procedure (See VI)	TLV (ACGIH) PPM (mg/M³)	NFPA 704M System Health	Fire	React.	Sp. Gr.	Vap. Dens. (Air=1)	Fl. Pt. °C (°F)	Ignit. Temp. °C (°F)	Flam. Limits %	B.P. °C (°F)	M.P. °C (°F)	Sol. in H₂O g/100g	Other Solvents	Misc. Ref.
711. 1-Naphthylamine $C_{10}H_7NH_2$	5	-	2	1	0	1.12	4.93	157 (315)	-	-	301 (574)	50 (122)	sl. sol.	alc.	BDH
712. 2-Naphthylamine $C_{10}H_7NH_2$	5	CAR		2	-	1.06					306 (583)	112 (234)	sol.	alc.	BGH, MCA
713. 1-Naphthylisothiocyanate $C_{10}H_7NCS$	7b			-	-	1.81					-	58 (136)	insol.	bz., alc., ace.	
714. Natural Gas (85%CH_4, 10%C_2H_6, N_2, C_3H_8, C_4H_{10})	18	-	1	4	0	-			482-632 (900-1170)	3.8-17		-	-	-	
715. Neon Ne	26	-		-	-	-	.69				-246 (-411)	-249 (-416)	1.522		
Neopentane, see 2, 2-Dimethylpropane															
716. Nickel Ni	27a	(1)	-	2	-	8.90	-				2900 (5252)	1452 (2646)	insol.	-	
717. Nickel Carbonyl $Ni(CO)_4$	18	.001	4	3	-	1.32	~6		60exp. (140)	2-	43 (109)	-25 (-13)	sl. sol.	alc., bz.	MGB
718. Nickel Nitrate $Ni(NO_3)_2 \cdot 6H_2O$	11	(1)	1	0	1	2.05					137 (279)	58 (136)	238°	alc.	
719. Nickel Sulfate $NiSO_4$	11	(1)	1	0	-	3.68					848dec. (1558)	840 (1544)	29°	-	
720. Nicotine $C_{10}H_{14}N_2$	5	(.5)	4	1	0	1.01	5.61		244 (471)	.7-4.0	247 (477)	<-80 (<-112)	∞	alc.	
721. Niobium Nb	27k	-	-	2	-	8.57					4930 (8906)	2468 (4474)	insol.	-	

SUBSTANCE/FORMULA	Waste Disposal Procedure (See VI)	TLV (ACGIH) PPM (mg/M³)	NFPA 704M System			Sp. Gr.	Vap. Dens. (Air=1)	Fl. Pt. °C (°F)	Ignit. Temp. °C (°F)	Flam. Limits %	B.P. °C (°F)	M.P. °C (°F)	Sol. in H₂O g/100g	Other Solvents	Misc. Ref.
			Health	Fire	React.										
722a. Nitrate-Nitrite Salt Baths (53% KNO₃,40%NaNO₂,7%NaNO₃)	11	-	-	-	-	-	-	-	-	-	-	140 (285)	-	-	NSC
722b. Nitrate-Nitrite Salt Baths (50%KNO₃,50%NaNO₂)	11	-	-	-	-	-	-	-	-	-	-	220 (428)	-	-	NSC
Nitrating Acid, see Mixed Acids															
723. Nitric Acid HNO₃	24b	2	2	0	1	1.50	-	-	-	-	86 (187)	-42 (-44)	∞	dec., alc.	AIA, BDH, NTTC, MCA
724. Nitric Oxide NO	21	25	3	-	3	1.34	1.04	-	-	-	-153 (-243)	-164 (-263)	7.3⁰	alc.	AIA, MGB
2,2',2''-Nitrilotriethanol, see Triethanolamine															
725. 3-Nitroacetophenone NO₂C₆H₄COCH₃	6	-	-	-	-	-	-	-	-	-	202 (396)	81 (178)	v. sol.	alc.	
726. m-Nitroaniline NO₂C₆H₄NH₂	6	-	3	1	1	1.18	-	-	-	-	306 (583)	114 (237)	sl. sol.	alc., bz. MeOH	BDH
727. o-Nitroaniline NO₂C₆H₄NH₂	6	-	3	1	1	1.44	-	168oc (335)	521 (970)	-	284 (543)	71 (160)	sl. sol.	alc., bz.	BDH
728. p-Nitroaniline NO₂C₆H₄NH₂	6	1 Skin	3	1	1	1.44	-	199 (390)	-	-	336 (637)	146 (295)	insol.	alc., MeOH	BDH, MCA
729. Nitrobenzene C₆H₅NO₂	6	1 Skin	3	2	0	1.20	4.24	88 (190)	482 (900)	1.8-@200°F.	211 (412)	5 (41)	sl. sol.	alc., bz.	AIA, BDH, MCA
730. o-Nitrobiphenyl C₆H₅C₆H₄NO₂	6	CAR	2	1	0	1.44	6.9s	143 (290)	180 (356)	-	330 (626)	35 (95)	insol.	alc.	

SUBSTANCE/FORMULA	Waste Disposal Procedure (See VI)	TLV (ACGIH) PPM (mg/M³)	NFPA 704M System Health	Fire	React.	Sp. Gr.	Vap. Dens. (Air=1)	Fl. Pt. °C (°F)	Ignit. Temp. °C (°F)	Flam. Limits %	B.P. °C (°F)	M.P. °C (°F)	Sol. in H₂O g/100g	Other Solvents	Misc. Ref.
Nitrochlorobenzene, see Chloronitrobenzene															
731. Nitroethane CH₃CH₂NO₂	4a	100	1	3	3	1.05	2.58	28 (82)	415 (779)	3.4–	114 (237)	–90 (–130)	sol.	alc.	AIA, BDH
732. 2-Nitrofluorene C₁₃H₉NO₂	6	-	CAR	-	-	-	-	-	-	-	-	160 (320)	insol.	hot bz.	
733. Nitrogen N₂	26	-	0	-	-	1.25	-	-	-	-	–196 (–321)	–210 (–346)	2.3O	sl. sol. alc.	CGA
734. Nitrogen Dioxide NO₂	21	5C	3	0	1	1.49	1.59	-	-	-	20 (68)	–11 (12)	dec.	chl., CS₂	AIA, MCA, MGB, NASA
735. Nitrogen Fertilizer Solutions	26		-	-	-	-				-	-		-	-	NTTC
Nitrogen Tetroxide, see Nitrogen Dioxide			-	-	-	-									
736. Nitrogen Trifluoride NF₃	12a	10	3	3	-	1.59	-	-	-	-	–129 (–200)	–209 (–344)	sl. sol	-	
737. Nitrogen Trioxide NO₃	26	-	3	-	-	1.45	-	-	-	-	3.5 (38)	–102 (–152)	-	eth.	MGB
738. Nitromethane CH₃NO₂	4a	100	1	3	4	1.14	2.11	35 (95)	418 (785)	7.3–	101 (214)	–28 (–18)	sol.	alc.	AIA
739. α-Nitronaphthalene C₁₀H₇NO₂	6	-	1	1	0	1.14	5.96	164 (327)	-	-	304 (579)	60 (140)	insol.	alc., CS₂	
740. m-Nitrophenol NO₂C₆H₄OH	6	-	3	-	-	1.28	-	-	-	-	194 (381) @70mm	97 (207)	sl. sol.	alc.	BDH

SUBSTANCE/FORMULA	Waste Disposal Procedure (See VI)	TLV (ACGIH) PPM (mg/M³)	NFPA 704M System			Sp. Gr.	Vap. Dens. (Air=1)	Fl. Pt. °C (°F)	Ignit. Temp. °C (°F)	Flam. Limits %	B.P. °C (°F)	M.P. °C (°F)	Sol. in H₂O g/100g	Other Solvents	Misc. Ref.
			Health	Fire	React.										
741. o-Nitrophenol $NO_2C_6H_4OH$	6	-	3	-	-	1.48	-	-	-	-	215 (419)	45 (113)	sl. sol.	alc.	BDH
742. p-Nitrophenol $NO_2C_6H_4OH$	6	-	3	-	-	1.48	-	-	-	-	279 (534)	113 (235)	sl. sol.	alc.	BDH
743. 1-Nitropropane $CH_3CH_2CH_2NO_2$	4a	25	1	2	3	.99	3.06	49oc (120)	421 (789)	2.6—	131 (268)	−108 (−162)	sl. sol.	alc.	
744. 2-Nitropropane $CH_3(NO_2)CHCH_3$	4a	25	1	2	3	.99	3.06	39oc (103)	428 (802)	2.6—	120 (248)	−93 (−135)	sl. sol.	-	
745. 4-Nitroquinoline-N-Oxide $NO_2C_9H_6NO$	6	-	-	-	-	-	-	-	-	-	-	153 (307)	-	-	
746. N-Nitrosodiethylamine $C_4H_{10}N_2O$	7a	-	-	-	-	-	-	-	-	-	-	-	-	-	
747. N-Nitrosodimethylamine $C_2H_6N_2O$	7a	CAR	-	-	-	1.01	-	-	-	-	152 (306)	-	sol.	alc.	
748. N-Nitroso-N-Methylaniline $C_6H_5N(NO)CH_3$	6	-	-	-	-	1.13	-	-	-	-	255dec. (491)	15 (59)	insol.	alc.	
749. 4-Nitrosomorpholine $C_4H_9N_2O_2$	6	-	-	-	-	-	-	-	-	-	-	-	-	-	
750. 1-Nitrosopiperazine $C_4H_9N_2O$	6	-	-	-	-	-	-	-	-	-	-	-	-	-	
751. N-Nitrosopiperidine $C_5H_{10}N_2O$	6	-	-	-	-	1.06	-	-	-	-	217 (423)	-	sol.	-	
752. Nitrosyl Chloride $NOCl$	12a	-	3	-	-	3.0	2.3	-	-	-	−5.8 (21)	−64 (−83)	dec.	-	MGB

SUBSTANCE/FORMULA	Waste Disposal Procedure (See VI)	TLV (ACGIH) PPM (mg/M³)	NFPA 704M System			Sp. Gr.	Vap. Dens. (Air=1)	Fl. Pt. °C (°F)	Ignit. Temp. °C (°F)	Flam. Limits %	B.P. °C (°F)	M.P. °C (°F)	Sol. in H₂O g/100g	Other Solvents	Misc. Ref.
			Health	Fire	React.										
753. m-Nitrotoluene $CH_3C_6H_4NO_2$	6	5 Skin	2	2	-	1.15	4.7	106 (223)	-	-	232 (450)	15 (59)	insol.	alc., bz.	
754. o-Nitrotoluene $CH_3C_6H_4NO_2$	6	5	2	2	-	1.16	4.72	106 (223)	-	-	220 (428)	-4.1 (24)	insol.	alc.	BDH
755. p-Nitrotoluene $CH_3C_6H_4NO_2$	6	5	2	1	3	1.29	4.72	106 (223)	-	-	238 (460)	52 (126)	insol.	alc., bz., CS_2	-
756. Nonyl Phenol $C_9H_{19}C_6H_4OH$	18	-	2	1	0	.95	7.6	141 (285)	-	-	290-301 (554-74)	-	-	-	
757. Octachloronaphthalene $C_{10}Cl_8$	4b	(.1)	-	-	-	1 53	-	-	-	-	440dec. (824)	197 (387)	-	bz.	
758. Octafluoro-2-Butene $F_3CFC=CFCF_3$	4b	-	-	-	-	-	-	-	-	-	1.2 (34)	-136 (-213)	-	-	
759. Octafluorocyclobutane C_4F_8 (Freon-C318)	26	-	-	-	-	1.65	-	-	-	-	-4 (25)	-41 (-42)	-	eth.	
760. Octafluoropropane C_3F_8	4b	-	-	-	-	1.29	-	-	-	-	-37 (-35)	-160 (-256)	-	-	
761. Octane $CH_3(CH_2)_6CH_3$	18	500	0	3	0	.70	3.86	13 (56)	220 (428)	1.0-4.66	125 (257)	-57 (-71)	insol.	sl. sol. alc.	
762. Octanoic Acid $CH_3(CH_2)_6COOH$	24a	-	-	-	-	.91	5.0	132oc (270)	-	-	220 (428)	16 (61)	sl. sol.	alc.	
763. 1-Octanol $CH_3(CH_2)_7OH$	18	-	1	2	0	.83	4.5	81 (178)	-	-	194 (381)	-17 (1.4)	sol.	alc.	
764. 2-Octanol $CH_3OHCH(CH_2)_5CH_3$	18	-	1	2	0	.82	4.48	88 (190)	-	-	178 (352)	-39 (-38)	sl. sol.	alc.	

| SUBSTANCE/FORMULA | Waste Disposal Procedure (See VI) | TLV (ACGIH) PPM (mg/M³) | NFPA 704M System | | | Sp. Gr. | Vap. Dens. (Air=1) | Fl. Pt. °C (°F) | Ignit. Temp. °C (°F) | Flam. Limits % | B.P. °C (°F) | M.P. °C (°F) | Sol. in H₂O g/100g | Other Solvents | Misc. Ref. |
			Health	Fire	React.										
765. Oil, Cocoanut	18	-	0	1	0	.9	-	216 (420)	-	-	-	23 (72)	-	-	-
766a. Oil, Fuel #1	18	-	0	2	0	<1	-	38oc (100)	229 (444)	.7-5	-	<−46 (<−50)	-	-	NTTC
766b. Oil, Fuel #2	18	-	0	2	0	<1	-	38oc (100)	257 (494)	-	-	<−46 (<−50)	-	-	NTTC
766c. Oil, Fuel #4	18	-	0	2	0	<1	-	54oc (130)	263 (505)	-	-	<−46 (<−50)	-	-	NTTC
766d. Oil, Fuel #5	18	-	0	2	0	<1	-	54oc (130)	-	-	-	<−46 (<−50)	-	-	NTTC
766e. Oil, Fuel #6	18	-	0	2	0	<1	-	71oc (160)	407 (765)	-	-	<−46 (<−50)	-	-	NTTC
767. Oil, Lubricating	18	-	0	1	0	-	-	>149oc (>300)	260-371 (500-700)	-	360 (680)	<−46 (<−50)	-	-	NTTC
768. Oil, Mineral Oil Mist	18	(5)	0	1	0	.81	-	193oc (380)	-	-	360 (680)	-	-	-	
769. Oil, Olive	18	-	0	1	0	.9	-	225 (437)	343 (650)	-	-	−6 (21)	-	-	
770. Oil, Peanut	18	-	0	1	0	.9	-	282 (540)	445 (833)	-	-	3 (37)	-	-	
771. Oil, Soybean	18	-	0	1	0	.9	-	282 (540)	445 (833)	-	-	22 (72)	-	-	

Oil of Vitriol, see Sulfuric Acid

SUBSTANCE/FORMULA	Waste Disposal Procedure (See VI)	TLV (ACGIH) PPM (mg/M³)	NFPA 704M System			Sp. Gr.	Vap. Dens. (Air=1)	Fl. Pt. °C (°F)	Ignit. Temp. °C (°F)	Flam. Limits %	B.P. °C (°F)	M.P. °C (°F)	Sol. in H₂O g/100g	Other Solvents	Misc. Ref.
			Health	Fire	React.										
772. Oil, Vegetable	18	-	0	1	0	<1	-	321 (610)	-	-	-	-9–1 (15-30)	-	-	NTTC
773. Oleic Acid $C_{17}H_{33}COOH$	24a	-	0	1	0	.89	-	189 (372)	363 (685)	-	360 (680)	14 (57)	insol.	alc., bz	NTTC
Oleum, see Sulfuric Acid															
774. Osmium Os	27k	-	1	2	-	22.5	-	-	-	-	>5300 (>9572)	2700 (4892)	insol.	-	
Onion Oil, see Allyl Propyl Disulfide															
775. Osmium Tetroxide OsO_4	27k	(.002)	3	-	-	4.91	-	None	-	-	130 (266)	41 (106)	5.710	alc.	
776. Oxalic Acid $(COOH)_2 2H_2O$	24a	(1)	3	-	-	1.65	-	-	-	-	157 (315)	102 (215)	sol.	alc.	NSC
777. Oxalyl Chloride $(COCl)_2$	1a	-	3	-	-	1.49	-	-	-	-	63 (145)	-16 (3.2)	dec.	dec., alc.	
Oxirane, see Ethylene Oxide															
778. Oxygen O_2	26	-	3	0	0	1.43	-	-	-	-	-183 (-297)	-218 (-360)	sol.	alc.	CGA
779. Oxygen Difluoride OF_2	21	.05	-	-	-	1.90	-	-	-	-	-145 (-229)	-224 (-371)	dec.		
780. Ozone O_3	26	.1	3	3	-	2.14	1.7	-	-	-	-112 (-170)	-193 (-315)	sol.	-	AIA, MGB, NASA

SUBSTANCE/FORMULA	Waste Dis-posal Pro-cedure (See VI)	TLV (ACGIH) PPM (mg/M³)	NFPA 704M System			Sp. Gr.	Vap. Dens. (Air=1)	Fl. Pt. °C (°F)	Ignit. Temp. °C (°F)	Flam. Limits %	B.P. °C (°F)	M.P. °C (°F)	Sol. in H₂O g/100g	Other Solvents	Misc. Ref.
			Health	Fire	React.										
781. Paint	26	-	3	3	-	-	-	<27 (<80)	-	-	-	-36 (-32)	-	-	NTTC
781a. Paladium	27a	-	-	-	-	-	-	-	-	-	-	-	-	-	
782. Paraffin	18	-	0	1	-	-	-	199 (390)	245 (473)	-	>370 (>698)	42-60 (108-40)	-	-	
783. Paraformaldehyde (HCHO)x	2	-	2	2	2	-	-	70 (158)	300 (572)	-	-	120-170 (248-338)	-	-	BDH, MCA
784. Paraldehyde (CH₃CHO)₃	2	-	2	3	1	.99	4.55	36oc (96)	238 (460)	1.3-	128 (262)	12 (54)	v. sol.	alc.	BDH
785. Parathion C₁₀H₁₄NO₅PS	7b	(.1) Skin	4	1	0	1.26	-	-	-	-	375 (707)	6.1 (43)	insol.	alc., ace.	
786. Pentaborane B₅H₉	17	.005	3	4	3	.66	2.2	30 (86)	-	.4-	58 (136)	-47 (-53)	dec.	-	
787. Pentachloroethane CHCl₂CCl₃	4b	-	3	2	-	1.67	7.0	-	-	-	162 (324)	-29 (-20)	insol.	alc.	BDH
788. Pentachloronaphthalene C₁₀H₃Cl₅	4b	(.5) Skin	3	-	-	-	-	-	-	-	-	-	-	-	
789. Pentachlorophenol C₆Cl₅OH	4b	(.5) Skin	3	-	-	1.98	-	-	-	-	310 (590)	190 (374)	sl. sol.	alc.	AIA, BDH
790. n-Pentane CH₃(CH₂)₃CH₃	18	1000	1	4	0	.63	2.48	-49 (-56)	309 (588)	1.4-8	36 (97)	-130 (-202)	v. sol.	alc.	BDH
791. iso-Pentane (CH₃)₂CHCH₂CH₃	18	1000	1	4	0	.62	2.48	-51 (-60)	420 (788)	1.4-7.6	28 (82)	-161 (-258)	insol.	alc.	BDH
792. 1,5-Pentanediol C₅H₁₀(OH)₂	18	-	1	1	0	.99	3.59	129oc (264)	-	-	260 (500)	-16 (3.2)	sol.	alc.	

SUBSTANCE/FORMULA	Waste Disposal Procedure (See VI)	TLV (ACGIH) PPM (mg/M³)	NFPA 704M System Health	Fire	React.	Sp. Gr.	Vap. Dens. (Air=1)	Fl. Pt. °C (°F)	Ignit. Temp. °C (°F)	Flam. Limits %	B.P. °C (°F)	M.P. °C (°F)	Sol. in H2O g/100g	Other Solvents	Misc. Ref.
793. 2, 4-Pentanedione $CH_3C(O)CH_2C(O)CH_3$	18	-	2	2	0	.98	3.45	41oc (106)	-	-	136-40 (277-84)	-23 (-9.4)	v. sol.	alc.	BDH
n-Pentanol, see n-Amyl Alcohol															
794. 2-Pentanol $C_5H_{11}OH$	18	-	1	2	0	.81	-	39 (103)	347 (657)	1.2-9.0	119 (246)	-	v. sol.	alc.	
2-Pentanone, see Methyl n-Propyl Ketone															
3-Pentanone, see Diethyl Ketone															
Pentene, see Amylene															
n-Pentyl Acetate, see n-Amyl Acetate															
sec-Pentyl Acetate, see sec-Amyl Acetate															
Pentyl Alcohol, see n-Amyl Alcohol															
Pentyl Amine, see n-Amylamine															
iso-Pentyl Nitrite, see Amyl Nitrite															
795. Peracetic Acid (60% Acetic Acid Solution) CH_3COOOH	12a	-	3	2	4	1.23	-	41 (105)	110exp. (230)	-	105 (221)	-30 (-22)	v. sol.	alc.	NFPA

SUBSTANCE/FORMULA	Waste Disposal Procedure (See VI)	TLV (ACGIH) PPM (mg/M³)	NFPA 704M System Health	Fire	React.	Sp. Gr.	Vap. Dens. (Air=1)	Fl. Pt. °C (°F)	Ignit. Temp. °C (°F)	Flam. Limits %	B.P. °C (°F)	M.P. °C (°F)	Sol. in H₂O g/100g	Other Solvents	Misc. Ref.
796. Perchloric Acid Solution $HClO_4 \cdot 2H_2O$	12a	-	3	0	3	1.70	-	-	-	-	203 (397)	-18 (0)	v. sol.	alc.	AIA, BDH, MCA, NSC
Perchloroethylene, see Tetrachloroethylene															
797. Perchloromethyl Mercaptan $ClSCCl_3$	13	0.1	3	-	-	1.70	6.4	-	-	-	149 (300)	-	-	-	
798. Perchloryl Fluoride ClO_3F	21	3	2	2	3	1.43	-	-	-	-	-47 (-53)	-146 (-231)	-	-	MCA
799. Petroleum Ethers	18	500	1	4	0	.6	2.50	-57 (-70)	288 (550)	1-6	30-60 (86-140)	<-73 (<-101)	-	alc., CS₂	BDH
800. Phenanthrene $(C_6H_4CH)_2$	18	-	-	1	-	1.18	6.14	-	-	-	340 (644)	101 (214)	insol.	alc., CS₂	
801. 2-Phenanthreneacetamide $C_{16}H_{13}NO$	20	-	CAR	-	-	-	-	-	-	-	-	-	-	-	
802. 3-Phenanthreneacetamide $C_{16}H_{13}NO$	20	-	CAR	-	-	-	-	-	-	-	-	-	-	-	
803. Phenol C_6H_5OH	18	5 Skin	3	2	0	1.07	3.24	79 (175)	715 (1319)	1.5-	181 (358)	40 (104)	sol.	alc., CS₂	BDH, NTTC, MCA
804. Phenyl Acetate $CH_3CO_2C_6H_5$	18	-	1	2	0	1.09	4.7	80 (176)	-	-	196 (385)	-	sl. sol.	alc.	
Phenyl Acetonitrile, see Benzyl Cyanide															
805. 1-Phenylazo-2-Naphthol $C_{16}H_{12}N_2O$	8	-	CAR	-	-	-					-	102-4 (216-19)	-	-	

SUBSTANCE/FORMULA	Waste Dis-posal Pro-cedure (See VI)	TLV (ACGIH) PPM (mg/M³)	NFPA 704M System Health	NFPA 704M System Fire	NFPA 704M System React.	Sp. Gr.	Vap. Dens. (Air=1)	Fl. Pt. °C (°F)	Ignit. Temp. °C (°F)	Flam. Limits %	B.P. °C (°F)	M.P. °C (°F)	Sol. in H₂O g/100g	Other Solvents	Misc. Ref.
Phenylcyclohexane, see Cyclohexylbenzene															
806. **p-Phenylenediamine** $H_2N(C_6H_4)NH_2$	5	(.1)	2	-	-	-	3.7	156 (312)	-	-	267 (513)	140 (284)	sol.	alc.	
807. **Phenylethanolamine** $C_6H_5NHCH_2CH_2OH$	5	-	2	1	0	1.09	-	152 (305)	-	-	285 (545)	35 (95)	4.6^{20}	alc.	
808. **Phenyl Ether** $C_6H_5OC_6H_5$	15	1	-	-	-	1.09	5.86	96oc (205)	646 (1195)	-	258 (496)	27 (81)	insol.	alc., bz.	
809. **Phenyl Ether** - Biphenyl Mixture	15	1	-	-	-	1.06	-	124oc (255)	610 (1130)	-	257 (494)	12 (54)	-	-	
Phenyl Glycidyl Ether, see 1, 2 - Epoxy-3-Phenoxy propane															
810. **Phenylhydrazine** $C_6H_5NHNH_2$	16	5 Skin	3	2	0	1.09	3.7	89 (192)	174 (345)	-	243dec. (469)	20 (68)	sol.	alc., bz.	BDH
811. **Phenyl Isocyanate** C_6H_5NCO	18	-	-	-	-	1.1	-	-	-	-	165 (329)	−30 (−22)	dec.	dec., alc.	
812. **Phenyl-2-Naphthylamine** $C_{10}H_7NHC_6H_5$	5	-	-	2	-	1.20	-	-	-	-	400 (752)	108 (226)	insol.	hot alc.	
813. **o-Phenyl Phenol** $C_6H_5C_6H_4OH$	18	-	2	1	0	1.21	4.8	124 (255)	-	-	286 (547)	57 (135)	insol.	alc., ace, bz.	
814. **Phorone** $CH_3(CH_3)C=CHC(O)CH= C(CH_3)CH_3$	18	-	2	2	0	.88	-	85 (185)	-	-	197 (387)	28 (82)	sl. sol.	alc.	
815. **iso-Phorone** $C_9H_{14}O$	18	-	2	1	0	.93	4.77	96oc (205)	462 (864)	.8-3.8	215 (419)	−8 (16)	sl. sol.	-	BDH

SUBSTANCE/FORMULA	Waste Disposal Procedure (See VI)	TLV (ACGIH) PPM (mg/M³)	NFPA 704M System Health	NFPA 704M System Fire	NFPA 704M System React.	Sp. Gr.	Vap. Dens. (Air=1)	Fl. Pt. °C (°F)	Ignit. Temp. °C (°F)	Flam. Limits %	B.P. °C (°F)	M.P. °C (°F)	Sol. in H₂O g/100g	Other Solvents	Misc. Ref.
816. Phosdrin $C_7H_{13}O_6P$	7b	(.1) Skin	3	-	-	1.23	-	79oc (175)	-	-	107 (225) @1mm	-	-	-	
817.															
818. Phosgene $COCl_2$	21	.1	3	-	-	1.39	3.4	-	-	-	8 (46)	-128 (-198)	dec.	dec., alc.	MCA MCB, MGB
819. Phosgene Solutions in Benzene	18	-	3	-	-	-	-	<22 (<72)	-	-	-	-	-	-	BDH
820. Phosphoric Acid H_3PO_4	24b	(1)	2	-	-	1.69	-	-	-	-	260 (500)	42 (108)	v. sol.	alc.	BHD, NTTC
821. Phosphoric Anhydride P_2O_5	24b	-	3	2	-	2.38	-	-	-	-	-	300 subl. (572)	dec.	dec., alc.	BDH, MCA
822. Phosphorus (White & Yellow) P_4	27c	(.1)	3	3	1	1.82	4.42	-	30 (86) spontaneous ignition in dry air	-	280 (535)	44 (111)	sl. sol.	bz.	BDH, MCA, NSC
823. Phosphorus (Red) P_4	27c	-	0	1	1	2.34	4.77	-	260 (500)	-	280	590 (1094) @43atm	v. sol	-	BDH
824. Phosphorus Oxychloride $POCl_3$	21	-	3	-	-	1.69	5.30	-	-	-	107 (225)	1.2 (34)	dec.	dec., alc.	BDH, MCA
Phosphorus Pentoxide, see Phosphoric Anhydride															
825. Phosphorus Pentachloride PCl_5	21	(1)	3	2	-	4.65	-	-	-	-	-	162 subl. (324)	dec.	CS_2	BDH

SUBSTANCE/FORMULA	Waste Dis-posal Pro-cedure (See VI)	TLV (ACGIH) PPM (mg/M³)	NFPA 704M System			Sp. Gr.	Vap. Dens. (Air=1)	Fl. Pt. °C (°F)	Ignit. Temp. °C (°F)	Flam. Limits %	B.P. °C (°F)	M.P. °C (°F)	Sol. in H₂O g/100g	Other Solvents	Misc. Ref.
			Health	Fire	React.										
826. Phosphorus Pentasulfide P_2S_5	21	(1)	3	1	2	2.03	-	-	142 (287)	-	514 (957)	276 (529)	insol.	sl. sol. CS_2	MCA
827. Phosphorus Tribromide P_4S_3	21	-	2	1	1	2.03	-	-	100 (212)	-	407 (765)	173' (343)	insol.	bz, CS_2	
828. Phosphorus Sesquisulfide PBr_3	21	-	3	-	-	2.85	-	-	-	-	173 (343)	-40 (-40)	dec.	dec., alc.	BDH
829. Phosphorus Trichloride PCl_3	21	.5	3	0	2	1.59	4.75	-	-	-	75 (167)	-112 (-169)	dec.	bz, CS_2	BDH, MCA
830. Phthalic Anhydride $C_6H_4(CO)_2O$	24a	2	2	1	0	1.53	5.10	151 (304)	584 (1083)	1.7-10.4	284 (544)	131 (268)	sl. sol.	alc.	AIA, MCA
831. 2-Picoline $CH_3C_5H_4N$	5	-	2	2	0	.95	3.2	39oc (102)	538 (1000)	-	129 (264)	-70 (-94)	v. sol.	alc.	BDH
832. 4-Picoline $CH_3C_5H_4N$	5	-	2	2	0	.96	3.21	57 (134)	-	-	143 (289)	4 (39)	∞	alc.	BDH
833. Picric Acid $(NO_2)_3C_6H_2OH$	6	(.1) Skin	2	4	4	1.76	7.9	150 (302)	300 (572)	-	>300exp. (△672)	122	sol.	alc.	BDH
834. Pimelic Acid $HOOC(CH_2)_5COOH$	24a	-	-	-	-	1.33	-	-	-	-	272 (522) @100mm	106 (223)	sl. sol.	alc.	
835. Pindone (Pival) $C_{14}H_{14}O_3$	18	(.1)	-	-	-	-	-	-	-	-	-	109 (228)	-	-	
836. 2-Pinene $C_{10}H_{16}$	18	-	1	3	0	.86	4.7	33 (91)	-	-	156 (313)	-55 (-67)	sl. sol.	alc.	
837. Piperidine $C_5H_{10}NH$	6	-	2	3	3	.86	3.0	16 (61)	-	-	106 (223)	-7 (19)	∞	alc., bz.	

SUBSTANCE/FORMULA	Waste Disposal Procedure (See VI)	TLV (ACGIH) PPM (mg/M³)	NFPA 704M System			Sp. Gr.	Vap. Dens. (Air=1)	Fl. Pt. °C (°F)	Ignit. Temp. °C (°F)	Flam. Limits %	B.P. °C (°F)	M.P. °C (°F)	Sol. in H₂O g/100g	Other Solvents	Misc. Ref.
			Health	Fire	React.										
838. Piperylene $CH_3CH=CHCH=CH_2$	18	-	1	3	1	.68	2.4	-43 (-45)	-	2-8.3	42 (108)	-141 (-222)	insol.	alc.	
Pival, see Pindone															
839. Platinum Pt	27a		1	0	0	21.5	-	-	-	-	3827 (6920)	1774 (3225)	insol.	aq. reg.	
840. Polytetrafluoroethylene $(F_2C:CF_2)n$	26	-	1	0	0	2.2	-	-	-	-	subl.	>260 (>500)	insol.	-	
841. Polyvinyl Acetate Emulsion	18	-	-	-	-	-	-	-	-	-	-	-36 (-32)	-	acet.	NTTC
842. Potassium K	3	-	3	1	2	.86	1.4	-	-	-	774 (1425)	64 (147)	dec.	dec., alc.	BHD
843. Potassium Acetate CH_3COOK	11	-	1	1	0	1.8	3.4	-	-	-	-	292 (558)	253²⁰	alc., MeOH	
844. Potassium Borohydride KBH_4	17	-	3	3	2	1.18	1.9	-	-	-	-	500dec. (932)	19²⁰	sl. sol., alc.	BDH
845. Potassium Carbonate K_2CO_3	11	-	3	0	0	2.43	-	-	-	-	dec.	891 (1636)	112²⁰	-	
846. Potassium Chlorate $KClO_3$	12a	-	1	0	2	2.32	-	-	-	-	400dec. (752)	368 (694)	7.1²⁰	alc.	AIA, BDH
847. Potassium Cyanide KCN	14	5	3	0	0	1.52	-	-	-	-	-	635 (1175)	50	sl. sol., alc.	BDH
848. Potassium Dichromate $K_2Cr_2O_7$	12a	(.5)	3	0	0	2.68	-	-	-	-	<500dec. (<932)	398 (748)	4.9²⁰	-	

SUBSTANCE/FORMULA	Waste Dis-posal Pro-cedure (See VI)	TLV (ACGIH) PPM (mg/M³)	NFPA 704M System			Sp. Gr.	Vap. Dens. (Air=1)	Fl. Pt. °C (°F)	Ignit. Temp. °C (°F)	Flam. Limits %	B.P. °C (°F)	M.P. °C (°F)	Sol. in H₂O g/100g	Other Solvents	Misc. Ref.
			Health	Fire	React.										
849. Potassium Ferrocyanide $K_4Fe(CN)_6 \cdot 3H_2O$	11	-	1	0	0	1.85	-	-	-	-	dec.	70 (158)	27.8¹²	ace.	
850. Potassium Fluoride KF	11	(2.5)	3	0	0	2.48	-	-	-	-	1505 (2741)	880 (1616)	v. sol.		
850a. Potassium Hydride KH	17	-	-	3	3	1.43	-	-	-	-	-	dec.	sol.		
851. Potassium Hydrogen Difluoride KHF_2	11	(2.5)	3	0	0	2.37	-	-	-	-	-	225dec. (437)	41²¹	-	BDH
Potassium Hydroxide, see Caustic Potash															
852. Potassium Nitrate KNO_3	11	-	1	0	1	2.11	-	-	-	-	<400dec. (752)	334 (633)	13⁰	-	
853. Potassium Perchlorate $KClO_4$	12a	-	1	0	2	2.52	-	-	-	-	-	610 (1130)	.75⁰	-	
854. Potassium Permanganate $KMnO_4$	12a	-	0	0	1	2.70	-	-	-	-	-	<240dec. (<464)	6.4²⁰	dec., alc. sol, MeOH	
855. Potassium Persulfate $K_2S_2O_8$	12a	-	1	0	1	2.48	-	-	-	-	-	<100dec. (<212)	1.7⁰	-	
856. Potassium Peroxide K_2O_2	22a	-	3	0	2	-	-	-	-	-	dec.	490 (914)	-	dec., alc.	
857. Potassium Sulfide K_2S	23	-	2	1	0	1.81	3.8	-	-	dust explosive	-	840 (1544)	sol.	alc.	
858. Praseodymium Pr	27k	-	1	2	0	6.64	4.9	-	-	-	3127 (5661)	935 (1715)	dec.	-	
859. Propane $CH_3CH_2CH_3$	18	1000	1	4	0	.58	1.56	(-104) (-156)	468 (874)	2.2-9.5	-45 (-49)	-187 (-305)	insol.	alc., bz.	NTTC

SUBSTANCE/FORMULA	Waste Dis-posal Pro-cedure (See VI)	TLV (ACGIH) PPM (mg/M³)	NFPA 704M System			Sp. Gr.	Vap. Dens. (Air=1)	Fl. Pt. °C (°F)	Ignit. Temp. °C (°F)	Flam. Limits %	B.P. °C (°F)	M.P. °C (°F)	Sol. in H₂O g/100g	Other Solvents	Misc. Ref.
			Health	Fire	React.										
860. **1,3 - Propanediamine** $NH_2(CH_3)_3NH_2$	7a	-	2	3	0	.86	2.56	24oc (75)	-	-	136 (276)	-24 (-10)	v. sol.	alc.	
861. **1, 2 - Propanediol** $CH_3CHOHCH_2OH$	18	-	0	1	0	1.04	2.62	99 (210)	371 (700)	2.6-12.5	189 (372)	-59 (-74)	∞	alc.	NTTC
n-Propanol, see n-Propyl Alcohol															
862. **Propargyl Alcohol** $HC{\equiv}CCH_2OH$	18	1	3	3	3	.96	1.93	36 (97)	-	3.4-	115 (239)	-17 (1.4)	sol.	alc.	
863. **Propargyl Bromide** $HC{\equiv}CCH_2Br$	4b	-	4	3	4	1.56	4.1	10 (50)	324 (615)	3.0-	90 (194)	-61 (-78)	-	-	
Propene, see Propylene															
864. **iso-Propenyl Acetate** $CH_3COOC(CH_3){:}CH_2$	18	-	2	3	0	.91	3.45	16 (60)	-	1.9-	93 (199)	-93 (-135)	sl. sol.	-	
β-Propiolactone, see Hydracrylic Acid-β-Lactone															
865. **Propionaldehyde** C_2H_5CHO	2	-	2	3	0	.81	2.0	-9-7oc (15-19)	207 (405)	2.9-17	48 (120)	-81 (-114)	sol.	alc.	BDH
866. **Propionic Acid** CH_3CH_2COOH	24a	-	2	2	0	.99	2.56	54 (130)	513 (955)	2.9-	141 (286)	-22 (-8)	∞	alc.	BHD
867. **Propionitrile** CH_3CH_2CN	14	-	4	3	1	.77	1.9	2 (36)	-	3.1-	97 (207)	-93 (-134)	v. sol.	alc.	
868. **Propionyl Chloride** CH_3CH_2COCl	1a	-	3	3	1	1.06	3.2	12 (54)	-	-	80 (176)	-94 (-137)	dec.	alc.	BDH

SUBSTANCE/FORMULA	Waste Disposal Procedure (See VI)	TLV (ACGIH) PPM (mg/M³)	NFPA 704M System Health	Fire	React.	Sp. Gr.	Vap. Dens. (Air=1)	Fl. Pt. °C (°F)	Ignit. Temp. °C (°F)	Flam. Limits %	B.P. °C (°F)	M.P. °C (°F)	Sol. in H₂O g/100g	Other Solvents	Misc. Ref.
869. iso-Propyl Acetate $CH_3COOCH(CH_3)_2$	18	250	1	3	0	.87	3.52	4 (40)	460 (860)	1.8-7.8	93 (199)	−73 (−99)	sol.	alc.	BDH
870. n-Propyl Acetate $CH_3COOCH_2CH_2CH_3$	1B	200	1	3	0	.89	3.5	14 (58)	450 (842)	2-8	102 (216)	−95 (−139)	sl. sol.	alc.	BDH
871. n-Propyl Alcohol $CH_3CH_2CH_2OH$	18	200	1	3	0	.78	2.07	25 (77)	433 (812)	2.1-13.5	97 (207)	−127 (−197)	v. sol.	alc., bz.	BDH
872. iso-Propyl Alcohol $CH_3CHOHCH_3$	18	400	1	3	0	.79	2.07	12 (53)	399 (750)	2.3-12.7	82 (180)	−89 (−128)	∞	alc.	BDH, MCA
873. Propylamine $CH_3CH_2CH_2NH_2$	7a	5	3	3	0	.72	2.0	−37 (−35)	318 (604)	2.0-10.4	49 (120)	−83 (−116)	sol.	alc.	BDH
874. iso-Propylamine $(CH_3)_2CHNH_2$	7a	5	3	4	0	.69	2.03	−37oc (−35)	402 (756)	2.3-10.4	32 (90)	−101 (−150)	∞	alc.	BDH, NTTC, MCA
875. Propyl Benzene $C_3H_7C_6H_5$	18	-	2	3	0	.86	4.14	30 (86)	450 (842)	0.8-6	159 (318)	−100 (−148)	insol.	alc., bz.	
iso-Propyl Benzene, see Cumene															
876. iso-Propyl Benzoate $C_6H_5COOCH(CH_3)_2$	18	-	1	1	0	1.01	5.67	99 (210)	-	-	218 (424)	−26 (−15)	insol.	alc.	
877. Propyl Bromide $CH_3CH_2CH_2Br$	4b	-	2	3	0	1.35	4.3	-	490 (914)	-	71 (160)	−110 (−166)	sl. sol.	alc.	BDH
878. Propyl Chloride $CH_3CH_2CH_2Cl$	4b	-	2	3	0	.89	2.71	<−18 (<0)	520 (968)	2.6-11.1	47 (117)	−123 (−189)	sl. sol.	alc.	BDH
n-Propyl Cyanide, see Butyronitrile															

SUBSTANCE/FORMULA	Waste Disposal Procedure (See VI)	TLV (ACGIH) PPM (mg/M^3)	NFPA 704M System Health	Fire	React.	Sp. Gr.	Vap. Dens. (Air=1)	Fl. Pt. °C (°F)	Ignit. Temp. °C (°F)	Flam. Limits %	B.P. °C (°F)	M.P. °C (°F)	Sol. in H$_2$O g/100g	Other Solvents	Misc. Ref.
879. Propylene C$_3$H$_6$	18	4000	1	4	1	.51	1.5	−108 (−162)	460 (860)	2-11.1	−48 (−54)	−185 (−301)	v. sol.	alc.	MCA
880. Propylene Carbonate C$_3$H$_6$CO$_3$	18	-	1	1	0	1.21	3.5	135oc (275)	-	-	242 (468)	−49 (−56)	v. sol.	bz.	
Propylene Dichloride, see 1,2-Dichloropropane															
881. Propylene Disulfate C$_3$H$_8$S$_2$O$_8$	4b	-	CAR	-	-	-	-	-	-	-	-	-	-	-	
Propylene Glycol, see 1,2-Propanediol															
882. Propylene Imine C$_3$H$_6$NH	7a	2 Skin	2	2	2	-	2.0	-	-	-	63 (145)	-	-	-	
Propylene Oxide, see 1,2-Epoxy-Propane															
883. iso-Propyl Ether [(CH$_3$)$_2$CH]$_2$O	15	500	2	3	1	.72	3.5	−28 (−18)	443 (830)	1.4-7.9	69 (155)	−60 (−76)	sl. sol.	alc.	MCA
884. n-Propyl Formate C$_3$H$_7$OOCH	18	-	2	3	0	.91	3.03	−3 (27)	455 (851)	2.3-	81 (178)	−93 (−135)	sl. sol.	alc.	
885. iso-propyl formate (CH$_3$)$_2$CHOOCH	18	-	2	3	0	.88	3.0	−6 (22)	485 (905)	-	68 (154)	-	sl. sol.	alc.	
886. iso-propyl glycidyl ether H$_3$COCH$_2$CH$_2$-O-HC(CH$_3$)$_2$	15	50	3	-	-	.92	4.15	-	-	-	137 (279)	-	-	-	
887. propyl nitrate CH$_3$CH$_2$CH$_2$NO$_3$	4a	25	2	3	3	1.06	-	20 (68)	177 (350)	2-100	111 (232)	<−100 (<−148)	sl. sol.	alc.	

SUBSTANCE/FORMULA	Waste Disposal Procedure (See VI)	TLV (ACGIH) PPM (mg/M³)	NFPA 704M System Health	Fire	React.	Sp. Gr.	Vap. Dens. (Air=1)	Fl. Pt. °C (°F)	Ignit. Temp. °C (°F)	Flam. Limits %	B.P. °C (°F)	M.P. °C (°F)	Sol. in H₂O g/100g	Other Solvents	Misc. Ref.
iso-propyl toluene, see cymene															
888. Propyne HC≡CCH$_3$	18	1000	2	4	2	0.58	1.38	-	-	1.7	−23 (−9)	−105 (−157)	sl. sol.	alc.	
889. Propyne-allene mixture (MAPP)	18	1000	2	4	2	-	>1	-	-	forms expl. with Cu & Ag	-	-	-	-	
890. Pyrethrum	26	(5)	2	1	0	-	>1				-	-	-	-	
891. Pyridine C$_5$H$_5$N	5	5	2	3	0	.99	2.7	20 (68)	482 (900)	1.8-12.4	115 (240)	−42 (−43)	∞	alc.	BDH, NSC
892. Pyrogallic acid C$_6$H$_3$(OH)$_3$	24a	-	3	1	0	1.45	4.4	-	-	-	309 (588)	133 (271)	v. sol.	alc.	
893. Pyrrolidine C$_4$H$_9$N	7a	-	2	3	1	.85	2.45	3 (37)	-	-	89 (192)	−63 (−81)	∞	alc.	
894. Pyruvic acid CH$_3$COCOOH	24a	-	-	-	-	1.23	-	3.0	-	-	165 (329)	14 (57)	∞	alc.	BDH
Quicklime, see Calcium oxide															
895. Quinaldine CH$_3$C$_6$H$_3$NCHCHCH	5	-	3	1	0	1.06	4.9	-	-	-	247 (477)	−2 (28)	insol.	ether	
896. Quinoline C$_6$H$_4$NCHCHCH	5	-	2	1	0	1.09	4.45	-	480 (896)	1.2-	238 (460)	−20 (−4)	sol.	alc., CS$_2$	BDH
897. 8-Quinolinol HOC$_6$H$_3$NCHCHCH	5	-	2	1	0	-	-	-	-	-	267 (513)	76 (169)	insol.	alc.	

SUBSTANCE/FORMULA	Waste Disposal Procedure (See VI)	TLV (ACGIH) PPM (mg/M³)	NFPA 704M System			Sp. Gr.	Vap. Dens. (Air=1)	Fl. Pt. °C (°F)	Ignit. Temp. °C (°F)	Flam. Limits %	B.P. °C (°F)	M.P. °C (°F)	Sol. in H₂O g/100g	Other Solvents	Misc. Ref.
			Health	Fire	React.										
898. Quinone OC$_6$H$_4$O	18	.1	3	-	-	1.32	-	293 (560)	-	-	-	116subl. (241)	sl. sol.	alc.	
Red fuming nitric acid, see nitric acid, anhydrous															
899. Refrigerant gases	26	-	-			-	-	-	-		-	<38 (<100)	-	-	NTTC
900. Resins	26	-	-			-									
901. Resorcinol C$_6$H$_4$(OH)$_2$	18	-	2	1	-	1.27	3.79	127 (261)	608 (1126)	1.4– @392ºF	281 (538)	110 (230)	sol.	alc.	
902. Rhodium Rh	27a	(.1)	-	2	-	12.4	-	-	-	-	3707 (6705)	1985 (3605)	insol.	-	
903. Ronnel C$_8$H$_8$Cl$_3$O$_3$PS	7b	(15)	-	-	-	-	-	-	-	-	-	41 (106)	-	org. solv.	
904. Rotenone C$_{23}$H$_{22}$O$_6$	18	(5)	2	1	-	1.27	-	-	-	-	210-20 (410-28) @5mm	163 (325)	insol.	org. solv.	
905. Rubber (tubing, etc.)	26	-	-	-	-	-	-	-	-	-	-	-	-	-	
906. Rubidium Rb	27k	-	2	3	-	1.53	-	-	-	Ignites in air	700 (1292)	38 (102)	dec.	dec. alc.	
907. Ruthenium Ru	27k	-	-	2	-	12.30	-	-	-	-	4150 (7502)	2450 (4442)	insol.	-	
908. Salicylaldehyde HOC$_6$H$_4$CHO	2	-	-	2	0	1.15	-	78 (172)	-	-	197 (387)	-10 (14)	sl. sol.	alc.	

SUBSTANCE/FORMULA	Waste Disposal Procedure (See VI)	TLV (ACGIH) PPM (mg/M³)	NFPA 704M System			Sp. Gr.	Vap. Dens. (Air=1)	Fl. Pt. °C (°F)	Ignit. Temp. °C (°F)	Flam. Limits %	B.P. °C (°F)	M.P. °C (°F)	Sol. in H₂O g/100g	Other Solvents	Misc. Ref.
			Health	Fire	React.										
909. Salicylic acid HOC_6H_4COOH	24a	-	1	1	-	1.44	4.8	157 (315)	545 (1013)	-	211 (412)	159 (318)	sl. sol.	alc., ace.	
910. Samarium Sm	27k	-	-	2	-	7.54	-	-	~150 (302)	-	1900 (3452)	1072 (1962)	insol.	-	
911. Selenic acid H_2SeO_4	27e	-	3	2	-	3.0	-	-	-	-	260dec. (500)	58 (136)	∞	dec. alc.	AIA, BDH
912. Selenium Se	27e	(.2)	3	2	-	~4.5	-	-	-	-	688 (1270)	217 (423)	insol.	chl.	AIA, NSC
913. Selenium dioxide SeO_2	27e	(.2)	3	-	-	3.95	-	-	-	-	-	315subl. (599)	38^{14}	alc., bz.	AIA, BDH
914. Selenium Hexafluoride SeF_6	27e	.05	3	-	-	3.25	-	-	-	-	-	-46 subl. (-51)	dec.	-	NSC
915. Selenium oxychloride $SeOCl_2$	27e	(.2)	3	-	-	2.42	-	-	-	-	176 (349)	9 (48)	dec.	bz., CS_2	AIA, NSC
Selenous acid, see selenium dioxide															
916. Sevin $C_{12}H_{11}NO_2$	14	(5)	-	-	-	-	-	-	-	-	-	-	-	-	
917. Silane SiH_4	17	-	3	3	-	.68	-	-	Spontaneously flammable in air	-	-112 (-170)	-185 (-301)	insol.	-	MGB
918. Silica (cristobalite) SiO_2	26	250 ($\overline{\%SiO_2+5}$)	2	-	-	2.32	-	-	-	-	2230 (4046)	1715 (3119)	insol.	-	
919. Silica (free) SiO_2	26	-	2	-	-	-	-	-	-	-	-	-	-	-	

SUBSTANCE/FORMULA	Waste Disposal Procedure (See VI)	TLV (ACGIH) PPM (mg/M³)	NFPA 704M System			Sp. Gr.	Vap. Dens. (Air=1)	Fl. Pt. °C (°F)	Ignit. Temp. °C (°F)	Flam. Limits %	B.P. °C (°F)	M.P. °C (°F)	Sol. in H₂O g/100g	Other Solvents	Misc. Ref.
			Health	Fire	React.										
920. Silica (quartz) SiO_2	26	250 (%SiO_2+5)	2	-	-	2.64	-	-	-	-	2230 (4046)	1610 (2930)	insol.	-	
921. Silica (amorphous) SiO_2	26	20	2	-	-	2.19	-	-	-	-	2230 (4046)	-	insol.	-	
922. Silica gel H_2SiO_3	26	-	1	-	-	-	-	-	-	-	-	150dec. (302)	insol.	-	
Silicic acid, see silica gel															
923. Silicon tetrachloride $SiCl_4$	1b	-	3	-	-	1.48	9.7	-	-	-	58 (136)	−70 (−94)	dec.	dec. alc.	BDH
924. Silicon tetrafluoride SiF_4	21	-	3	-	-	4.67	-	-	-	-	−65 (−85) @1810mm	−90 (−130) @1318mm	dec.	alc.	MGB
925. Silver Ag	27a	(.01)	1	2	-	10.5	-	-	-	-	2212 (4014)	961 (1762)	insol.	-	
926. Silver Nitrate $AgNO_3$	27k	(.01)	1	0	1	4.35	-	-	-	-	444dec. (831)	212 (414)	122⁰	eth.	
927. Sodium Na	3	-	3	1	2	.97	-	-	>115 (239) spontaneous ignition in dry air	-	892 (1638)	98 (208)	dec.	dec. alc.	BDH, MCA, NSC
928. Sodium acetate $NaOOCCH_3$	11	-	-	-	-	1.53	-	-	607 (1125)	-	-	324 (615)	119⁰	sl. sol. alc.	
929. Sodium amide $NaNH_2$	19	-	3	2	-	-	-	-	-	-	400 (752)	210 (410)	dec.	-	BDH
930. Sodium azide NaN_3	8	-	3	2	2	2.85	-	-	-	-	dec.	-	4217	sl. sol. alc.	BDH

SUBSTANCE/FORMULA	Waste Disposal Procedure (See VI)	TLV (ACGIH) PPM (mg/M³)	NFPA 704M System			Sp. Gr.	Vap. Dens. (Air=1)	Fl. Pt. °C (°F)	Ignit. Temp. °C (°F)	Flam. Limits %	B.P. °C (°F)	M.P. °C (°F)	Sol. in H₂O g/100g	Other Solvents	Misc. Ref.
			Health	Fire	React.										
931. Sodium benzoate $NaOOCC_6H_5$	11	-	0	-	-	-	-	-	-	-	-	-	66[20]	sl. sol. alc.	
932. Sodium bicarbonate $NaHCO_3$	11	-	-	-	-	2.16	-	-	-	-	-	270 (518)	6.9[0]	sl. sol. alc.	
933. Sodium bisulfate $NaHSO_4$	12a	-	2	-	-	2.44	-	-	-	-	-	>315dec. (599)	29[25]	sl. sol. alc.	
934. Sodium bisulfite $NaHSO_3$	12b	-	-	-	-	1.48	-	-	-	-	-	dec.	v. sol.	sl. sol. alc.	
935. Sodium borohydride $NaBH_4$	17	-	2	-	-	1.07	-	-	-	-	-	>400 (>752)	55[25]	alc., MeOH	AIA, BDH
936. Sodium carbonate Na_2CO_3	11	-	2	2	-	2.53	-	none	none	-	dec.	851 (1564)	7.1[0]	sl. sol. alc.	
937. Sodium chlorate $NaClO_3$	12a	-	1	0	2	2.49	-	-	-	-	300dec. (572)	248 (478)	79[0]	alc.	AIA, BDH, MCA
938. Sodium chloride $NaCl$	11	-	0	-	-	2.17	-	-	-	-	1413 (2575)	801 (1474)	36[0]	sl. sol. alc.	
939. Sodium chlorite $NaClO_2$	12a	-	1	1	2	-	-	-	-	-	175dec. (347)	-	39[17]	-	
940. Sodium chromate Na_2CrO_4	12a	-	3	-	-	-	-	-	-	-	-	792 (1458)	sol.	-	MCA
941. Sodium cyanide $NaCN$	14	(5)	2	0	0	-	-	-	-	-	1496 (2725)	560 (1040)	48[10]	sl. sol. alc.	BDH, MCA
942. Sodium dichromate $Na_2Cr_2O_7$	12a	-	3	-	-	-	-	-	-	-	120 (248)	-	sol.	-	MCA

| SUBSTANCE/FORMULA | Waste Dis-posal Pro-cedure (See VI) | TLV (ACGIH) PPM (mg/M³) | NFPA 704M System | | | Sp. Gr. | Vap. Dens. (Air=1) | Fl. Pt. °C (°F) | Ignit. Temp. °C (°F) | Flam. Limits % | B.P. °C (°F) | M.P. °C (°F) | Sol. in H₂O g/100g | Other Solvents | Misc. Ref. |
			Health	Fire	React.										
943. Sodium ethoxide $NaOC_2H_5 \cdot 2C_2H_5OH$	3	-	3	3	-	-	-	-	-	-	dec.	200 (392)	dec.	alc.	BDH
944. Sodium fluoride NaF	11	-	3	-	-	2.56	-	-	-	-	1700 (3092)	980-97 (1796-1827)	4.2^{18}	-	
945. Sodium fluoroacetate $CH_2FCOONa$	4b	(.05) Skin	3	-	-	-	-	-	-	-	-	200 (392)	111^{25}	sl. sol. alc.	
946. Sodium formate NaOOCH	11	-	3	1	-	1.92	-	-	-	-	-	253 (487)	97^{20}	sl. sol. alc.	
947. Sodium hydride NaH	17	-	3	2	-	.92	-	Explodes with water				800dec. (1472)	dec.	-	NSC
948. Sodium hydrogen difluoride NaF-HF	11	-	-	-	-	2.08	-	-	-	-	-		-	-	BDH
Sodium hydroxide, see caustic soda															
949. Sodium hypochlorite NaOCl	12a	-	2	-	-		-	-	-	-	-	-	dec.	-	BDH
950. Sodium iodide NaI	11	-	2	-	-	3.67	-	-	-	-	1304 (2379)	651 (1204)	184^{25}	alc., ace.	
951. Sodium methoxide $NaOCH_3 \cdot 2CH_3OH$	3	-	-	-	-		-	-	-	-	-	dec.	dec.	MeOH	BDH
952. Sodium nitrate $NaNO_3$	11	-	1	0	2	2.26	-	-	-	-	380dec. (716)	307 (585)	92^{25}	alc., MeOH	
953. Sodium nitrite $NaNO_2$	12b	-	3	2	-	2.17	-	-	538exp. (1000)	-	320dec. (608)	271 (520)	81^{15}	sl. sol. alc.	

SUBSTANCE/FORMULA	Waste Disposal Procedure (See VI)	TLV (ACGIH) PPM (mg/M³)	NFPA 704M System			Sp. Gr.	Vap. Dens. (Air=1)	Fl. Pt. °C (°F)	Ignit. Temp. °C (°F)	Flam. Limits %	B.P. °C (°F)	M.P. °C (°F)	Sol. in H2O g/100g	Other Solvents	Misc, Ref.
			Health	Fire	React.										
954. Sodium pentachlorophenate C_6Cl_5ONa	4b	(.5)	3	-	-	-	-	-	-	-	-	-	-	-	-
955. Sodium perchlorate $NaClO_4$	12a	-	2	0	2	2.02	-	-	-	-	-	482dec. (900)	sol.	alc.	
956. Sodium peroxide Na_2O_2	22a	-	3	0	2	2.81	-	-	-	-	-	460dec. (860)	sol.	dec. alc.	BDH
957. Sodium o-phosphate $Na_2HPO_4 \cdot 7H_2O$	11	-	-	-	-	1.68	-	-	-	-	48 (118)	-	104^{40}	-	
958. Sodium-potassium alloys NaK	3	-	3	3	2	-	-	-	Ignite in air	-	-	-	-	-	
959. Sodium propionate $NaOOCC_2H_5$	11	-	-	-	-	-	-	-	-	-	-	-	sol.	alc.	
960. Sodium silicate (water glass) $Na_2O.xSiO_2$	11	-	1	-	-	-	-	-	-	-	-	-	sol.	-	NTTC
961. Sodium sulfide Na_2S	23	-	2	1	0	1.86	-	-	-	-	-	1180 (2156)	15^{10}	sl. sol. alc.	BDH
962. Sodium sulfite Na_2SO_3	12b	-	2	-	-	2.63	-	-	-	-	dec.	-	12^0	sl. sol. alc.	
963. Sodium tetraborate $Na_2B_4O_7$	11	-	3	-	-	2.37	-	-	-	-	157dec. (2867)	741 (1366)	1.06^0	-	
964. Sodium thiocyanate NaSCN	11	-	3	-	-	-	-	-	-	-	-	287 (549)	139^{21}	alc., ace.	
965. Sodium thiosulfate $Na_2S_2O_3$	12b	-	1	-	-	1.67	-	-	-	-	-	48 (118)	50	-	

SUBSTANCE/FORMULA	Waste Dis- posal Pro- cedure (See VI)	TLV (ACGIH) PPM (mg/M³)	NFPA 704M System Health	Fire	React.	Sp. Gr.	Vap. Dens. (Air=1)	Fl. Pt. °C (°F)	Ignit. Temp. °C (°F)	Flam. Limits %	B.P. °C (°F)	M.P. °C (°F)	Sol. in H₂O g/100g	Other Solvents	Misc. Ref.
966. Stannic Chloride $SnCl_4$	1b	-	3	1	-	2.23	-	-	-	-	114 (237)	-33 (-27)	sol.	eth.	BDH
Stannic Chloride hydrate	11														
967. Stearic acid $C_{17}H_{35}COOH$	24a	-	1	1	0	.95	9.8	196 (385)	395 (743)	-	358-83 (676-721)	69 (156)	insol.	CS_2	NTTC
968. Stibine SbH_3	27d	.1	3	2	-	5.30	-	-	-	-	-17 (1.4)	-88 (-126)	.40	CS_2	-
969. 4-stibenamine $C_{14}H_{14}$	5	-	CAR	-	-	-	-	-	-	-	-	90 (194)	-	-	
970. Stoddard solvent	18	500	1	2	-	1.0	-	38-43 (100-110)	227-60 (441-500)	.8-5	220-300 (428-572)	-	-	-	
971. Strontium Sr	27h	-	2	2	-	2.6	-	-	-	-	1384 (2523)	752 (1386)	dec.	alc.	
972. Strontium carbonate $SrCO_3$	27h	-	-	-	-	3.7	-	-	-	-	1340 (2444)	1497 (2727) @69atm	sl. sol.	-	
973. Strontium nitrate $Sr(NO_3)_2$	27h	-	1	0	1	2.99	-	-	-	-	645 (1193)	570 (1058)	71¹⁸	sl. sol. alc.	
974. Strontium peroxide SrO_2	27h	-	1	0	1	4.56	-	-	-	-	-	215dec. (419)	sl. sol.	alc.	
975. Strychnine $C_{21}H_{22}N_2O_2$	5	(.15)	3	-	-	1.36	-	-	-	-	270 (518) @5mm	268 (514)	sl. sol.	chl.	
976. Styrene $(C_6H_5CHCH_2)n$	18	-	2	3	-	.909	3.6	31 (88)	490 (914)	1.1-6.1	146 (295)	-33 (-27)	insol.	alc., CS_2	AIA, BDH, NTTC
977. Styrene monomer $C_6H_5CHCH_2$	18	100C	2	3	2	.905	1	31 (88)	490 (914)	1.1-6.1	145 (293)	-31 (-23)	insol.	alc., CS_2	BDH, NTTC, MCA

SUBSTANCE/FORMULA	Waste Disposal Procedure (See VI)	TLV (ACGIH) PPM (mg/M³)	NFPA 704M System Health	Fire	React.	Sp. Gr.	Vap. Dens. (Air=1)	Fl. Pt. °C (°F)	Ignit. Temp. °C (°F)	Flam. Limits %	B.P. °C (°F)	M.P. °C (°F)	Sol. in H₂O g/100g	Other Solvents	Misc. Ref.
978. Succinic Acid COOH(CH₂)₂COOH	24a	-	2	1	-	1.57	-	-	-	-	235dec. (455)	185 (365)	sl. sol.	alc., MeOH	
979. Succinic Anhydride C₄H₄O₃	24a	-	2	1	-	1.10	-	-	-	-	261 (502)	120 (248)	insol.	alc.	
980. Succinonitrile NCCH₂CH₂CN	14	-	3	1	-	.98	2.76	132 (270)	-	-	266 (511)	58 (136)	v. sol.	alc., bz.	
980a. Sulfamic Acid SO₂(NH₂)OH	19	-	2	-	3	2.03	-	-	-	-	dec.	200	dec.	-	
981. Sulfamide SO₂(NH₂)₂	19	-	-	-	-	1.61	-	-	-	-	250dec. (482)	92 (198)	sol.	alc.	
982. Sulfur S₈	26	-	2	1	0	2.07	-	207 (405)	232 (450)	-	444 (832)	119 (246)	insol.	CS₂	NTTC, MCA, NSC
983. Sulfur Decafluoride S₂F₁₀	21	-	3	-	-	2.08	-	-	-	-	29 (84)	-92 (-134)	-	-	
984. Sulfur Dichloride SCl₂	21	-	3	2	-	1.62	3.6	-	-	-	59dec. (138)	-78 (-108)	-	bz.	BDH, MCA
985. Sulfur Dioxide SO₂	12b	5	3	0	0	1.4	2.3	-	-	-	-10 (14)	-76 (-105)	23[0]	alc.	AIA, MCA CGA
986. Sulfur Hexafluoride SF₆	21	1000	-	-	-	1.67	6.2	-	-	-	-	-64subl (-83)	sl. sol.	alc.	
987. Sulfuric Acid H₂SO₄	24b	(1)	3	0	1	1.84	2.8	None	-	None	338 (640)	10 (50)	∞	dec., alc.	AIA, BDH, NTTC, MCA
988. Sulfur Monochloride S₂Cl₂	21	1	2	1	1	1.69	-	118 (245)	234 (453)	-	136 (277)	-80 (-112)	dec.	bz., CS₂	BDH, MCA
989. Sulfur Trioxide SO₃	24b	-	3	-	-	2.75	-	-	-	-	45 (113)	17 (63)	dec.	-	MCA

SUBSTANCE/FORMULA	Waste Disposal Procedure (See VI)	TLV (ACGIH) PPM (mg/M³)	NFPA 704M System			Sp. Gr.	Vap. Dens. (Air=1)	Fl. Pt. °C (°F)	Ignit. Temp. °C (°F)	Flam. Limits %	B.P. °C (°F)	M.P. °C (°F)	Sol. in H₂O g/100g	Other Solvents	Misc. Ref.
			Health	Fire	React.										
990. Sulfuryl Chloride SO$_2$Cl$_2$	21	-	3	-	-	1.67	4.7	-	-	-	69 (156)	-54 (-65)	dec.	bz.	BDH
991. Sulfuryl Fluoride SO$_2$F$_2$	21	5	3	-	-	3.72	-	-	-	-	-55 (-67)	-137 (-215)	10^9	alc.	-
992. Systox (C$_2$H$_5$O)$_2$PSOC$_2$H$_4$SC$_2$H$_5$	7b	(.1)	3	-	-	-	-	-	-	-	-	134 (273)	-	-	-
993. Tall Oil (Liquid Rosin)	18	-	-	1	-	-	-	182 (360)	-	-	-	-	-	-	NTTC
994. Tallow	18	-	0	1	0	.895	-	265 (509)	-	-	-	32 (90)	-	-	NTTC
995. Tannic Acid C$_{76}$H$_{52}$O$_6$	24a	-	2	1	-	-	-	199oc. (390)	527 (980)	-	-	210dec. (410)	sol.	alc., ace.	
996. Tantalum Ta	27a	(5)	-	2	-	16.7	-	-	-	-	5425 (9797)	2997 (5300)	insol.	-	-
997. Tar, Liquid	26	-	3	2	-	-	-	>93 (>200)	-	-	-	49 (120)	-	-	NTTC
TEDP, see Tetraethyl Dithionopyrophosphate															
998. Tellurium Te	27e	(.1)	2	2	-	6.25	-	-	-	-	1390 (2534)	452 (846)	insol.	-	
999. Tellurium Hexafluoride TeF$_6$	27e	0.02	2	-	-	1.18	-	-	-	-	35 (95)	-36 (-33)	dec.	-	
TEPP, see Tetraethyl Pyrophosphate															

SUBSTANCE/FORMULA	Waste Dis-posal Pro-cedure (See VI)	TLV (ACGIH) PPM (mg/M^3)	NFPA 704M System			Sp. Gr.	Vap. Dens. (Air=1)	Fl. Pt. °C (°F)	Ignit. Temp. °C (°F)	Flam. Limits %	B.P. °C (°F)	M.P. °C (°F)	Sol. in H$_2$O g/100g	Other Solvents	Misc. Ref.
			Health	Fire	React.										
1000. Terbium Tb	27k	-	-	2	-	8.77	-	-	-	-	2800 (5072)	1356 (2473)	insol.	-	
1001. m-Terphenyl $C_{18}H_{14}$	18	1C	-	1	0	1.16	-	135oc. (275)	-	-	365 (689)	87 (189)	insol.	alc., bz.	
1002. o-Terphenyl $C_{18}H_{14}$	18	1C	-	1	0	1.14	7.9	163oc. (325)	-	-	332 (630)	57 (135)	insol.	bz., MeOH	
1003. p-Terphenyl $C_{18}H_{14}$	18	1C	-	1	-	1.24	-	-	-	-	405 (761)	213 (415)	-	-	
1004. Tetrabromoethane $Br_2CHCH\ Br_2$	26	1	3	0	1	2.97	-	-	-	-	239dec. (374)	-1 (30)	insol.	alc.	BDH
1005. 1,2,4,5-Tetrachlorobenzene $C_6H_2Cl_4$	4b	-	-	1	0	1.86	-	155 (311)	-	-	243 (469)	139 (282)	insol.	bz., CS$_2$	
1006. 1,1,1,2-Tetrachloro-2,2-Difluoro Ethane $F_2CCl\ CCl_3$	26	500	2	-	-	1.64	7.03	-	-	-	92 (198)	41 (106)	insol.	alc.	
1007. 1,1,2,2-Tetrachloro-1,2, Difluoro Ethane $Cl_2FCCFCl_2$	26	500	-	-	-	1.64	7.03	-	-	-	93 (199)	25 (77)	insol.	alc.	
1008. 1,1,2,2-Tetrachloroethane $Cl_2HCCHCl_2$	4b	5 Skin	3	-	-	1.59	5.8	None	-	-	146 (295)	-43 (-45)	sl. sol.	alc.	AIA, BDH, MCA
1009. Tetrachloroethylene $Cl_2C=CCl_2$	27j	100	3	-	-	1.62	5.8	None	-	-	121 (250)	-24 (-11)	insol.	alc., bz.	AIA, BDH, NTTC, MCA
1010. Tetrachloronaphthalene $C_{10}H_8Cl_4$	6	(2) Skin	2	-	-	-	-	-	-	-	-	182 (360)	-	-	
1011. Tetradecane $C_{14}H_{30}$	18	-	-	1	0	.76	6.8	100 (212)	202 (396)	.5—	254 (489)	6 (43)	insol.	alc.	

| SUBSTANCE/FORMULA | Waste Dis-posal Pro-cedure (See VI) | TLV (ACGIH) PPM (mg/M³) | NFPA 704M System | | | Sp. Gr. | Vap. Dens. (Air=1) | Fl. Pt. °C (°F) | Ignit. Temp. °C (°F) | Flam. Limits % | B.P. °C (°F) | M.P. °C (°F) | Sol. in H₂O g/100g | Other Solvents | Misc. Ref. |
			Health	Fire	React.										
1012. Tetraethyldithionopyro-phosphate $(C_2H_5O)_4P_2OS_2$	7b	-	3	-	-	1.19	-	-	-	-	138 (280)	-	insol.	alc.	
1013. Tetraethylenepentamine $H_2N(CH_2CH_2NH)_3$ $CH_2CH_2NH_2$	7a	-	2	1	0	.99	-	163oc. (325)	-	-	333 (631)	-	-	-	
1014. Tetraethyl Lead $Pb(C_2H_5)_4$	4b	(.075) Skin	3	2	3	1.66	8.6	93 (206)	-	-	170exp. (338)	-137 (-215)	insol.	alc., bz.	MCA
1015. Tetraethyl Pyrophosphate $(C_2H_5O)_4P_2O_3$	7b	-	3	-	-	1.20	-	-	-	-	155 (311)	-	∞	alc., ace.	
1016. Tetrahydrofuran C_4H_8O	15	200	2	3	1	.89	2.5	-14 (6)	321 (610)	2-11.8	65 (151)	-65 (-85)	v. sol.	org., sol.	BDH
3,4,7,7a-Tetrahydro-4,7-Methanoindene, see Dicyclopentadiene															
1017. Tetrahydronaphthalene $C_{10}H_{12}$	18	-	1	2	0	.97	4.55	71 (160)	384 (723)	.8-5 @302°F	207 (405)	-30 (-22)	insol.	alc.	
1018. Terralin, see Tetrahydronaphthalene															
1019. N,N,N',N'-Tetramethyl-3,3'-Dimethoxybenzidine $C_{18}H_{24}O_2N_2$	6	-	CAR	-	-	-	-	-	-	-	-	-	-	-	
1020. N,N,N',N'-Tetramethylethy-lenediamine $(CH_3)_2NCH=$ $CHN(CH_3)_2$	6	-	-	2	-	.78	-	-	-	-	121 (250)	-	sol.	-	
1021. Tetramethyl Lead $Pb(CH_3)_4$	4b	(.075) Skin	3	3	3	1.99	6.5	38 (100)	-	-	110 (230)	-28 (-18)	insol.	alc., bz.	
1022. Tetramethyl Silane $Si(CH_3)_4$	26	-	3	-	-	.65	-	-	-	-	27 (81)	-	-	eth.	

SUBSTANCE/FORMULA	Waste Disposal Procedure (See VI)	TLV (ACGIH) PPM (mg/M³)	NFPA 704M System			Sp. Gr.	Vap. Dens. (Air=1)	Fl. Pt. °C (°F)	Ignit. Temp. °C (°F)	Flam. Limits %	B.P. °C (°F)	M.P. °C (°F)	Sol. in H₂O g/100g	Other Solvents	Misc., Ref.
			Health	Fire	React.										
1023. Tetramethyl Succinonitrile $NCC(CH_3)_2C(CH_3)_2CN$	14	.5	3	-	-	1.30	-	-	-	-	-	190subl. (374)	sl. sol.	sl. sol., alc.	
1024. Tetramethylthiuram Disulfide (Thiram) $(H_2NCSCH_3)_4S_2$	13	(5)	2	-	-	1.3	-	-	-	-	-	70 (158)	-	-	
1025. Tetranitromethane $C(NO_2)_4$	4a	1	3	3	-	1.64	-	-	-	-	126 (259)	13 (55)	insol.	alc.	
Tetryl, see N-Methyl-N-2, 4,6-Tetranitroaniline															
1026. Thallium Tl	27k	(.1) Skin	3	-	-	11.9	-	-	-	-	1460 (2660)	304 (579)	insol.	-	
1027. Thallium, soluble compounds	27k	(.1) Skin	3	-	-	-	-	-	-	-			-	-	
1028. Thallous Sulfate Tl_2SO_4	27k	-	3	-	-	6.77	-	-	-	-	dec.	632 (1170)	20 4.9	-	
1029. Thioacetamide CH_3CSNH_2	13	-	2	-	-	-	-	-	-	-	-	115 (239)	v. sol.	alc.	
1030. 2,2′-Thiodiethanol $(CH_2CH_2OH)_2S$	13	-	1	1	0	1.18	4.2	160oc. (320)	-	-	28 (82)	-11 (12)	∞	alc.	
1031. Thioglycollic Acid $HSCH_2-COOH$	13	-	3	-	-	1.35	-	-	-	-	105 (221) @11mm	-17 (1)	∞	alc.	BDH
1032. Thionyl Chloride $SOCl_2$	21	-	3	-	-	1.64	-	-	-	-	76 (169)	-105 (-157)	dec.	bz.	BDH
1033. Thionyl Fluoride SOF_2	21	-	3	-	-	2.93	-	-	-	-	-44 (-47)	-110 (-166)	dec.	bz., ace.	

SUBSTANCE/FORMULA	Waste Disposal Procedure (See VI)	TLV (ACGIH) PPM (mg/M³)	NFPA 704M System			Sp. Gr.	Vap. Dens. (Air=1)	Fl. Pt. °C (°F)	Ignit. Temp. °C (°F)	Flam. Limits %	B.P. °C (°F)	M.P. °C (°F)	Sol. in H₂O g/100g	Other Solvents	Misc. Ref.
			Health	Fire	React.										
1034. Thiophene C4H4S	13	-	2	3	-	1.06	2.9	-1 (30)	-	-	84 (183)	-38 (-36)	-	org., sol.	BDH
1035. Thiourea NH2CSNH2	13	-	1	-	-	1.41	-	-	-	-	dec.	181 (358)	sol.	alc.	
Thiram, see Tetramethyl Thiuram Disulfide															
1036. Thorium Th	27k	-	-	2	-	11.7	-	-	-	-	4500 (8100)	1827 (3321)	insol.	-	AIA
1037. Thorium Nitrate Th(NO3)4	27k	-	1	0	1	-	-	-	-	-	-	500dec. (932)	v. sol.	alc.	
1038. Tin Sn	27a	-	0	1	-	5.75	-	-	-	-	2260 (4100)	232 (450)	insol.	-	
1039. Tin, Inorganic compounds	11	(2)	variable, oxides toxic	-	-	-	-	-	-	-	-	-	most soluble	-	
1040. Tin, Organic compounds	26	(.1)	3	-	-	-	-	-	-	-	-	-	-	-	
1041. Titanium Ti	27a	(15)	0	-	-	4.5	-	-	700-800 (1292-1472)	-	3262 (5432)	1800 (3146)	insol.	-	AIA
1042. Titanium Dioxide (rutile) TiO2	26	(15)	1	-	-	4.26	-	-	-	-	~2700 (4890)	1840 (3344)	insol.	-	
1043. Titanium Tetrachloride TiCl4	1b	5	3	0	1	1.73	-	-	-	-	136 (277)	-30 (-22)	sol.	alc.	BDH, NSC
1044. Toluene C6H5CH3	18	200	2	3	0	.87	3.1	4.4 (40)	536 (997)	1.4-6.7	111 (231)	-95 (-139)	insol.	alc., bz., CS2	AIA, BDH, NTTC, MCA

SUBSTANCE/FORMULA	Waste Disposal Procedure (See VI)	TLV (ACGIH) PPM (mg/M³)	NFPA 704M System Health	Fire	React.	Sp. Gr.	Vap. Dens. (Air=1)	Fl. Pt. °C (°F)	Ignit. Temp. °C (°F)	Flam. Limits %	B.P. °C (°F)	M.P. °C (°F)	Sol. in H₂O g/100g	Other Solvents	Misc. Ref.
1045. Toluene-2,4-diisocyanate $CH_3C_6H_3(NCO)_2$	6	.02C	2	1	2	1.2	6.0	132 (270)oc	-	.9-9.5	251 (484)	20 (68)	-	-	AIA, MCA
α-Toluenethiol, see Benzyl Mercaptan															
1046. 1-o-Tolylazo-2-Naphthol $C_{17}H_{14}N_2O$	8	-	CAR	-	-	-	-	-	-	-	-	-	-	-	
Tolylene Diisocyanate, see Toluene-2,4-Diisocyanate															
o-Tolyl Phosphate, see Tritolyl Phosphate															
1047. m-Toluidine $CH_3C_6H_4 \cdot NH_2$	5	-	3	2	0	.99	3.9	86 (187)	482 (900)	-	203 (398)	-31 (-24)	sl. sol.	alc.	BDH, MCA
1048. o-Toluidine $CH_3 \cdot C_6H_4 \cdot NH_2$	5	5 Skin	3	2	0	1.004	3.7	85 (185)	482 (900)	-	200 (392)	-16 (3)	sl. sol.	alc.	BDH, MCA
1049. p-Toluidine $CH_3C_6H_4 \cdot NH_2$	5	-	3	2	0	1.046	3.9	87 (189)	482 (900)	-	200 (392)	44 (111)	sl. sol.	alc.	BDH, MCA
1049a. Transite	26														
1049b. Toxaphene $C_{10}H_{10}Cl_8$	4b	(0.5) Skin	3	1	0	1.66	-	-	-	-	-	65-90	insol.	org. sol.	
1050. Tremolite $Ca_2Mg_5Si_8O_{24}H_2$	26	-	-	-	-	-	-	-	-	-	-	-	-	-	
1051. Triamylamine $(C_5H_{11})_3N$	7a	-	2	1	0	.8	7.8	102 (215)	-	-	232 (450)	-	-	alc., ace.	
Tribromomethane, see Bromoform															

SUBSTANCE/FORMULA	Waste Disposal Procedure (See VI)	TLV (ACGIH) PPM (mg/M³)	NFPA 704M System			Sp. Gr.	Vap. Dens. (Air=1)	Fl. Pt. °C (°F)	Ignit. Temp. °C (°F)	Flam. Limits %	B.P. °C (°F)	M.P. °C (°F)	Sol. in H₂O g/100g	Other Solvents	Misc. Ref.
			Health	Fire	React.										
1052. Tri-n-Butyl Amine (C₄H₉)₃N	7a	-	2	2	0	.8	6.38	86 (187)	-	-	216 (421)	−70 (−94)	sl. sol.	alc.	BDH
1053. Tributylchlorotin C₁₂H₂₇SnCl	4b	-	-	-	-	1.13	-	-	-	-	174 (345)	30 (86)	-	-	
1054. Tributyl Phosphate (C₄H₉O)₃PO	7b	(5)	2	1	0	.97	9.2	146 (295)	-	-	292 (558)	<−80 (<−112)	sol.	alc., bz., cs2	
1055. Trichloroacetic Acid C Cl₃COOH	4c	-	3	-	-	1.63	-	-	-	-	198 (388)	58 (136)	v. sol.	alc.	BDH
Trichloroacetaldehyde, see Chloral															
1056. Trichloroacetonitrile CCl₃ CN	14	-	-	-	-	1.44	-	-	-	-	85 (185)	−44 (−47)	-	-	
1057. Trichloroacetyl Chloride CCl₃COCl	1a	-	3	-	-	1.63	-	-	-	-	118 (244)	-	dec.	dec., alc.	
1058. 1,2,4-Trichlorobenzene C₆H₃Cl₃	4b	-	2	1	0	1.45	6.3	99 (210)	-	-	214 (417)	17 (63)	insol.	sl. sol., alc	
1059. 1,1,1-Trichloroethane CH₃CCl₃	27j	350	2	-	-	1.34	4.6	none	-	-	74 (165)	−38 (−36)	insol.	alc.	AIA, NTTC, MCA
1060. Trichloroethylene ClCHCCl₂	27j	100	2	-	-	1.46	4.54	-	410 (770)	-	87 (188)	−73 (−99)	sl. sol.	alc.	AIA, BDH, NTTC, MCA, NSC
Trichlorofluoromethane, see Fluorotrichloromethane															
Trichloromethanethiol, see Perchloromethyl Mercaptan															

SUBSTANCE/FORMULA	Waste Disposal Procedure (See VI)	TLV (ACGIH) PPM (mg/M³)	NFPA 704M System Health	Fire	React.	Sp. Gr.	Vap. Dens. (Air=1)	Fl. Pt. °C (°F)	Ignit. Temp. °C (°F)	Flam. Limits %	B.P. °C (°F)	M.P. °C (°F)	Sol. in H₂O g/100g	Other Solvents	Misc. Ref.
1061. Trichloronaphthalene $C_{10}H_5Cl_3$	4b	(5) Skin	2	-	-	-	-	-	-	-	-	133 (271)	-	alc.h	
Trichloronitromethane, see Chloropicrin															
1062. 2,4,5-Trichlorophenoxy Acetic Acid $Cl_3C_2H_2OCH_2COOH$	4c	(10)	-	-	-	-	-	-	-	-	-	157 (315)	sl.sol.	-	
1063. 1,2,3-Trichloropropane $C_3H_5Cl_3$	4b	50	3	2	0	1.39	5.0	82 (180)oc	304 (579)	3.2-12.6	156 (313)	-15 (5)	sl. sol.	alc.	
1064. 1,1,2-Trichloro-1,2,2-Trifluoro Ethane $Cl_2FCC\,Cl\,F_2$	26	1000	-	1	-	1.56	-	-	680 (1256)	-	48 (118)	-36 (-33)	insol.	alc., bz.	
Tricresyl Phosphate, see Tritolyl Phosphate															
1065. Tridecanol $CH_3(CH_2)_{11}CH_2OH$	18	-	-	-	-	.82	6.9	121 (250)	-	-	274 (525)	31 (88)	insol.	alc.	
1066. Triethyl Aluminum $(C_2H_5)_3Al$	3	-	3	4	3	.84	-	<-53 (<-63)	<-53 (<-63)	-	194 (381)	-53 (-63)	exp.H₂O	-	AIA
1067. Triethyl Amine $(C_2H_5)_3N$	7a	25	2	3	0	.73	3.48	<-7 (<-20loc)	-	1.2-8.0	89 (192)	-115 (-175)	sol.	alc.	BDH
1068. Triethanolamine $(CH_2OHCH_2)_3N$	7a	-	1	1	1	1.13	-	179 (355)	371 (700)	-	360 (680)	20 (68)	∞	alc.	
1069. Triethylene Glycol Triethy $(CH_2OCH_2CH_2OH)_2$	18	-	1	1	0	1.13	5.17	177 (350)	-	.9-9.2	276 (529)	-4 (25)	∞	alc.	
1070. Triethylene-Tetramine $H_2NCH_2(CH_2NHCH_2)_2CH_2 NH_2$	7a	-	3	1	0	.98	-	135 (275)	338 (640)	-	267 (513)	12 (54)	sol.	alc.	BDH

SUBSTANCE/FORMULA	Waste Disposal Procedure (See VI)	TLV (ACGIH) PPM (mg/M³)	NFPA 704M System Health	Fire	React.	Sp. Gr.	Vap. Dens. (Air=1)	Fl. Pt. °C (°F)	Ignit. Temp. °C (°F)	Flam. Limits %	B.P. °C (°F)	M.P. °C (°F)	Sol. in H₂O g/100g	Other Solvents	Misc, Ref.
1071. Triethyl o-Formate (C2H5O)3CH	18	-	3	2	-	.89	5.1	30 (86)	-	-	146 (295)	-	dec.	alc.	BDH
1072. Trifluoroacetic Acid CF3COOH	4c	-	3	-	-	1.54	3.92	-	-	-	72 (162)	-15 (5)	sol.	-	BDH
1073. Trifluoromethane CHF3	26	-	2	-	-	1.52	-	-	-	-	-84 (-119)	-163 (-261)	sl. sol.	alc.	
1074. Triisobutyl Aluminum (C4H9)3Al	3	-	3	3	-	.79	-	<0 (<32)	<4 (<39)	-	114 (237)	4 (39)	-	-	AIA
1075. Trimethyl Amine (CH3)3N	7a	-	3	4	0	.66	2.0	-13-8 (8-18)	190 (374)	2-11.6	4 (39)	-117 (-179)	v. sol.	alc., bz.	BDH, MCA, MGB
1076. 5,9,10-Trimethyl-1,2-Benzanthracene (CH3)3C18H13	18	-	CAR	-	-	-	-	-	-	-	-	-	-	-	
1077. 6,9,10-Trimethyl-1,2-Benzanthracene (CH3)3C18H13	18	-	CAR	-	-	-	-	-	-	-	-	-	-	-	
1078. Trimethyl Borate B(OCH3)3	18	-	2	3	1	.92	3.6	<27 (<80)	-	-	67 (153)	-29 (-20)	dec.	MeOH	
3,5,5-Trimethyl-2-Cyclohexenone, see iso-Phorone															
1079. 2,2,4-Trimethyl Pentane (CH3)3C CH2 CH(CH3)2	18	-	2	3	0	.69	3.9	-12 (10)	418 (784)	1.1-6.0	99 (210)	-107 (-161)	insol.	sl. sol., alc.	BDH
1080. 2,4,4-Trimethyl-2-Pentane (CH3)2C=CH C(CH3)3	18	-	-	3	0	.72	3.9	2 (35)	-	-	112 (234)	-107 (-161)	insol.	-	
1081. 1,3,5-Trinitrobenzene C6H3(NO2)3	6	-	2	4	4	1.69	-	-	-	exp.	-	121 (250)	sl. sol.	ace., bz.	

SUBSTANCE/FORMULA	Waste Disposal Procedure (See VI)	TLV (ACGIH) PPM (mg/M³)	NFPA 704M System			Sp. Gr.	Vap. Dens. (Air=1)	Fl. Pt. °C (°F)	Ignit. Temp. °C (°F)	Flam. Limits %	B.P. °C (°F)	M.P. °C (°F)	Sol. in H₂O g/100g	Other Solvents	Misc. Ref.
			Health	Fire	React.										
Trinitrophenol, see Picric Acid															
1082. 2,4,6-Trinitrotoluene $CH_3C_6H_2(NO_2)_3$	6	(1.5) Skin	2	4	4	1.65	-	-	-	exp.	240exp. (464)	81 (178)	insol.	ace., bz.	
1083. 1,3,5-Trioxane $OCH_2OCH_2OCH_2$	18	-	2	2	0	1.17	-	45 (113)oc	414 (777)	3.6-29	115 (239)	62 (144)	v. sol.	alc., bz. cs2	
1084. Triphenyl Phosphate $(C_6H_5O)_3 PO$	7b	(3)	2	1	0	1.21	-	220 (428)	-	-	245 (473) @11mm	49 (120)	insol.	alc., bz.	
1085. Triphenyl Phosphine $(C_6H_5)_3 P$	7b	-	3	2	-	1.19	-	180 (356)oc	-	-	>360 (680)	80 (176)	insol.	alc., bz.	
1086. Tripropylamine $(C_3H_7)_3 N$	7a	-	2	2	0	.75	4.9	41 (106)oc	-	-	156 (313)	-94 (-137)	sl. sol.	alc.	
1087. Trisodium Phosphate $Na_3PO_4 \cdot 12H_2O$	11	-	2	-	-	1.62	-	-	-	-	-	74dec. (165)	sl. sol.	-	
Triton, see Trinitrotoluene															
1088. Tritolyl Phosphate $(CH_3C_6H_4)_3PO_4$	7b	-	2	1	0	1.17	12.7	225 (437)	385 (725)	-	410 (770)	11 (52)	-	-	AIA, BDH
1089. Turpentine $C_{10}H_{16}$	18	100	1	3	0	.87	4.6	35-39 (95-102)	253 (487)	.8-	153-75 (307-47)	-	-	-	AIA
1090. Unsymmetrical Dimethyl Hydrazine $(CH_3)_2NNH_2$	16	.5	3	3	1	.79	1.94	~-15 (~5)	249 (480)	2-95	63 (145)	-58 (-72)	v. sol.	alc., MeOH	MCA
1091. Uranium U	27k	(.25)	3	3	-	19.05	-	-	-	-	3818 (6905)	1130 (2066)	insol.	-	AIA

SUBSTANCE/FORMULA	Waste Disposal Procedure (See VI)	TLV (ACGIH) PPM (mg/M³)	NFPA 704M System Health	NFPA 704M System Fire	NFPA 704M System React.	Sp. Gr.	Vap. Dens. (Air=1)	Fl. Pt. °C (°F)	Ignit. Temp. °C (°F)	Flam. Limits %	B.P. °C (°F)	M.P. °C (°F)	Sol. in H₂O g/100g	Other Solvents	Misc. Ref.
1092. Uranium, Insoluble Compounds	27k	.25	3	-	-	-	-	-	-	-	-	-	insol.	-	-
1093. Uranium, Soluble Compounds	27k	(.05)	3	-	-	-	-	-	-	-	-	-	sol.	-	-
1094. Uranium, Nitrate $UO_2(NO_3)_2 \cdot 6H_2O$	27k	(.05)	1	0	1	2.81	-	-	-	-	118 (244)	59 (138)	∞	alc., ace.	-
1095. Urea H_2NCONH_2	26	-	0	-	-	1.32	-	-	-	-	dec.	133	v. sol.	alc.	NTTC
1096. Valeraldehyde $CH_3CH_2CH_2CH_2CHO$	2	-	-	-	-	.81	3.0	12 (54)	-	-	102 (217)	-92 (-134)	sl. sol.	alc.	NTTC
1097. Valeric Acid C_4H_9COOH	24a	-	1	-	-	.94	3.5	96 (205)oc	-	-	186 (367)	-35 (-31)	sol.	alc.	
1098. Vanadium Dichloride VCl_2	27i	-	3	-	-	3.23	-	-	-	-	-	-	dec.	alc.	
1099. Vanadium Oxytrichloride $VOCl_3$	27i	(.1)	-	2	-	1.83	-	-	-	-	127 (261)	-77 (-107)	dec.	alc.	
1100. Vanadium Pentoxide V_2O_5	27i	(.1)	-	2	-	3.36	-	-	-	-	1750dec. (3182)	690 (1274)	sl. sol.	-	
1100a. Vanadyl Sulfate $VOSO_4$	27i														
1101. Varnish	18	-	-	3	0	-	-	176-248	-	-	-	-	-	-	NTTC
1102. Vinyl Acetate $CH_3COOCHCH_2$	18	-	2	3	2	.94	3.0	-8 (18)	427 (800)	2.6-13.4	73 (163)	-100 (-148)	insol.	org. sol.	NTTC, MCA
Vinyl Bromide, see Bromoethylene															
Vinyl Butyl Ether see Butyl Vinyl Ether															

SUBSTANCE/FORMULA	Waste Disposal Procedure (See VI)	TLV (ACGIH) PPM (mg/M³)	NFPA 704M System Health	Fire	React.	Sp. Gr.	Vap. Dens. (Air=1)	Fl. Pt. °C (°F)	Ignit. Temp. °C (°F)	Flam. Limits %	B.P. °C (°F)	M.P. °C (°F)	Sol. in H₂O g/100g	Other Solvents	Misc. Ref.
1103. Vinyl Chloride CH_2CHCl	4b	500C	2	4	2	.91	2.15	−78 (−108)	472 (882)	4-22	−14 (7)	−154 (−245)	sl. sol.	alc.	MCA
Vinyl Cyanide, see Acrylonitrile															
1104. Vinyl Ether $CH_2CHOCHCH_2$	15	-	2	3	2	.77	2.4	<−30 (<−22)	360 (680)	1.7-27	39 (102)	-	-	alc. ace.	
1105. Vinylidene Chloride CH_2CCl_2	4b	5	2	4	2	1.3	3.4	−15oc (5)	458 (856)	5.6-11.4	37 (99)	−122 (−188)	insol.	-	MCA
Vinyl Toluene, see Methyl Styrene															
1106. Warfarin $C_{19}H_{16}O_4$	18	(.1)	3	-	-	-		-	-		-	161 (322)	insol.	alc., ace.	
1107. Xenon Xe	26	-	0	-	-	5.89	-	-	-	-	−108 (−162)	−112 (−170)	24⁰	-	
Xenylamine, see 4-Biphenylamine															
1108. m-Xylene $C_6H_4(CH_3)_2$	18	100	2	3	0	.87	-	29 (84)	528 (982)	1.1-7.0	139 (282)	−48 (−54)	insol.	org. sol.	AIA, BOH, NTTC, MCA
1109. o-Xylene $C_6H_4(CH_3)_2$	18	100	2	3	0	.90	1.1	32 (90)	464 (867)	1.0-6.0	144 (291)	−26 (−15)	insol.	org. sol.	AIA, BDH, NTTC, MCA
1110. p-Xylene $C_6H_4(CH_3)_2$	18	100	2	3	0	.86	-	27 (81)	529 (984)	1.1-7.0	138 (280)	13 (55)	insol.	org. sol.	AIA, BDH, NTTC, MCA
1111. Xylenol $C_6H_3(CH_3)_2$ OH	18	-	3	1	-	-	-	-	-	-	218 (424)	75 (167)	sl. sol.	alc.	BDH

| SUBSTANCE/FORMULA | Waste Disposal Procedure (See VI) | TLV (ACGIH) PPM (mg/M³) | NFPA 704M System | | | Sp. Gr. | Vap. Dens. (Air=1) | Fl. Pt. °C (°F) | Ignit. Temp. °C (°F) | Flam. Limits % | B.P. °C (°F) | M.P. °C (°F) | Sol. in H₂O g/100g | Other Solvents | Misc. Ref. |
			Health	Fire	React.										
1112. Xylidine $C_6H_3(CH_3)_2NH_2$	5	5 Skin	3	1	0	.99	4.2	97 (206)	-	-	224 (435)	<−15 (<5)	sl. sol.	alc.	BDH
1113. Yttrium Y	26	(1)	-	2	-	4.34	-	-	-	-	2927 (5300)	1500 (2732)	sl. dec.	-	
1114. Zinc Zn	27a	(15)	0	1	1	7.14	-	-	-	-	905 (1661)	419 (787)	insol.	-	
1115. Zinc Acetate $Zn(C_2H_3O_2)_2$	11	-	0	-	-	1.84	-	-	-	-	-	200 dec. (392)	30²⁰	alc.	
1116. Zinc Chlorate $Zn(ClO_3)_2 \cdot 4H_2O$	12a	-	2	0	2	2.15	-	-	-	-	-	60 dec. (140)	262²⁰	alc., ace.	AIA
1117. Zinc Chloride $Zn\,Cl_2$	11	(1)	2	0	2	2.91	-	-	-	-	732 (1350)	283 (541)	432²⁵	alc.	
1118. Zinc Oxide ZnO	26	(5)	3	-	-	5.47	-	-	-	-	-	1975 (3587)	sl. sol.	-	
1119. Zirconium Zr	26	(5)	1	4	1	6.5	-	-	260 (500)	-	3578 (5252)	1850 (3362)	insol.	-	AIA, NSC
1120. Zirconium Compounds	11	(5)	1	-	-	-	-	-	-	-	-	-	-		
1121. Zirconium Hafnium Powder	26	(5)	1	4	1	-	-	-	20 (68)	-	-	-	insol.		MCA, NSC

443

SECTION VI.

WASTE DISPOSAL PROCEDURES

ORGANIC ACID HALIDES

Wear:

Rubber gloves, self-contained breathing apparatus (or work in an effective fume hood with full face shield), laboratory coat.

Spills:

Cover with sodium bicarbonate. If a small quantity is involved, scoop the mixture into a large beaker of water and let stand for a few minutes. Slowly pour into the drain with copious amounts of water. If a large quantity is involved, scoop the resulting bicarbonate mixture into a plastic bag, cardboard box or small fiber drum. This material can then be burned in an incinerator or spread on the ground and flooded with water. The site of the spill should be washed with soapy water.

Package lots:

Slowly sift or pour into a large glass or plastic vessel containing a layer of sodium bicarbonate. Mix thoroughly and add slowly to a large container of water with stirring. Slowly pour this mix down the drain with copious amounts of water.

Examples:

Acetyl bromide
Acetyl chloride
Benzene sulfonyl chloride
Benzoyl chloride
Butyryl chloride
Chloroacetyl chloride
3-Chloropropionyl chloride
o-Chlorobenzoyl chloride
Crotonyl chloride
Dichloroacetyl chloride
Oxalyl chloride
Propionyl chloride
Trichloroacetyl chloride

1b DISPOSAL PROCEDURE
INORGANIC HALIDES

Wear:

Rubber gloves, self-contained breathing apparatus (or work in fume hood), laboratory coat. For the more active compounds work from behind a body shield.

Spills:

Cover with excess sodium bicarbonate. If a small quantity is involved, scoop the mixture into a large beaker of water and let stand for a few minutes. Slowly pour into the drain with copious amounts of water. If a large quantity is involved, scoop the resulting bicarbonate mixture into a plastic bag, cardboard box or small fiber drum. This material can then be burned in an incinerator or spread on the ground and flooded with water. The site of the spill should be washed with soapy water.

Package lots:

Sift or pour onto a dry layer of sodium bicarbonate in a large evaporating dish. After mixing thoroughly spray with 6M-NH$_4$OH while stirring. Cover with a layer of crushed ice and stir. Continue spraying with 6M-NH$_4$OH. When the smoke of NH$_4$Cl has partly subsided add iced water and stir. Dump this slurry into a large container. Repeat until all has been treated. Neutralize* and slowly siphon the suspension into the drain with excess running water.

Examples:

Aluminum bromide, anhydrous
Aluminum chloride, anhydrous
Chlorosulfonic acid
Ferric chloride, hexahydrate
Germanium tetrachloride
Silicon tetrachloride
Stannic chloride
Tin tetrachloride
Titanium tetrachloride

* If excess of 6M-NH$_4$OH has been used, neutralize with 6M-HCl (use litmus paper indicator). If acidic, neutralize with 6M-NH$_4$OH.

ALDEHYDES

Wear:

Rubber gloves, self-contained breathing apparatus, laboratory coat.

Spills:

Eliminate all sources of ignition and flammables.

Small—Absorb on paper towel. Evaporate in fume hood and burn the paper.

Large—Cover with sodium bisulfite ($NaHSO_3$). Add small amount of water and mix. Scoop into large beaker. After one hour wash down the drain with a large excess of water. Wash site with soap solution.

Package lots:

(CHOICE OF PROCEDURES)

1 • Absorb on vermiculite. Burn in an open pit or open incinerator.

2 • Dissolve in a flammable solvent (such as acetone or benzene). Spray into the fire-box of an incinerator equipped with an afterburner.

Examples:

Acetaldehyde
Acrolein
Acrolein dimer
o-Anisaldehyde
Benzaldehyde
Butyraldehyde
Caprylaldehyde
Chloral
Chloralhydrate
Chloroacetaldehyde
Cinnamaldehyde
Crotonaldehyde
iso-Decaldehyde
2-Ethyl-3-propylacrolein
Formaldehyde
Formalin (MeOH free)
Formalin (15% MeOH)
Furfural
Glutaraldehyde
Glyoxal
Paraformaldehyde
Paraldehyde
Propionaldehyde
Salicylaldehyde
Valeraldehyde

3 DISPOSAL PROCEDURE

ALKALI AND ALKALINE EARTH METALS, METAL ALKYLS, AND ALKOXIDES

Wear:

Leather gloves, large face shield, laboratory coat. (Class D fire extinguisher should be available)

Spills:

Small—Cover with excess dry soda ash. Mix and add slowly to butyl alcohol. After 24 hours dilute and add to drain with large excess of water.

Package lots:

OR Large spill—Mix with dry soda ash. Scoop in a **dry** bucket. In a remote area spread onto a large iron pan. Cover with scrap wood, paper and ignite with an excelsior train.

OR—Burn in an open pit incinerator.

OR—Direct "dry" steam onto the waste, spread on an iron pan. Beware of splatter.

Examples:

Aluminum alkyls
Aluminum ethoxide
n-Butyllithium
Calcium
Chlorodiethylaluminum
Diethyl zinc
Lithium
Potassium
Sodium
Sodium ethoxide
Sodium methoxide
Sodium-potassium alloys
Triethyl aluminum
Triisobutyl aluminum

WARNING: Possible Violent Reaction with Water

DISPOSAL PROCEDURE 4a

CHLOROHYDRINS, NITROPARAFFINS

Wear:

Neoprene gloves, plastic laboratory coat, self-contained breathing apparatus. Provide good ventilation.

Spills:

Eliminate all sources of ignition.

On skin—Wash immediately with soap solution. Rinse thoroughly.

On clothing—Remove clothing immediately and place in a fume hood. Wash clothing before wearing again. Shoes are difficult to decontaminate and may have to be discarded and burned.

On bench and floor—Cover with soda ash. Mix and spray with water. Scoop into a bucket of water. Let stand two hours. Neutralize* and wash into sewer with large excess of water. Wash site with soap solution.

* Neutralize with 6M-HCl.

Package lots:

(CHOICE OF PROCEDURES)

1 • Pour or sift over soda ash. Mix and wash slowly into large tank. Neutralize and pass to sewer with excess water.

2 • Absorb on vermiculite. Mix and shovel into paper boxes. Drop into incinerator with afterburner and scrubber.

** **Cyclonite-RDX:** Burning not recommended except in an area equipped to contain a detonation. Decompose cyclonite-RDX by adding slowly to 25 times its weight of boiling 5% sodium hydroxide. Boil for 1/2 hour.

Examples:

iso-Amyl nitrate
iso-Amyl nitrite
1-Chloro-1-nitropropane
Chloropicrin
Cyclonite-RDX**
1,1-Dichloro-1-nitroethane
EPN
Ethylene chlorohydrin
Ethylene nitrate
Ethyl nitrite
Methyl ethylnitrosocarbamate
Nitroethane
Nitromethane
Nitropropanes
Propyl nitrate
Tetranitromethane

4b DISPOSAL PROCEDURE

ORGANIC HALOGEN AND RELATED COMPOUNDS

Wear:

Rubber gloves, self-contained breathing apparatus, laboratory coat.

Spills:

Eliminate all sources of ignition. Absorb on paper towels or with vermiculite. Place on an iron, glass or plastic dish in a hood. Allow to evaporate. Burn the paper or vermiculite. Wash site with soap solution.

Package lots:

(CHOICE OF PROCEDURES)

1 • Pour onto vermiculite, sodium bicarbonate or a sand-soda ash mixture (90-10). (If a fluoride is present, add slaked lime to the mixture.) Mix and shovel into paper boxes. Place in an open incinerator. Cover with scrap wood and paper. Ignite with an excelsior train; stay on upwind side. **Or** dump into a closed incinerator with afterburner.

2 • Dissolve in a flammable solvent. Spray into the fire box of an incinerator equipped with afterburner and scrubber (alkali).

Examples:

Aldrin
Allyl bromide
Allyl chloride
Allyl chloroformate
Allyl iodide
Amyl bromide
Benzal chloride
Benzotrichloride
Benzotrifluoride
Benzyl bromide
Benzyl chloride
Benzyl chloroformate
Bromobenzene
Bromoethane
Bromoethylene
o-Bromotoluene
n-Butyl bromide
n-Butyl chloride
t-Butyl chloride
Chlordane
Chloroacetophenone
Chlorobenzene
Chlorocresols
Chlorodiphenyl
2-Chloro-2-methyl propene
Chloronaphthalene

EXAMPLES—Continued

Chloroprene
Chlorotrifluoroethylene
Dibutyl dichlorotin
1,4-Dichlorobutane
1,3-Dichloro-2-butene
1,1-Dichloroethane
1,2-Dichloroethylene
1,2-Dichloropropane
1,3-Dichloropropene
Dieldrin
Diethyl aluminum chloride
Diethyl sulfate
Dimethyl sulfate
Dimethyl sulfoxide
Dodecyl sodium sulfate
Epichlorohydrin
Ethyl bromoacetate
Ethyl chloride
Ethyl chloroacetate

Ethyl chloroformate
Ethylene dibromide
Ethylene dichloride
Ethyl fluoride
Ethyl fluoroacetate
Ethyl iodide
Fluoroethylene
Heptachlor
Hexachlorobenzene
Hexachloronaphthalene
Lindane
Methoxychlor
Methyl bromide
Methyl chloride
Methyl chloroformate
Methyl iodide
Octachloronaphthalene
Octafluoro-2-butene
Octafluoropropane

Pentachloroethane
Pentachloronaphthalene
Pentachlorophenol
Propargyl bromide
Propyl bromide
Propyl chloride
Propylene disulfate
Sodium fluoroacetate
Sodium pentachlorophenate
1,2,4,5-Tetrachlorobenzene
Tetrachloroethane
Tetraethyl lead
Tetramethyl lead
Tributyl chlorotin
1,2,4-Trichlorobenzene
Trichloronaphthalene
1,2,3-Trichloropropane
Vinyl chloride
Vinylidene chloride

ORGANIC HALOGEN AND RELATED COMPOUNDS

4c DISPOSAL PROCEDURE
SUBSTITUTED ORGANIC ACIDS

Wear:

Rubber gloves, self-contained breathing apparatus or all-purpose canister respirator, laboratory coat.

Spills:

Eliminate all sources of ignition. Turn on the fume hood if acid is volatile. Cover the spill on bench and floor with excess sodium bicarbonate and vermiculite. Mix and scoop into a large beaker of water. When reaction is complete, pour down the drain with a large excess of water. Wash site with soap solution.

Package lots:

(CHOICE OF PROCEDURES)

1 • Pour onto excess sodium bicarbonate. Mix and scoop into a bucket. Dump into a 55-gal. drum and fill with water. After 24 hours slowly pour into drain with large excess of water.

2 • Pour onto vermiculite in an open incinerator. Cover with scrap wood and paper. Pour waste alcohol over all and ignite with an excelsior train. Stay on upwind side.

3 • Dissolve in a flammable solvent such as waste alcohol. Spray into an incinerator with an afterburner and scrubber.

Examples:

Benzene sulfonic acid
Bromoacetic acid
Chloroacetic acid
Dichloroacetic acid
Ethylenediamine tetracetic acid (EDTA)
Fluoroacetic acid
Iodoacetic acid
Methanesulfonic acid
Trichloroacetic acid
2, 4, 5-trichlorophenoxy acetic acid
Trifluoroacetic acid

AROMATIC AMINES

Wear:

Butyl rubber gloves, plastic laboratory coat, self-contained breathing apparatus.

Spills:

On skin and clothing—Wash skin with strong soap solution immediately. Rinse thoroughly. Contaminated clothing should be removed, dried, and washed with strong soap solution—or destroyed. It may be necessary to destroy shoes by burning.

Small spills—Absorb liquids on paper towels. Brush solids onto paper. Place in an iron pan and allow evaporation in the fume hood. Add crumpled paper and burn. Wash site with strong soap solution.

Large spills—Cover large spills with sand and soda ash mixture (90-10). Mix and shovel into a cardboard box. Pack with much excess crumpled paper. Burn in an open pit or in an incinerator with afterburners and scrubber.

Package lots:

(CHOICE OF PROCEDURES)

1 • Pour or sift onto a thick layer of sand and soda ash mixture (90-10). Mix and shovel into a heavy paper box with much paper packing. Burn in incinerator. Fire may be augmented by adding excelsior and scrap wood. Stay on upwind side.

2 • Waste may be dissolved in flammable solvent (alcohols, benzene, etc.) and sprayed into fire box of an incinerator with afterburner and scrubber.

Examples:

Acridine
2-aminodiphenylene oxide
2-aminopyridine
Aniline
Anisidines
2-anthramine
Auramine.
Aziridine
Benzidine*
Benzyl amine
N-4-biphenyl acetohydroxamic acid
2-biphenyl amine
Chloroanilines
1, 2, 5, 6-Dibenzacridine
1, 2, 7, 8-Dibenzacridine
1, 2, 5, 6-Dibenzcarbazole
3, 4, 5, 6-Dibenzcarbazole
Dibenzyl amine

N, N-diethyl aniline
3, 3'-dimethoxy benzidine*
N, N-dimethyl aniline
Diphenyl amine
N-ethyl aniline
Ethyl morpholine
1-methyl-2-naphthylamine
Morpholine
1-naphthylamine
2-naphthylamine*
Nicotine
p-phenylenediamine
Phenylethanolamine
Phenyl-2-napthylamine
Picolines
Pyridine
Quinaldine
Quinoline
8-Quinolinol
4-Stilbenamine
Strychnine
Toluidines
Xylidines

---

*Carcinogenic

6 DISPOSAL PROCEDURE

AROMATIC HALOGENATED AMINES and NITRO COMPOUNDS

Wear:

Butyl rubber gloves, protective laboratory coat, self-contained breathing apparatus, protective shoes.

Spills:

On skin—Wash with strong soap solution **immediately**. Rinse well.

Contaminated gloves, clothing, shoes—Remove and clean **at once** or destroy by burning.

Small spills on tables or floor—Absorb liquid spills on paper towels or vermiculite; sweep solid spills onto paper. Put on an iron pan in the fume hood and allow to evaporate. Burn the paper or vermiculite in the absence of other flammables. Wash the site thoroughly with strong soap solution.

Large spills—Absorb or mix with vermiculite, sodium bicarbonate or sand. Package this in a paper carton and burn in an open pit. Use fuel such as crumpled paper and wood splinters. Wash site thoroughly as above.

Package lots:

(CHOICE OF PROCEDURES)***

1 • Pour or sift onto sodium bicarbonate or a sand-soda ash mixture (90-10). Mix and package in heavy paper cartons with plenty of paper packing to serve as fuel. Burn in an incinerator. Fire may be augmented with scrap wood.

2 • The packages of #1 may be burned more effectively in an incinerator with afterburner and scrubber (alkaline).

3 • The waste may be mixed with a flammable solvent (alcohol, benzene, etc.) and sprayed into the fire chamber of an incinerator with afterburners and scrubber.

—————
*** See page 133.

Examples:

Brucine
Carbazole
1-Chloro-2, 4-dinitrobenzene
Chloro-2-naphthylamine
Chloro-nitroanilines
Chloro-nitrobenzene
Chlorophenols
2, 6-Dibromo-N-chloro-p-benzoquinonimine
2, 5-Dichloroaniline
Dichlorobenzene
3, 3'-Dichlorobenzidine*
1, 3-Dichloro 5, 5-dimethylhydantoin
2, 4-Dichlorophenol
2-Dimethylaminofluorene
9, 10-Dimethyl-1, 2-benzanthracene
N, N-Dimethyl-4-biphenylamine
Dinitroaniline
m-Dinitrobenzene**
o-Dinitrobenzene**
p-Dinitrobenzene**
4, 6-Dinitro-o-cresol
2, 7-Dinitrofluorene
2, 4-Dinitrophenol**

—————
* Carcinogenic.
** Explosive. Other nitro compounds may also be unstable.

AROMATIC HALOGENATED AMINES and NITRO COMPOUNDS

EXAMPLES—Continued

1, 4-Dinitropiperazine
2, 4-Dinitrotoluene**
Elon
Endrin
4-Ethoxy-2-nitroaniline
N-2-Fluorenylacetamide
4-Fluoro-4-biphenylamine
2'-Fluoro-4'-phenylacetanilide
4'''-Fluoro-4'-phenylacetanilide
Methylene bis-phenylisocyanate
Methyl isothiocyanate
N-methyl-N-nitrosoacetamide
N-methyl-N-nitrosoallylamine
N-methyl-N-nitrosoaniline
N-methyl-N-nitrosobenzylamine

1-Methyl-1-nitrosourea
N-Methyl-N-nitrosovinylamine
1-Methyl pyrole
N-Methyl-N-2, 4, 6-tetranitroaniline
3-Nitroacetophenone
m-, o-, p-Nitroanilines
Nitrobenzene
Nitrobiphenyl
2-Nitrofluorene
∝-Nitronaphthalene
Nitrophenols
4-Nitroquinoline-N-oxide
N-Nitroso-N-methylaniline
4-Nitrosomorpholine

1-Nitrosopiperazine
N-Nitrosopiperidine
m-, o-, p-Nitrotoluenes
Picric acid**
Piperidine
Tetrachloronaphthalene
N,N,N′,N′-tetramethyl-3,
   3′-dimethoxybenzidine
N, N, N′, N′-tetramethylenediamine
Toluene-2, 4-diisocyanate
Trinitrobenzene**
Trinitrotoluene**

*** Destruction by chemical decomposition is recommended for dinitro, trinitro and other compounds with explosive potential.

Add the material slowly, while stirring, to 30
times its weight of a solution prepared by dissolving 1 part sodium sulfide ($Na_2S.9H_2O$) in 6
parts of water.

For unstable acidic materials (e.g. Picric acid),
dissolve in 25 times its weight of a solution
made from 1 part sodium hydroxide and 21 parts
sodium sulfide in 200 parts of water. Some H_2S
and NH_3 is evolved.

** Explosive. Other nitro compounds may also be
unstable.

7a DISPOSAL PROCEDURE

ALIPHATIC AMINES

Wear:

Butyl rubber gloves, face shield or all-purpose canister respirator, laboratory coat.

Spills:

Liquid or solid—Cover with sodium bisulfate. Spray with water and wash into drain with large excess of water.

Package lots:

(CHOICE OF PROCEDURES)

1 • Add the contaminated amine to a layer of sodium bisulfate in a large evaporating dish. Spray with water. Make neutral and wash into the drain with large excess of water.

2 • Dissolve in a flammable solvent (e.g., waste alcohols). Burn in an open pit by means of an excelsior train. Stay on the upwind side.

3 • Solution of #2 may be sprayed into the fire box of an incinerator with afterburner and scrubber.

Examples:

Allyl amine
Amyl amine
Aminoethylethanola-
  mine
Butyl amine
iso-Butyl amine
Tert-Butyl amine
Cyclohexylamine
Dibutyl amine
Dicyclohexylamine
Diethanolamine
Diethylamine
2-Diethyl
  aminoethanol
Diethylenetriamine
Diisopropylamine
Dimethylamine
Ethanolamine
Ethylamine
Ethylene diamine
N-Ethyl-N-nitroso-
  N-butylamine
N-Ethyl-N-N
  nitrosovinylamine
n-Heptylamine

Hexamethylenetetra-
  mine
1-6-Hexanediamine
Hydroxylamine
Hydroxylamine
  hydrochloride
Lutidine
N-Methylbutylamine
Monomethylamine
N-Nitrosodiethanola-
  mine
N-Nitrosodimethyla-
  mine*
1,3-Propanediamine
Propyl amines
Propylene imine
Pyrrolidine
Tetraethylenepenta-
  mine
Triamylamine
Tri-n-butylamine
Tri-ethylamine
Triethanolamine
Triethylene tetramine
Trimethylamine
Tripropylamine

---

* Carcinogenic.

DISPOSAL PROCEDURE 7b

ORGANIC PHOSPHATES AND RELATED COMPOUNDS

Wear:

Rubber gloves, self-contained breathing apparatus, laboratory coat.

Spills:

Absorb with vermiculite or paper towels. Scoop the mixture into a plastic bag. Take bag outside to incinerator or pad and burn. If an incinerator is not available, set the bag in a pan of waste flammable solvent and burn.

Package lots:

(CHOICE OF PROCEDURES)

1 • Take packages to an open incinerator. Stay on upwind side and mix with equal parts of sand and pulverized limestone. Wet down with a flammable solvent (benzene or alcohol). Ignite from a safe distance with an excelsior train.

2 • Shovel mixture of #1 into a paper box and drop into an incinerator with an efficient afterburner. Alkaline scrubbing will prevent escape of any oxides of phosphorous and arsenic.

Examples:

o-Chlorophenyl diphenyl phosphate
Dibutyl phosphate
Dibutyl phosphite
Dichlorophenyl phosphine
Diethyl ethyl phosphate
Diisopropyl fluorophosphate
Dimethyl-1, 2-dibromo-2, 2-dichloro-ethyl phosphate
Grain fumigants
Hydroxy dimethyl arsine oxide
Malathion
Methyl parathion
1-Naphthylisothiocyanate
Parathion
Phosdrin
Ronnel
Systox
Tetra ethyl dithiono pyrophosphate
Tetra ethyl pyrophosphate
Tributyl phosphate
Triphenyl phosphate
Triphenyl phosphine
Tritolyl phosphate

8 DISPOSAL PROCEDURE

AZIDES and AZO-COMPOUNDS

The organic azides and heavy metal azides are explosive. Alkali and alkaline earth azides are not considered explosive under normal laboratory conditions.

Keep stock of all azides very low. Stamp date and receipt on package.

Wear:

Leather gloves, heavy face shield, laboratory coat. Work from behind a barricade (body shield or wall). Avoid unnecessary heat, friction or impact.

Small spills:

Absorb the liquid on paper or with vermiculite. If it is a solid, dampen and brush onto paper with great care. Place in plastic bag and take outside for burning.

OR

Sponge up with water, followed by de-

contamination with a 10% ceric ammonium nitrate solution.

Large spills or Package lots:
(CHOICE OF PROCEDURES)

"Kill" by adding to a greater-than-stoichiometric amount of ceric ammonium nitrate solution with agitation sufficient to provide suspension of all solids. Cool the reaction.

Examples:

2,2'-Azonaphthalene
Azoxybenzene
Diazo methane
Dimethyl amino azo benzene-2-naphthalene
2,3-Dimethyl-azo benzene
Hydrazoic acid
3-Methyl-4-dimethyl amino-azobenzene
1-Phenylazo-2-naphthol
1-Ortho-tolylazo-2-naphthol
Sodium azide

DISPOSAL PROCEDURE 9

CARBON DISULFIDE

Wear:

Rubber gloves, safety glasses, laboratory coat. If hood is not available wear self-contained breathing apparatus. Carbon dioxide fire extinguisher should be available.

Spills:

Eliminate flammables and all sources of ignition. Allow to evaporate or absorb with paper towels and evaporate in hood on an iron pan or glass dish. Burn the paper.

Package lots:

All equipment or contact surfaces should be grounded to avoid ignition by static charge. Absorb on vermiculite, sand, or ashes and cover with water. Transfer under water in buckets to an open area. Ignite from a distance with an excelsior train. If quantity is large, carbon disulfide may be recovered by distillation and repackaged for use.

10 DISPOSAL PROCEDURE
CAUSTIC ALKALI AND AMMONIA

Wear:

Rubber gloves, large face shield (wear all-purpose or special canister respirator for NH_3*), Laboratory coat.

Spills:

Solid—Sweep up, dilute and neutralize with 6M-HCl in a large bucket. Wash down drain with large excess of water.

Solution—Neutralize and mop up—or use water-vac. Discharge to sewer with large excess of water.

Package lots:

Pour into large tank of water and neutralize. Transfer to sewer with large excess of water.

Examples:

Ammonia, anhydrous
Ammonia, aqua
Calcium hydroxide (slaked lime)
Calcium oxide (quick lime)
Potassium hydroxide (caustic potash)
Sodium hydroxide (caustic soda)

DISPOSAL PROCEDURE 11

INORGANIC SALTS

Wear:

Rubber gloves, safety glasses, laboratory coat.

Spills:

1 • Solutions—Cover with soda ash, mix and scoop into a beaker of water. Neutralize with 6M-HCl and wash down drain with excess water.

2 • Solids—Collect in a beaker. Dissolve in large amount of water. Add soda ash, mix and treat as above.

If spill contains a fluoride, add slaked lime in addition to the above treatment.

Package lots:

Add slowly to a large container of water. Stir in slight excess of soda ash. If fluoride is present add slaked lime also. Let stand 24 hours. Decant or siphon into another container and **neutralize** with 6M-HCl before washing down drain with large excess of water. The sludge may be added to land fill.

Examples:

Alums
Aluminum chloride, hydrate
Aluminum nitrate, hydrate
Aluminum sulfate, hydrate
Ammonium fluoride
Ammonium nitrate
Ammonium thiocyanate
Chromic (III) salts, hydrates
Cobaltous nitrate
Copper nitrate
Cuprous chloride
Ferrous Ammonium sulfate
Ferrous chloride
Ferrous sulfate
Lithium carbonate
Magnesium nitrate
Manganese sulfate
Molybdenum compounds (sol.)
Nickel nitrate
Nickel sulfate

Potassium acetate
Potassium carbonate
Potassium ferrocyanide
Potassium fluoride
Potassium hydrogen difluoride
Potassium nitrate
Sodium acetate
Sodium benzoate
Sodium fluoride
Sodium formate
Sodium hydrogen difluoride
Sodium iodide
Sodium nitrate
Sodium propionate
Sodium silicate
Sodium tetraborate
Stannic chloride, hydrate
Zinc acetate
Zinc chloride

Wear:

Rubber gloves, face shield, laboratory coat. Body shield should be available for the more active agents. Replace face shield with self-contained breathing apparatus for such agents as chlorine and bromine.

Spills:

1 • Gas leak: If the valve is leaking because it cannot be closed (a common occurrence), the gas can be bubbled through a reducer (sodium sulfite) and excess sodium bicarbonate solution. Be sure to include a trap in the line to prevent the solution being sucked back into the cylinder. If this cannot be done, the cylinder should be placed in or adjacent to a fume hood and left to bleed off.

If the leak is in the valve assembly, a plastic bag can be fastened over the head of the cylinder which can then be taken outside or to a fume hood.

2 • If the oxidizer is a liquid or a solid—Cover with a reducer (hypo, a bisulfite, or a ferrous salt but not carbon, sulfur or strong reducing agents). Mix well and spray with water. A sulfite or a ferrous salt will require addition of some $3M-H_2SO_4$ to promote rapid reduction. Scoop slurry into a container of water and **neutralize** with soda ash. Wash down the drain with excess water. Wash site thoroughly with a soap solution containing some reducer.

Package lots:

Add to a large volume of concentrated solution of reducer (hypo, a bisulfite or a ferrous salt and acidify with $3M-H_2SO_4$). When reduction is complete add soda ash or dilute hydrochloric acid to neutralize the solution. Wash into drain with large excess of water.

Examples:

Ammonium dichromate

Ammonium perchlorate
Ammonium persulfate
Barium chlorate
Bromic acid
Bromine
tert.-Butyl chromate
Bromic acid
Bromine
Calcium chlorate
Calcium hypochlorite
Chloric acid
Chlorine
Chlorine dioxide
Cleaning solution (acid dichromate)
Chromium oxide
Chromium oxychloride
Cleaning solution (acid dichromate)
Fluorine
Iodine
Magnesium chlorate
Magnesium perchlorate
Nitrogen trifluoride
Nitrosyl chloride
Peracetic acid
Perchloric acid
Potassium and sodium perchlorates, chlorates, chlorites, dichromates, hypochlorites, permanganates, persulfates
Zinc chlorate

DISPOSAL PROCEDURE 12b

REDUCING SUBSTANCES

Wear:

Rubber gloves, safety glasses, laboratory coat. Work in hood or wear a respirator.

Spills:

Gas leak: Eliminate all sources of ignition. If the valve is leaking because it cannot be closed, the gas can be bubbled through a calcium hypochlorite solution. Be sure to include a trap in the line to prevent the solution being sucked back into the cylinder.

Solid: Cover spill with soda ash or sodium bicarbonate. Mix and spray with water. If effervescent wait until reaction is complete. Scoop into a large beaker and cautiously add equal volume of calcium hypochlorite (reaction may be vigorous). Add more water, stir, and allow to stand for one hour. Dilute and neutralize* the oxidized solution and transfer to the drain with excess of water.

Package lots:

If a **gas,**** bubble into soda ash solution. If a **solid,** mix with equal volume of soda ash and add water to form a slurry in a large container. In either case add calcium hypochlorite. Add more water if necessary and let stand two hours. Neutralize* the oxidized solution. Wash down drain with large excess of water.

Examples:

Chromous salts
Sodium bisulfite
Sodium nitrite
Sodium sulfite
Sodium thiosulfate
Stannous chloride
Sulfur dioxide

---

*Test with litmus. Neutralize with 6M-HCl or 6M-NaOH as required.

** If a tank of reducing gas has developed a permanent leak lower it upside down into a drum filled with water. Add a mixture of soda ash and calcium hypochlorite. Continue the treatment until the tank is empty and the drum contains a solution of stable element or compound.

13 DISPOSAL PROCEDURE
MERCAPTANS—and organic sulfides

Wear:

Rubber gloves, self-contained breathing apparatus, laboratory coat.

Spills:

Eliminate all sources of ignition. Cover with calcium hypochlorite and mix. Scoop into a large beaker. After 12 hours, neutralize* if necessary. Wash to sewer with excess water. Wash site of spill with strong soap solution to which has been added some hypochlorite.

Package lots:

1 • As for spills.

2 • Dissolve in waste alcohol or other flammable solvent. Burn in an incinerator with an afterburner and scrubber to neutralize the SO_2.

WARNING! Addition of dry calcium hypochlorite may cause violent reaction or flash fire. *Use a weak aqueous solution (up to 15%)* with stirring.

NOTE: Sodium hypochlorite available in 5% concentrations as household bleach is also suitable.

Examples:

Allyl propyl disulfide
Amyl mercaptan
Benzyl mercaptan
Butyl mercaptan
Carbonyl sulfide**
Crag 974
Dimethyl sulfide
Diphenyl sulfide
Ethanethiol
Ferbam
2-Mercaptoethanol
Methyl mercaptan**
Perchloromethyl mercaptan
Thioacetamide
2, 2-Thiodiethanol
Thioglycolic acid
Thiophene
Thiourea
Tetramethylthiuram disulfide (Thiram)

---

* Test with litmus. Neutralize with 6M-HCl or 6M-NH₄OH as required.
** If quantity is large, seal and return container to the supplier. If small, allow it to dissipate in a fume hood.

CYANIDES AND NITRILES

Wear:

Long rubber gloves, self-contained breathing apparatus, laboratory apron or coat.

Evacuate the laboratory and isolate the area during decontamination.

Spills:

Eliminate all sources of ignition and flammables.

1 • General treatment:

(a) Absorb liquid with vermiculite or on paper towel (sweep solid onto paper). Place on an iron or glass dish in a hood. Evaporate and burn paper.

(b) On skin—Wash away **immediately** with much soap and water.

2 • Hydrocyanic acid (HCN) leak:*

Turn on fume hood. Allow gas to leak into a container of sodium hydroxide solution while stopping leak.** Add excess calcium hypochlorite to the alkali cyanide. Discharge the cyanate into the drain with excess water.

3 • Cyanides:

(a) Scoop into a larger beaker and make alkaline with sodium hydroxide solution. Add to the slurry an excess of ferrous sulfate solution. After one hour, flush down the drain with excess water.

OR

(b) Add excess sulfur to the alkaline slurry of the cyanide. Heat to convert to thiocyanate. Flush down drain with excess water.

4 • Nitriles:

Add excess of sodium hydroxide and calcium hypochlorite solution to produce a cyanate. Scoop slurry into a large beaker. After one hour flush down the drain with excess water. Wash site with soap solution containing some hypochlorite.

(Continued on next page)

---

* If the leak cannot be stopped set the tank upside down in a drum filled with a strong solution of sodium hydroxide and calcium hypochlorite. Continue the treatment until the tank is empty.

** Avoid possibility of suck back of alkaline material into liquid HCN.

14 DISPOSAL PROCEDURE

Package lots:

(CHOICE OF PROCEDURES)

1 • Add the cyanide with stirring to strong alkaline solution of calcium hypochlorite. Maintain an excess sodium hydroxide and calcium hypochlorite. Let stand 24 hours. Flush the cyanate down drain with large excess of water.

2 • Nitriles are more effectively converted to soluble sodium cyanate by treating with excess alcoholic sodium hydroxide. After about one hour evaporate the alcohol and then add calcium hypochlorite. Maintain an excess of the hydroxide and hypochlorite. After 24 hours flush the cyanate down the drain with a large excess of water.

Examples:

Acetone cyanohydrin
Acetonitrile
Adiponitrile
Benzonitrile
Benzyl cyanide
Butyl nitrile
n-Butyronitrile
Calcium cyanide
Chloroacetonitrile
Crotononitrile
Cuprous cyanide
Cyanamide
Cyanoacetamide
Cyanogen
Cyanogen chloride
Ethyl cyanoacetate
Hydrocyanic acid
Lactonitrile
Potassium cyanide
Propionitrile
Sevin
Sodium cyanide
Succinonitrile
Tetramethyl-succinonitrile
Trichloroacetonitrile

DISPOSAL PROCEDURE 15

ETHERS

Wear:

Rubber gloves, large heavy face shield (if in doubt use body shield also). Self-contained breathing apparatus.

Spills:

Eliminate all sources of ignition and flammables. Absorb on paper towel. Evaporate from an iron pan in a hood. Allow time for vapors to completely escape the hood vents, then burn the paper.

If large spill, absorb on much more paper or vermiculite and allow complete evaporation from all surfaces. Use same precaution before burning paper.

Package lots:

1 • Pour on ground in open area. Allow evaporation or ignite from a distance by means of a long fuse or excelsior train.

2 • Dissolve waste in higher alcohol (e.g., butyl), benzene, or petroleum ether. Incinerate.

3 • PEROXIDE FORMATION. (See MCA Data Sheet SD-29 ethyl ether.) Ether of long standing in contact with air and exposed to light may contain peroxides, especially if stored in clear glass. **Explosions have occurred when caps or stoppers were turned.** (See MCA Accident Case History No. 603.)

Transport cans or bottles to an isolated area (e.g., deserted quarry). Each container should be wrapped in padding material or packed in sawdust. At site uncover containers and arrange an excelsior train. From a safe distance puncture the cans near bottom with rifle fire. Ignite excelsior train. Local regulations must be observed.

Examples:

Allyl glycidyl ether
n-Amyl ether
Anisole
n-Butyl glycidyl ether
Butyl vinyl ether
Di-n-butyl ether
Chloromethyl ether
2,2-Dichloroethyl ether
1,1-Diethoxyethane
Diethylene glycol monoethyl ether
Diglycidyl ether
Diisopropyl ether
Dimethoxy ethane
Dimethoxy methane
Dimethoxy propane
Dipropylene glycol methyl ether
1,2-Epoxy-3-phenoxy-propane
1,2-Epoxy-propane
Ethoxy acetylene
Ethyl ether
Ethyl vinyl ether
Ethylene oxide
Glycidol
Guaiacol
Hydroquinone monomethyl ether
Methyl ether
Methyl ethyl ether
Methyl vinyl ether
Phenyl ether
Phenyl ether-biphenyl mixture
iso-Propyl ether
iso- Propyl glycidyl ether
Tetrahydrofuran
Vinyl ether

16 DISPOSAL PROCEDURE

HYDRAZINES

Wear:

Rubber gloves, self-contained breathing apparatus. Impervious clothing recommended. Body shield should be available.

Spills:

Eliminate all sources of ignition and flammables.

1 • On skin or clothing—Wash skin immediately. Remove contaminated clothing at once.

2 • Absorb liquid with vermiculite or paper towels. Scoop mixture, paper or solids into a plastic bag and take to a burning pit or incinerator for burning.

3 • Large spills—Collect the liquid with an aspirator such as used for recovering spilled mercury. Empty into a large beaker and neutralize with dilute sulfuric acid. Wash to drain with excess water. Wash site with soap and water.

Package lots:

(CHOICE OF PROCEDURES)

1 • Dilute to at least 40% and neutralize with dilute sulfuric acid. Flush to sewer with excess water.

OR

2 • Dissolve in large volume of waste alcohol or other flammable solvent and burn in an open pit. Ignite from a distance with an excelsior train.

Examples:

1,1-Dimethyl hydrazine (UDMH)
Hydrazine
Hydrazine salts
Methyl hydrazine
Phenyl hydrazine

HYDRIDES

Wear:

Rubber gloves, fire proof clothing, face shield. Work from behind body shield where possible. Keep available pulverized dolomite or dry graphite for fire fighting.

Spills:

Eliminate all sources of ignition. Scoop in dry plastic bag which has first been purged to inert gas. Remove to the outside for burning. Flood the burned residues with water to ensure complete destruction of hydrides.

Package lots:

(CHOICE OF PROCEDURES)

1 • Mix with dry sand to avoid or stop fire. Scoop into bucket and remove to open area. Slowly spray with dry butyl alcohol. Later add water by fogging until last of hydride is destroyed. Scoop the solid into a large container. Neutralize* with 6M-HCl. Let settle. Decant and flush to sewer with excess of water. Send sand residue to land fill. If the hydride is a gas (e.g. diborane, silane, germanium hydride), dispose of it by controlled burning.

2 • Burn in iron pan or in open pit.

---

* Test with litmus—
neutralize with 6M HCl if necessary.

Examples:

Aluminum borohydride
Calcium cyanide
Decaborane
Diborane
Germanium hydride
Lithium aluminum hydride
Lithium borohydride
Lithium hydride
Pentaborane
Potassium borohydride
Potassium hydride
Silane
Sodium amide
Sodium borohydride
Sodium hydride

HYDROCARBONS, ALCOHOLS, KETONES, and ESTERS

Wear:

Rubber gloves, face shield, laboratory coat. Have all-purpose canister mask available.

Spills:

Eliminate all sources of ignition and flammables.

1 • A gas leak from a faulty tank— Keep concentration of gas below the explosive mixture range by forced ventilation. Remove tank to an open area and allow dissipation to the atmosphere. Attempt to cap the valve outlet and return tank to the supplier.

2 • A liquid—Absorb on paper. Evaporate on an iron pan in a hood. Burn the paper.

3 • A solid—Sweep onto paper and place in an iron pan in the hood. Burn the paper and compound.

Package lots:

1 • A gas—Pipe the gas into the incinerator. Or lower into a pit and allow it to burn away.

2 • A liquid—Atomize into an incinerator. Combustion may be improved by mixing with a more flammable solvent.

3 • A solid—Make up packages in paper or other flammable material. Burn in the incinerator. Or the solid may be dissolved in a flammable solvent and sprayed into the fire chamber.

Examples:

Acenaphthene
Acetone
Acetylene
Alizarin dye
Allene
Allyl acetate
Allyl alcohol
n-Amyl acetate
iso-Amyl acetate
sec-Amyl-acetate
n-Amyl-alcohol
iso-Amyl alcohol
tert-Amyl alcohol
Amylene
iso-Amyl formate
Anthracene
Anthraquinone
1, 2-Benzanthracene
Benzene
Benzyl acetate
Benzyl alcohol
Benzyl benzoate
Biphenyl
Borneol
Butadiene
n-Butane
iso-Butane
1-Butene
2-Butene
n-Butyl acetate
iso-Butyl acetate
sec-Butyl acetate
tert-Butyl acetate

n-Butyl alcohol
iso-Butyl alcohol
sec-Butyl alcohol
tert-Butyl alcohol
Butyl cellosolve
n-Butyl formate
iso-Butyl formate
Butyl methacrylate
iso-Butyl methyl
  ketone
p-tert-Butyl toluene
1-Butyne
2-Butyrolactone
Camphor
Carbon monoxide
Cellosolve
Cellosolve acetate
Cresols
Creosote
Croton oil
Crude oil
Cumene
Cycloheptanone
Cyclohexane
Cyclohexanol
Cyclohexanone
Cyclohexene
Cyclohexyl benzene
Cyclopentadiene
Cyclopentane
Cyclopentanone
Cyclopropane
p-Cymene

EXAMPLES—Continued

Decahydronaphthalene
n-Decane
n-Decyl alcohol
Diacetone alcohol
1, 2, 5,
6-Dibenzanthracene
1, 2, 5,
6-Dibenzofluorene
1, 2, 3,
4-Dibenzophenanthrene
3, 4, 8, 9-Dibenzpyrene
Dibutyl tin diluarate
Dibutyl oxalate
Dibutyl phthalate
Dicyclopentadiene
Diethyl adipate
Diethyl carbonate
Diethylene glycol
Diethyl ketone
Diethyl malonate
Diethyl phthalate
3, 4-Dihydropyran
Diisobutyl ketone
2, 2-Dimethyl butane
Dimethyl carbonate
1, 2 Dimethyl chrysene
Dimethyl fumarate
Dimethyl naphthalene
Dimethyl propane
Dimethyl phthalate
Di-n-octyl phthalate
Di-sec-octyl phthalate
1, 4-Dioxane
Dipentene
Dipentene monoxide
Diphenyl methane
Dodecane
Ethane
Ethanol
Ethyl acetate
Ethyl acetoacetate
Ethyl acrylate

Ethyl alcohol
Ethyl-sec-amyl ketone
Ethyl benzene
Ethyl benzoate
Ethyl butyl ketone
Ethyl butyrate
Ethyl crotonate
Ethylene
Ethylene glycol
Ethyl formate
2-Ethyl hexanol
Ethyl lactate
Ethyl oxalate
p-Ethyl phenol
Ethyl silicate
Furan
Furfuryl alcohol
Gasoline
Glycerol
Greases
Heptane
Hexane
n-Hexanol
1-Hexene
2-Hexene
sec-Hexyl acetate
p-Hydroquinone
5-indanol
Industrial gases (LHG)
Isoprene
Jet fuels
Kerosene
Ketene
Lacquer diluent
Ligroin
Liquefied petroleum gas
p-Mentha-1, 8-diene
Mesityl oxide
Methane
Methyl acetate
Methyl acrylate

Methyl alcohol
Methyl amyl alcohol
Methyl-m-amyl ketone
6 Methyl-1, 2-benzanthra-
cene
10 Methyl, 1, 2-benzanthra-
cene
Methyl benzoate
Methyl benzyl alcohol
2-Methyl-1-butene
2-Methyl-2-butene
3-Methyl-1-butene
Methyl butyl ketone
Methyl butyrate
Methyl cellosolve
Methyl cellosolve acetate
3-Methyl cholanthrene
Methyl cyclohexane
2-Methyl cyclohexanol
2-Methyl cyclohexanone
4-Methyl cyclohexene
Methyl ethyl ketone
Methyl formate
2-Methyl furan
Methyl isobutyl ketone
Methyl isobutyrate
Methyl isocyanate
Methyl methacrylate
1-Methyl naphthalene
Methyl-n-propyl ketone
Methyl salicylate
Methyl styrenes
Methyl-toluene sulfonate
Naphthas
Naphthalene
1-Naphthol
2-Naphthol
Natural gas
Nickel carbonyl
Nonyl phenol
Octane
1-Octanol

2-Octanol
Oil, cocoanut
Oil, fuel
Oil, lubricating
Oil, mineral mist
Oil, olive
Oil, peanut
Oil, soybean
Oil, vegetable
Paraffin
n-Pentane
iso-Pentane
1, 5-Pentanediol
2, 4-Pentanedione
2-Pentanol
Petroleum ether
Phenanthrene
Phenol
Phenyl acetate
Phenyl isocyanate
o-Phenyl phenol
Phorone
iso-Phorone
Phosgene solutions in
benzene
2-Pinene
Piperylene
Pival
Polyvinyl acetate emulsion
Propane
1, 2-Propanediol
Propargyl alcohol
iso-Propenyl acetate
iso-Propyl acetate
n-Propyl acetate
n-Propyl alcohol
iso-Propyl alcohol
Propyl benzene
iso-Propyl benzoate
Propylene
Propylene carbonate

Propyl formate
iso-Propyl formate
Propyne
Propyne-allene mixture
Quinone
Resorcinol
Rotenone
Stoddard solvent
Styrene
Tall oil
Tallow
Terphenyls
Tetradecane
Tetrahydronaphthalene
1-Tetralone
Toluene
Tridecanol
Triethylene glycol
Triethyl-ortho-formate
5, 9, 10-Trimethyl-1,
2-benzanthracene
6, 9, 10-Trimethyl-1,
2-benzanthracene
Trimethyl borate
2, 2, 4-Trimethyl pentane
2, 4, 4-Trimethyl-2-pentane
1, 3, 5-Trioxane
Turpentine
Varnish
Vinyl acetate
Warfarin
Xylenes
Xylenols

HYDROCARBONS, ALCOHOLS, KETONES, and ESTERS

19 DISPOSAL PROCEDURE
INORGANIC AMIDES and Derivatives

Wear:

Rubber gloves, large face shield, protective laboratory coat. A large body shield should be available.

Spills:

Eliminate all sources of ignition. Sweep up solid amide onto dry paper. Cautiously add to cold water in small portions with agitation. Neutralize* and discharge into drain with large excess of water.

Package lots:

Sift slowly into a large container of cold water, with agitation. When all has reacted, neutralize* and pour into drain with large excess of water.

Examples:

Ammonium sulfamate
Monochloroamine
Sulfamic acid
Sulfamide

---

*Test with litmus. Neutralize with 3M-HCl or 6M-NH₄OH as required.

DISPOSAL PROCEDURE 20

ORGANIC AMIDES

Wear:

Rubber gloves, safety glasses, laboratory coat.

Spills:

Scoop into a plastic bag or onto a paper towel. Remove to the outside, add alcohol and burn in a safe place.

Package lots:

Add to a flammable solvent (alcohol or benzene). Pour into an iron pan in an open pit. Ignite.

OR

Spray into an incinerator. Oxides of nitrogen may be scrubbed out with alkaline solution.

Examples:

Acetamide
Chloroacetamide
N, N-Dimethylacetamide
Dimethylformamide
Ethyl Acetanilide
Formamide
2-Phenanthreneacetamide
3-Phenanthreneacetamide

21 DISPOSAL PROCEDURE— INTER NON-METALLIC COMPOUNDS

Wear:

Long rubber gloves, safety glasses or goggles, self-contained breathing apparatus, laboratory coat.

Recommended that work be done in an effective hood from behind a body shield or in an open barricaded area out of doors. **NOTE**—Some suppliers will collect unwanted cylinders of compressed gases on request.

Spills:

Eliminate all sources of ignition.

Gas leak (e.g., boron trichloride, chlorine trifluoride). Allow gas to flow into a mixed solution of caustic soda and slaked lime. If possible, keep in a hood until cylinder is emptied.

Liquid or solid—Cover with vermiculite, sodium bicarbonate or a mixture of soda ash and slaked lime (50-50). Mix and spray water cautiously from an atomizer. Scoop up and add slowly to a large container of water (if too active continue spraying). When reaction is complete neutralize* and wash down the drain with a large excess of water. Wash site with soap solution.

Package lots:

Sprinkle or sift onto a thick layer of mixed dry soda ash and slaked lime (50-50) from behind a body shield. Mix and spray water cautiously with an atomizer. Scoop up and sift cautiously into a large volume of water. Neutralize* and wash down the drain with large excess of water.

Examples:

Boron fluoride-ethyl ether complex
Boron tribromide
Boron trichloride
Boron trifluoride
Boron trifluoride complexes with acetic acid and methanol
Bromine pentafluoride
Bromine trifluoride
Carbon tetrafluoride
Carbonyl fluoride
Chlorine trifluoride
Iodine chloride
Iodine pentafluoride
Iodine trichloride
Nitric oxide
Nitrogen dioxide
Oxygen difluoride
Perchloryl fluoride
Phosgene
Phosphorus oxychloride
Phosphorus pentachloride
Phosphorus pentasulfide
Phosphorus sesquisulfide
Phosphorus tribromide
Phophorus trichloride
Silicon tetrafluoride
Sulfur decafluoride
Sulfur dichloride
Sulfur hexafluoride
Sulfur monochloride
Sulfuryl chloride
Sulfuryl fluoride
Thionyl chloride
Thionyl fluoride

* Make litmus test. Neutralize with 6M-NH_4OH or 6M-HCl as required.

DISPOSAL PROCEDURE 22a
PEROXIDES, INORGANIC

Wear:

Rubber gloves, large face shield, laboratory coat. A body shield should be available.

Spills:

Cover with at least double volume of sand-soda ash mixture (90%-10%). Mix thoroughly and break up any lumps of peroxide. With a plastic scoop add slowly to a large beaker of sodium sulfite solution (3 or 4 liters) with stirring. Neutralize with dilute sulfuric acid. When settled decant the sulfate solution into drain with excess water. The sand can be sent to the landfill.

Hydrogen Peroxide (spill or package lots): Dilute and wash down drain with excess of water.

Package lots:

Use the above method for the encrusted impure peroxide. Repackage any recovered pure granular peroxide in glass containers with rubber stoppers.

Examples:

Hydrogen peroxide (3% to 90%)
Potassium peroxide
Sodium peroxide

22b DISPOSAL PROCEDURE
PEROXIDES, ORGANIC

CAUTION: Keep stock low and date each container as received. Never transfer to glass stoppered containers or screw cap bottles which can cause dangerous friction. See Procedure 15.

Wear:

Rubber gloves, large heavy face shield, laboratory coat. Work from behind heavy body shield in hood.

Spills:

Eliminate all sources of ignition.

1 • Liquid peroxide spills may be absorbed in large quantity of vermiculite or sand. Using a soft plastic scoop, carefully place the mixture in a plastic container. Spread on a steel pan or in a deep pit. Ignite from a distance with an excelsior train or a long torch. Wash the scoop and container with 20%

NaOH. Burn the original cartons and bags.

2 • Solid peroxide spills should be mixed with a large volume of vermiculite or sand. Cautiously transfer as above and burn.

3 • A very small quantity of #1 or #2 may be destroyed by adding 10 volumes of 20% NaOH. After 24 hours, neutralize* and pass into drain with large excess of water.

Package lots:

Absorb or mix in small portions on vermiculite or sand. Wet down with 10% NaOH. Scoop up with plastic scoops and take to open-incinerator pit. When dry, ignite from a distance with an excelsior train.

---

* Neutralize with 6M-HCl.

Examples:

Acetyl peroxide
Benzoyl peroxide
Butyl hydroperoxide
tert.-Butyl peracetate
tert.-Butyl perbenzoate
Butyl peroxy pivalate
Cumene hydro peroxide
3,4-dichlorobenzoyl peroxide
Dibutyl peroxide
Diisopropyl peroxy dicarbonate
Lauroyl peroxide
Methyl ethyl ketone peroxide

DISPOSAL PROCEDURE 23

SULFIDES, INORGANIC

Wear:

Rubber gloves, safety glasses. Work in hood or wear self-contained breathing apparatus, laboratory coat.

Spills:

Eliminate all sources of ignition. Add FeCl$_3$ solution. Stir until FeS formation is complete. Add slight excess of soda ash. Scoop up and wash into drain with excess water. Wash site with soap solution.

OR

Absorb with vermiculite or sodium bicarbonate and scoop into plastic bag or wide-mouthed glass jar. Close tightly, remove to safe place outside and add FeCl$_3$ solution. Stir until reaction is complete. Add slight excess sodium bicarbonate and wash into drain with excess water.

Package lots:

Add to a large volume of FeCl$_3$ solution with stirring. Add more FeCl$_3$ if necessary. Add soda ash with stirring until neutral. Scoop up and wash down the drain with excess water.

Note: If a gas (e.g., H$_2$S), seal the cylinder and return to supplier. If the valve is leaking the gas can be bubbled through a FeCl$_3$ solution. Be sure to include a trap in the line to prevent the solution being sucked back into the cylinder. If this cannot be done, the cylinder should be placed in or adjacent to a fume hood and left to bleed off.

Examples:

Ammonium polysulfide
Ammonium sulfide
Calcium sulfide
Hydrogen sulfide
Potassium sulfide
Sodium sulfide

DISPOSAL PROCEDURE
ACIDS, ORGANIC (Limited to C, H, and O Compositions)

Wear:

Rubber gloves, face shield, laboratory coat. Body shield and self-contained breathing apparatus should be available.

Spills:

Eliminate all sources of ignition.

1 • Cover contaminated surfaces with soda ash or sodium bicarbonate. Mix and add water if necessary. Scoop up slurry and wash neutral* waste down drain with excess water. Wash site with soda ash solution.

Package lots:

(CHOICE OF PROCEDURES)

1 • Liquid acid may be injected at base of incinerator or after mixing with a flammable solvent. Afterburner is suggested for complete combustion.

2 • A solid acid may be dissolved in a flammable solvent and burned as above.

3 • Solid acid may be packaged in paper or other flammable material and burned in an incinerator.

Examples:

Acetic acid
Acetic anhydride.
Acrylic acid
Adipic acid
Benzoic acid
n-Butyric acid
iso-Butyric acid
n-Butyric anhydride
Caproic acid
Citraconic anhydride
Citric acid
Formic acid
Fumaric acid
Gallic acid
Glutaric anhydride
Glycolic acid
Hydracrylic
   acid-B-lactone*

Lauric acid
Maleic acid
Maleic anhydride
Methacrylic acid
Octanoic acid
Oleic acid
Oxalic acid
Phthalic anhydride
Pimelic acid
Propionic acid
Pyrogallic acid
Pyruvic acid
Salicylic acid
Stearic acid
Succinic acid
Succinic anhydride
Tannic acid
Valeric Acid

* Carcinogenic.

---

* Make litmus test. Neutralize with 6M-NH₄OH or 6M-HCl as required.

DISPOSAL PROCEDURE 24b

ACIDS, INORGANIC

Wear:

Rubber gloves, self contained breathing apparatus, laboratory coat. Have body shield available.

Spills:

1 • Cover the contaminated surface with sodium bicarbonate or a soda ash—slaked lime mixture (50-50). Mix and add water if necessary to form a slurry. Scoop up slurry and wash down the drain with excess water. Wash site with soda ash solution.

Package lots:

Add slowly to large volume of agitated solution of soda ash and slaked lime. Add neutralized solution to excess running water. As an added precaution, the sink can be lined with protective matting and filled with coarse chipped marble.

Examples:

Boric acid
Boron oxide
Fluoroboric acid
Fluorosilicic acid
Hydriodic acid
Hydrobromic acid
Hydrochloric acid
Hydrofluoric acid
Iodic acid
Mixed acids
Nitric acid
Phosphoric acid
Phosphoric anhydride
Sulfuric acid
Sulfur trioxide

25 DISPOSAL PROCEDURE

CARBIDES

Wear:

Rubber gloves, safety glasses, laboratory coat.

Spills:

Cover with dry vermiculite, scoop into a dry bucket or plastic bag and transfer to a safe open area. Dispose of the material cautiously by adding it slowly to a large volume of water. Burn the hydrocarbon gas using a pilot burner. Allow to stand for 24 hours and run to sewer with excess water.

Package lots:

Take to a safe open area and add slowly to a large container of water. Burn off the hydrocarbon gas with a pilot flame. Allow to stand for 24 hours. Siphon off the liquid and transfer the precipitate to a landfill.

Examples:

Aluminum carbide
Calcium carbide

DISPOSAL PROCEDURE 26

WASTES TO BE DUMPED INTO LANDFILLS OR RELEASED TO AIR

Wear:

Heavy work gloves, safety glasses.

Properly assembled waste ready for pick-up trucks may be used as fill in reclaiming low areas or may be dumped into a landfill.

Examples:

Argon
Asphalt
Batteries, dry cell
Boron
Bromochloromethane
Bromoform
Bromotrifluoromethane
Calcium carbonate
Calcium oxide
Carbon black
Chlorobromomethane
1-Chloro-1,1-difluoroethane
Chloroform
Chloropentafluoroethane
Chlorotrifluoroethylene
Chromium
Crude lime
1,2-Dibromotetrafluoromethane
Dichlorodifluoromethane

Dichloromethane
Dichloromonofluoromethane
Dichlorotetrafluoroethane
Epoxy resin systems
Ferrosilicon
Ferrovanadium dust
Fluorotrichloromethane
Helium
Hexachloroethane
Hexafluoroethane
Hydrogen
Lamp bulbs
Latex
Magnesium oxide
Metal scrap
Molybdenum, insoluble compounds
Neon
Nitrogen
Nitrogen fertilizers
Nitrogen trioxide
Octafluorocyclobutane
Osmium tetroxide
Oxygen
Ozone
Paint
Pyrethrum
Resins
Rubber
Scrap glass
Scrap stoneware
Silica
Sludges
Stone, alberine
Sulfur
Tar
Tetrabromoethane

1,1,1,2-Tetrachloro-2,2-difluoroethane
1,1,2,2-Tetrachloro-1,2-difluoroethane
Tetramethyl silane
Transite
Tin, organic compounds
Titanium oxide
Tremolite
1,1,1-Trichloroethane
1,1,2-Trichloro-1,2,2-trifluoroethane
Trifluoromethane
Urea
Xenon
Yttrium
Zinc oxide
*Zirconium
*Zirconium-hafnium powders

---

* Clean, dry material may be burned on steel plate. Ignite from distance.
To dispose of 2 lbs of moist or contaminated material, add, with water, to 10 lbs. cement mix. Allow to set for 2 days and dump into landfill.

481

27a DISPOSAL PROCEDURE

RECOVERY

Scrap metal in the form of sheets, rods, wire, tubes

Wear:

Heavy gloves, safety glasses, laboratory coat.

1 • Larger pieces of less expensive metals can be salvaged profitably for use in local shops, or can be sold as scrap metal—e.g., aluminum, copper, brass.

2 • The more expensive metals are worth salvaging, even in small pieces—e.g., platinum wire, silver foil.

Metal should be sorted, classified, and placed in boxes, properly labelled. Turnings, shot, cuttings of the cheaper metals may be assigned to landfill. (See 26.)

Examples:

Sheet, rods, wire, tubes, foil, etc.

Aluminum
Antimony
Batteries, wet cells
Beryllium
Bismuth
Bronze
Cadmium
Cobalt
Copper
Gold
Lead
Magnesium
Manganese
Molybdenum
Nickel
Paladium
Platinum
Rhodium
Silver
Steel, carbon
Steel, stainless
Tantalum
Tin
Titanium
Zinc

DISPOSAL PROCEDURE 27b

RECOVERY*—MERCURY

Wear:

Rubber gloves, self-contained breathing apparatus, laboratory coat.

CAUTION: The toxicity of mercury is such that the element and its compounds should not be allowed to contaminate air or water.

Spills and package lots:

Metal—Collect all droplets and pools at once by means of suction pump and aspirator bottle with a long capillary tube. Cover fine droplets in non-accessible cracks with calcium polysulfide and excess sulfur. Combine all contaminated mercury in a tightly stoppered bottle. Hold it for purification or sale.

Compounds—Dissolve all water soluble contaminated compounds. Convert other contaminated compounds to the soluble nitrates. Adjust the acidity and precipitate as mercuric sulfide. Wash and dry the precipitate. Ship to the supplier.

Examples:

Mercuric nitrate, chloride and thiocyanate
Mercurous nitrate and chloride
Mercury
Mercury fulminate**
Organic mercury compounds

---

* In procedures 27b to 27m, recovery is essential because of high toxicity of all elements and compounds—e.g. the American Conference of Governmental Industrial Hygienists' threshold limit values (1968) expressed as mg/M³ (milligrams per meter cubed) for mercury = 0.1, for arsenic = 0.5, antimony = 0.5, selenium = 0.2, beryllium 0.002, lead 0.2, cadmium 0.2, barium 0.5, uranium 0.05, vanadium 0.1—as opposed to hydrogen cyanide 11, and hydrogen sulfide 15.

** Readily destroyed by reaction with 20% sodium thiosulfate. Ventilate process to avoid cyanogen exposure.

27c DISPOSAL PROCEDURE
RECOVERY—PHOSPHORUS
YELLOW AND RED

Wear:

Rubber gloves, large face shield.

YELLOW PHOSPHORUS

Spills:

Cover with wet sand. Spray with water to keep sand wet. Scoop into a bucket of water. After standing overnight, recover and repackage.

OR

If quantity is very small (e.g., fragments of sticks of yellow phosphorus), cover with water and remove to an open area. Pour onto the ground or a steel pan. The water will evaporate and dry yellow phosphorus will ignite spontaneously in air and burn away.

Defective package:

Submerge defective package in a large container of water. Repackage under water and return to the suppliers.

RED PHOSPHORUS

Spills:

Sweep up and burn on an iron pan in the hood.

Defective package:

Repackage and return to the shelf—or to the supplier.

DISPOSAL PROCEDURE 27d

RECOVERY—ARSENIC, ANTIMONY AND BISMUTH*

Wear:

Rubber gloves, safety glasses, respirator, laboratory coat. Work in fume hood.

Waste:

Dissolve in minimum hydrochloric acid (concentrated, reagent). Filter if necessary. Dilute with water until white precipitates form (SbOCl and BiOCl). Add just enough 6M-MCl to redissolve. Saturate with hydrogen sulfide. Filter, wash the precipitate, dry, package and ship to the supplier.

OR

If the waste is of very little value, use Procedure 11.

Examples:

Antimony pentasulfide
Antimony trioxide
Arsenic acid
Arsenic trichloride
Arsenic trioxide
Arsenous acid
Calcium arsenate
Lead arsenate
Metals (arsenic, antimony and bismuth)
Nitrates and chlorides of arsenic, antimony and bismuth
Stibine

---

* Although bismuth is much less toxic than arsenic or antimony it is included here because the recovery procedure is similar.

27e DISPOSAL PROCEDURE
RECOVERY—SELENIUM AND TELLURIUM

Wear:

Rubber gloves, safety glasses, respirator (or work in hood), laboratory coat.

Spills:

Absorb on paper and place in wide mouth stoppered bottle for later recovery. Wash site with soap solution.

Package lots and recovered spills:

Liquid or **solid**—Make a solution strongly acidic with hydrochloric acid. Slowly add sodium sulfite to the cold solution with stirring, thus producing sulfur dioxide, the reducer.

Upon heating dark grey selenium and black tellurium form. Let stand overnight. Filter and dry. Ship to supplier.

Examples:

Hydrogen selenide
Selenium hexafluoride
Selenium oxides
Selenium oxychloride
Selenium tetrachloride
Tellurium hexafluoride
Tellurium oxides

DISPOSAL PROCEDURE 27f

RECOVERY—
LEAD AND CADMIUM COMPOUNDS

Wear:

Rubber gloves, safety glasses, respirator (or work in hood).

Package lots and recovered spills:

Convert to nitrates with a minimum of nitric acid (concentrated, reagent). Evaporate in a fume hood to a thin paste. Add about 500 ml. water and saturate with hydrogen sulfide. Filter, wash, and dry the precipitate. Package and ship to the supplier.

OR

If the waste is of small volume, use Procedure 11.

Examples:

Cadmium oxide
Cadmium salts
Lead oxides
Lead salts

27g DISPOSAL PROCEDURE

RECOVERY— BERYLLIUM COMPOUNDS

Wear:

Rubber gloves, self-contained respirator, laboratory coat (wash after each job).

Spills:

Absorb on paper towels and place in large stoppered wide-mouth bottle. Save for recovery. Wash site with soap solution.

Package lots and recovered spills:

Dissolve in minimum of 6M-HCl. Filter and treat filtrate with slight excess of 6M-NH$_4$OH (use litmus). Boil and allow coagulated precipitate to settle for about 12 hours. Filter and dry. Package and ship to the supplier.

Examples:

Beryllium salts

DISPOSAL PROCEDURE 27h

RECOVERY—STRONTIUM AND BARIUM COMPOUNDS

Wear:

Rubber gloves, safety glasses, laboratory coat.

Package lots and recovered spills:

Dissolve waste in 6M-HCl and filter. Neutralize the filtrate with 6M-NH₄OH (use litmus) and precipitate with excess sodium carbonate. Filter, wash, and dry the precipitate. Package and ship to the supplier.

OR

If the waste is of small volume, use Procedure 11.

Examples:

Oxides and salts of strontium and barium

27i DISPOSAL PROCEDURE

RECOVERY—VANADIUM COMPOUNDS

Wear:

Rubber gloves, large face shield, laboratory coat. (Wear self-contained respirator if spill is large.)

Spills:

Cover with powdered ammonium carbonate. Add a layer of crushed ice and spray with 6M-NH₄OH while stirring. Scoop slurry into a wide mouth bottle, save for salvage. Wash site with soap water.

Package lots:

(Work in a fume hood).

Add (e.g., VOCl₃) slowly to a thick layer of powdered ammonium carbonate in a large evaporating dish. Spray with 6M-NH₄OH while stirring. Add a layer of crushed ice and continue stirring and spraying. Add more 6M-NH₄OH if necessary. May add more waste vanadium compound with stirring. Pour into large beaker and let stand overnight. Filter off the crude ammonium vanadate, dry and package for shipment to supplier for reprocessing.

Examples:

Ammonium vanadate
Sodium vanadate
Vanadium chlorides
Vanadium oxychlorides
Vanadyl sulfate

DISPOSAL PROCEDURE 27j

RECOVERY—HALOGENATED SOLVENTS

Wear:

Rubber gloves, self-contained respirator (or work in hood), laboratory coat.

Spills:

Absorb on paper towels and allow to evaporate in the fume hood. Burn the paper. Wash site with soap solution.

Package lots:

The toxic liquid compounds concerned here are insoluble in water and cannot be burned. Purify the contaminated liquids by distillation and place the pure distillate back on the shelf.

OR

Return to the supplier.

Examples:

Bromoform
Carbon tetrabromide
Carbon tetrachloride
Chloroform
Tetrachloroethylene
1,1,1-Trichloroethane
Trichloroethylene

27k DISPOSAL PROCEDURE—RECOVERY—MISCELLANEOUS

The elements and their compounds listed below should be dealt with separately. If the quantity justifies recovery get in touch with the supplier for special instructions.

Examples:

Deuterium
Erbium
Gadolinium
Gallium
Germanium
Germanium dioxide
Hafnium
Holmium
Indium
Lanthanum
Lutecium
Niobium
Osmium
Osmium oxide
Praseodymium
Rubidium
Ruthenium
Samarium
Silver nitrate
Terbium
Thallium
Thallous sulfate
Thorium*
Uranium*

* These materials as well as all other naturally radioactive materials, elements or substances must be disposed of in accordance with regulations of the U.S. Atomic Energy Commission or those of the local or state departments of health. This may usually be done by turning them over to the radiation safety officer or a commercial disposal agency handling radioactive materials.

CELLULOSE NITRATE, COLLODION, CELLULOID

Wear:

Rubber gloves and face shield.

Note: Outside disposal is preferred but can be done in the hood.

Spills:

CELLULOSE NITRATE

Eliminate all sources of ignition. Gather up the nitrocellulose and dampen with an alcohol. Spread out in a thin layer (2 inches deep max.) in an outside open area on top of papers or other combustible material. Ignite from a distance with an excelsior train or a long torch. When disposal by burning must be done in a hood, the quantity of nitrocellulose burned at any one time should be limited to 250 ml or less. This small quantity of nitrocellulose can be placed in a shallow pyrex dish and ignited after it has been dampened with an alcohol.

OR

Eliminate all sources of ignition. Gather up the nitrocellulose and dampen with water. Squeeze out excess water and place nitrocellulose (250 ml or less) in a large stainless steel beaker (2500 ml or larger), and place beaker in a large pyrex dish. Add an equal amount of 10% caustic (**no stronger**) to the beaker. Most of the nitrocellulose should be consumed in 20 minutes. Remove beaker (**Caution:** beaker will be hot from the heat of reaction) and pour remaining contents into drain with large excess of water.

Package lots:

Small quantities of collodion (250 ml or less) may be disposed of in a hood by igniting it in a pyrex dish. Larger quantities should be disposed of in an outside open area by pouring into a steel pan and igniting with an excelsior train or a long torch.

CELLULOID (Nitrocellulose Plastic)

Celluloid disposal should be done outside in an open area. Eliminate sources of ignition. Spread the waste celluloid on top of paper or other combustible material. Ignite from a distance with an excelsior train or a long torch.

COLLODION (Nitrocellulose Lacquer)

Eliminate sources of ignition. Wipe up spilled material with paper towels or rags. Remove to an outside open area, spread on the ground and ignite with an excelsior train or a long torch.

WARNING: Whenever cellulose nitrate, collodion or celluloid is burned or decomposed, oxides of nitrogen are released. **Avoid direct breathing of vapors.**

29 DISPOSAL PROCEDURE

DIRECT BURNING

Wear:

Heavy leather gloves, safety glasses.

Disposal:

Materials should be properly packed for safe handling, and placed in a pre-scribed site for daily pick-up. Burning can be done in an open or closed incin-erator with afterburners.

Examples:

Materials:

Bags, paper and cloth
Biological wastes—animal remains, bedding, feed wastes
Clothing, discarded or contaminated
Excelsior
Paper wastes, packing boxes, etc.
Wood scraps, packing boxes, etc.
Bags, paper and cloth

Disposal Materials – Minimum Requirements

(a) CHEMICALS

Acetone, waste (2 gal.)
Alcohol, butyl (2 gal.)
Alcohol, denatured (waste) (2 gal.)
*Ammonium hydroxide (concentrated, Reagent) (2 liters)
Ammonium hydroxide (6M) (4 liters)
Ammonium carbonate (2 x 1 lb.)
Benzene, waste (2 gal.)
Calcium hypochlorite (2 x lb.)
**Calcium polysulfide (2 x 1 liter)
Excelsior (box)
Fuels — scrap wood
　　　　 paper
　　　　 organic flammable solvents
Ferrous sulfate (2 x 1 lb.)
Ferrous sulfate, 30% (4 x 1 liter)
Ferric chloride (2 x 1 lb.)
Graphite powder (5 lb.)
Hydrochloric acid (concentrated, Reagent) (2 liters)
***Hydrochloric acid, (6M) (4 liters)
Hydrogen sulfide (small cylinder)
Ice, chipped
Kaolin (2 x 5 lb.)
Kaolin — Soda ash (50-50) (2 x 5 lb.)
Limestone, pulverized (2 x 5 lb.)

Litmus paper, red and blue (1 doz. vials each)
Marble chips (2 x 5 lb.)
Nitric acid (centrated, Reagent) (2 liters)
Paper boxes, waste
Potassium iodide, 10% (peroxide test)
Sand, dry (2 x 10 lb.)
Sand-Soda ash (90-10) (2 x 10 lb.)
Slaked lime (calcium hydroxide) (2 x 5 lb.)
Soap power (6 x 1 lb.)
Soda ash (sodium carbonate) (2 x 5 lb.)
Soda ash-Slaked lime (50-50) (2 x 10 lb.)
Sodium bicarbonate (2 x 5 lb.)
Sodium hydroxide, 10% (4 liters)
Sodium hydroxide, 20% (4 liters)
Sodium bisulfite (5 lb.)
Sodium sulfite (5 lb.)
Sodium thiosulfate (hypo) (2 x 5 lb.)
Sulfur, sublimed (2 x 1 lb.)
Sulfuric acid (concentrated, Reagent) (2 liters)
****Sulfuric acid (3M) (4 liters)
Towels, paper (1 doz. rolls)
Vermiculite (or Oil-Dri, Sol-Speedi-Dri, etc.) (10 lb.)

*Requires approximately 400 ml. of concentrated lime water with hydrogen sulfide.
　 per liter of solution.
**Saturate one liter of concentrated lime water with hydrogen sulfide.
　 Add excess sublimed sulfur. Shake.
***Requires 516 ml. of concentrated HCl per liter of solution.
****Requires 336 ml. of concentrated reagent H_2SO_4 per liter of solution.

(See over for EQUIPMENT listing)

Disposal Materials—Minimum Requirements

(b) EQUIPMENT

Aspirator bottle (mercury collector) (2)

Atomizer spray bottle (2)

Beakers, 3 or 4 liter size (2)

Brush, hand (6)

Brush, small paint

Bucket, plastic (3 gal.)

Burners, Bunsen

Condenser, Liebig

Dish, evaporating, 12″ (2)

Extinguisher (Class D fires) (30 lb. size)

Extinguisher (CO_2) (10 lb. size)

Flask, Distilling — 1 liter (2)

Gloves, heavy work

Gloves, leather

Gloves, neoprene

Gloves, butyl rubber

Iron pan (for fume hood) (approx. 18″ x 24″)

Laboratory coat, cloth

Laboratory coat, plastic

Matting, open mesh, rubber, to fit sink

Mop and bucket

Pipettes 10 ml. (peroxide test) (4)

Respirator, canisters (1 of each)

Respirator (self-contained breathing apparatus)

Safety glasses or goggles (2 prs.)

Scoop, plastic

Shield, body

Shield, face, large and heavy, 0.050″, to cover ears and neck

Spray jar (insecticide type), stainless steel

Suction pump (mercury collector)

Tank (oil drum, 55 gal.)

Tools — hammer
pliers
wrench, crescent
screwdrivers

Water-vac

BIBLIOGRAPHY

ACGIH—"Threshold Limit Values," American Conference of Governmental Industrial Hygienists, 1014 Broadway, Cincinnati, Ohio 45202

AIA—"Nitroparaffins and their hazards," Report #12, and "Chemical Hazards Bulletin," American Insurance Association, 85 John St., New York, N.Y. 10038

ASTM—"Fire and Explosion Hazards of Peroxy Compounds," Tech. Bul. #394, American Society for Testing and Materials, 1916 Race St., Philadelphia, Pa. 19103

Audrieth, L. F. & Ogg, B.A.—"The Chemistry of Hydrazine," John Wiley & Sons Inc., 605 3rd Ave., New York, N.Y. 10016

BDH—"Spillages of Hazardous Chemicals," British Drug Houses Ltd., Poole, Dorset, England. Also available from Gallard-Schlesinger Chemical Mfg. Corp., 584 Mineola Ave., Carle Place, Long Island, New York, N.Y. 11514 or from British Drug Houses (Canada) Ltd., Barclay Ave., Toronto 18, Ontario, Canada

Brookes and Alyea—"Poisons," Van Nostrand Reinhold Company, 430 Park Avenue, New York, N.Y. 10022

Campbell, Neil—"Schmidts Organic Chemistry," Oliver and Boyd Ltd., Edinburgh, Scotland

CGA—"Safe Handling of Compressed Gases" Pamphlet P-1, Compressed Gas Association Inc., 500 5th Ave., New York, N.Y. 10036

Cloyd, D. R. & Murphy, W. J.—"Handling Hazardous Materials—N.A.S.A. SP-5032," National Aeronautics and Space Administration, Washington, D. C. 20546

Elkins, H. B.—"Chemistry of Industrial Toxicology," 1963, John Wiley & Sons, 605 3rd Ave., New York, N.Y. 10016

Fairhall, L. T.—"Industrial Toxicology," Williams and Wilkins, Baltimore, Md.

Gaston, P. J.—"The care, handling and disposal of dangerous chemicals," 1965, Institute of Science Technology, Northern Publishers (Aberdeen Ltd.), Aberdeen, Scotland

Gonzales, Vance, Helpern—"Legal Medicine and Toxicology," 1940, Appleton-Century Trade Books, 250 Park Ave., New York, N.Y. 10017

Hurd—"Chemistry of Hydrides," 1952, John Wiley & Sons, Inc., 605 3rd Ave., New York, N.Y. 10016

IBM—"Freon" solvent data book, IBM Corp., Yorktown Heights, New York 10598

Kirk & Othmer—"Encyclopedia of Chemical Technology," Vol. 8, p.489 2nd Ed., Wiley-Interscience, 605 3rd Ave., New York, N.Y. 10016

Los Angeles Fire Dept.—"Dangerous Chemicals Code," 1951, Parker & Son, Inc., Los Angeles, Calif.

McElroy, F. E.—"Accident Prevention Manual for Industrial Operations," 5th Ed., National Safety Council, 425 No. Michigan Ave., Chicago, Ill. 60611

MCA—"Chemical Safety Data Sheets," "Safety Guides" and "Chemical Card Manual," Manufacturing Chemists Association, 1825 Connecticut Ave., Washington, D. C. 20009

Pennsylvania Department of Revenue—"Highway Transportation of Hazardous Substances," 1968, Hazardous Substances Transportation Board, Harrisburg, Pa. 17127

Remy, H.—"Treatise on Inorganic Chemistry," Vol. I & II, 1956, American Elsevier Publishing Co., Inc., 52 Vanderbilt Ave., New York, N.Y. 10017

Ross, R. D.—"Industrial Waste Disposal," Van Nostrand Reinhold Company, 430 Park Avenue, New York, N.Y. 10022

Sax, N. I.—"Dangerous Properties of Industrial Materials," 1968, Van Nostrand Reinhold Company, 430 Park Avenue, New York, N.Y. 10022

Sidgwick—"Chemical Elements and their Compounds," Vol. I & II, 1950, Oxford University Press, Inc., 200 Madison Ave., New York, N.Y. 10016

Sittig, Marshall—"Sodium, its manufacture, properties and uses," 1956, Van Nostrand Reinhold Company, 430 Park Avenue, New York, N.Y. 10022

Steere, N. V.—"Handbook of Laboratory Safety," Waste disposal pp. 34-40, 1967, Chemical Rubber Co., 18901 Cranwood Parkway, Cleveland, Ohio 44128

Voeglein, Joseph F.—"Storage and Disposal of Dangerous Chemicals," J. Chem. Educ. 43, A151 (Feb. 1966).

Walsh, John—"Pollution—The Wake of the Torrey Canyon" (news comment), Science 160, 167 (April 12, 1968)

Weast, R. C. & Selby, S. M.—"Handbook of Chemistry and Physics," 1967, Chemical Rubber Co., 18901 Cranwood Parkway, Cleveland, Ohio 44128

World Health Organization—"Treatment and Disposal of Wastes," WHO Tech. Report Series No. 367 (1967)

MCB—"Safety Procedures: Laboratory Chemical Catalog," pp. 9-31, Matheson, Coleman & Bell, 2909 Highland Ave., Norwood, Ohio 45212

MGB—"Gas Data Book," The Matheson Co., Inc., P.O. Box 85, East Rutherford, N.J. 07073

N.F.P.A.—No. 49, "Hazardous Chemicals Data"; No. 325, "Fire Hazard Properties of Flammable Liquids, Gases and Volatile Solids"; No. 491, "Manual of Hazardous Chemical Reactions"; No. 704M, "Fire Hazards of Materials," National Fire Protection Association, 60 Batterymarch St, Boston, Mass. 02110

Nat'l Research Council—"Evaluation of the Hazard of Bulk Water Transportation of Industrial Chemicals," 1966, National Academy of Science, 2101 Constitution Ave., N.W., Washington, D. C. 20418

Nicholson, H. Page—"Pesticide Pollution Control," Science 158, 871 (Nov. 17, 1967)

Noller, D. C. & Bolton, D. F.—"Safe Handling and Storage of Organic Peroxides in the Laboratory," Anal. Chem. 35, 887 (June 1963).

NSC—"Chemical Safety References," National Safety News 97, 47 (Apr. 1968), and "Industrial Safety Data Sheets," National Safety Council, 425 No. Michigan Avenue, Chicago, Ill. 60611

NTTC—"Commodity and Equipment Data Sheets," National Tank Truck Carriers, Inc., 1616 P St., N.W., Washington, D. C. 20006

O'Brien, Richard D.—"Toxic Phosphorous Esters," 1960, Academic Press Inc., 111 5th Ave., New York, N.Y. 10003

Patty, Frank A.—"Industrial Hygiene and Toxicology," Vol. II, 1963, Wiley-Interscience, 605 3rd Ave., New York, N.Y. 10016

Pennsylvania Department of Health—"Short Term Limits for Exposure to Airborne Contaminants," "Hygienic Information Guides," "Industrial Waste Manual," Division of Sanitary Engineering, Harrisburg, Pa.

INDEX